MACHINE INTELLIGENCE AND THE IMAGINAL REALM

MACHINE INTELLIGENCE AND THE IMAGINAL REALM

Spiritual Freedom and the Re-animation of Matter

A Sacred Planet Book

LUKE LAFITTE

Inner Traditions
Rochester, Vermont

Inner Traditions
One Park Street
Rochester, Vermont 05767
www.InnerTraditions.com

Text stock is SFI certified

Sacred Planet Books are curated by Richard Grossinger, Inner Traditions editorial board member and cofounder and former publisher of North Atlantic Books. The Sacred Planet collection, published under the umbrella of the Inner Traditions family of imprints, is composed of works on the themes of consciousness, cosmology, alternative medicine, dreams, climate, permaculture, alchemy, shamanic studies, oracles, astrology, crystals, hyperobjects, locutions, and subtle bodies.

Cataloging-in-Publication Data for this title is available from the Library of Congress

ISBN 978-1-64411-406-3 (print)
ISBN 978-1-64411-407-0 (ebook)

Printed and bound in the United States by Lake Book Manufacturing, Inc.
The text stock is SFI certified. The Sustainable Forestry Initiative® program promotes sustainable forest management.

10 9 8 7 6 5 4 3 2 1

Text design by Debbie Glogover and layout by Priscilla Harris Baker
This book was typeset in Garamond Premier Pro with Supermolot used as a display typeface

To send correspondence to the author of this book, mail a first-class letter to the author c/o Inner Traditions • Bear & Company, One Park Street, Rochester, VT 05767, and we will forward the communication, or contact the author directly at **drlucaslafitte@gmail.com**.

Contents

Editor's Note to the Reader

It is our practice to put any author asides in footnotes directly on the text page, which we have done in this book as well. However, we encourage all readers to also refer to the endnotes at the back of the book when citations are given, as these often contain fascinating historical and contextual comments from the author on the specific work being cited—as opposed to simply being a page reference for the material quoted.

FOREWORD

What Science and Invention Might Yet Become

Jeffrey J. Kripal

This mega merger of super-sciences may transform who and what we are. It could transplant our senses into other entities, and then convert those entities into something else. It may alter millions of years of evolution that imbue us with wonderful and terrible traits.

RICHARD FEYNMAN

The theme of co-evolution between technology and humanity was advancing now at a rapid pace. . . . Feynman knew that our world and reality are merely mental constructions from the same book of spells.

LUKE LAFITTE

Who writes like that? Well, Richard Feynman did. And Luke Lafitte does.

The following pages are "spellings" in both the linguistic and the entrancing sense of that *double entendre:* a vast arrangement of letters on the page to create words and sentences and so meaning, but also to conjure new future realities. *Caveat lector.* Reader beware. You are about to be spelled, hexed. But you are about to be spelled to get you out of a spell, hexed to get you out of a hex.

The sharp division that we pretend between the fictive and the real, between the present and the future, between the possible and the impossible, and, most of all, between mind and machine, wobble and waver in these pages before they finally threaten to disappear altogether. At the very end of the book, one gets this haunting, uncomfortable, and yet somehow ecstatic sense that things are not at all what they seem to be (the uncomfortable sensibility), and that we—as individuals, yes, but especially as community and culture—have this astonishing power to change it all, alter some of its most intimate workings (the ecstatic sensibility).

But how?

Through technology.

It's not what you think. Lafitte is not arguing that you need another smart phone or shiny gadget. Nor is he arguing that future silicon chips implanted into our bodies will transform our collective into some kind of posthuman Borg. That's precisely what he is writing against (as, by the way, were the writers who wrote the Borg). He is calling us instead to understand, *to really understand,* that each and every piece of human technology—from the prehistoric sharpened rock and hand-painted cave wall, through the ancient writing stylus and leaf palm, to the most recent promise of AI, the nanotechnology of the contemporary university lab, and the most powerful mathematical formula—can become so many indirect pathways to a near or far future enlightenment. That enlightenment comes down to this: it is *mind* that has produced the machine, not the other way around.

It turns out that we have it all backward, or upside down, or inside out. Consciousness is not an accidental by-product of random cosmological and now accidental biological processes, as we are asked to believe over and over again, as if the ideologues are not so sure of themselves, as if the evidence does not add up (it doesn't). Consciousness appears here rather as a force, field, or frequency, as fundamental to the cosmos as gravity. The brain does not produce consciousness. Consciousness produces the brain. There is no ghost in the machine, as Gilbert Ryle had it in the middle of the last century, but there is a machine in the ghost, if you can call consciousness a "ghost" (and you can't).

Indeed, for Lafitte, consciousness is not just a force of the cosmos. It *is* the Force. Think *Star Wars*. And Luke Lafitte is our Yoda, clinging to our backs as we run, sweat, and somersault through the dark forest, chattering away in our ear in oddly formed sentences to flip us, turn us inside out, wake us up from our long cultural dream so that we can get in touch with that Force, not to "master" it (that's more machine talk), but to *become* it.

Lafitte is certainly clear enough about what he is doing in this book. It would be all too easy to get lost in the stories, biographies, and movies he so lovingly recounts here, including the *Star Wars* scenes. But getting lost in them is precisely what this book is *not* about. Lafitte is telling us stories to remind us, to help us remember who we really are, to free us from our own self-created illusions and begin to take back the ultimate modern projection, the projection of mind into matter and machine, including and especially that most intimate of material machines—the human body. The end result is a three-hundred-year American story of loss and recovery, of invention and realization, of material gain and spiritual awakening, of a new rebalancing or reconciliation of materialism and idealism. Matter and machine are not left behind or denied here. They are recognized for what they have always been: us.

It is all very gnostic for the author, "gnostic" in the sense that the American science fiction writer Philip K. Dick used that most powerful of ancient and now modern words, particularly in his late novels after he (and his poor cat) were irradiated by some kind of radioactive pink light or palpable Cosmic Mind in the winter of 1974. Such a literally glowing gnosis for Dick was not really about his dead cat, who died quickly from cancer (are you freaked out yet?). It was about a direct "knowing" (*gnosis*) that such an illumination bestows, a spiritual-physical light that really and truly saves us.

Wait, what? *Saves* us? From what? And how?

Such a gnosis saves us, not by any bloody sacrifice or unbelievable *deus ex machina* (more machine talk!), but by shocking us into the memory of who and what we actually are. Salvation, in other words, does not and cannot come from outside us. It can only come from inside us. Dick, being a science fiction writer (there is that collapse of the fictive and the real again), imagined this saving secret identity as an extraterrestrial being from the stars, with three eyes no less, which he doodled in his private journals (not very well). Dick also wrote about this saving gnosis philosophically—for thousands of obsessed pages in his private diaries and published novels, in fact—as the Mind of all creation lying ahead of us, speaking back to us in dream, vision, and mystical experience, pulling us, haunting us, inspiring us forward into a grand future like some kind of giant Magnet mesmerizing everything at the end of time.

He was perfectly serious. In his most stunning lines, Dick even suggested that we are always mistakenly relating to this future being, who is really present here and now, as someone external, as "God." I assume that he thought we did this because we will not allow ourselves to imagine that this superbeing is in fact our own future presence that we are presently worshipping as not us. To make the same point, Lafitte highlights a key Philip K. Dick passage in his own Phil Dickian book. Here is the passage:

> I did not think I should tell Fat that I thought his encounter with God was in fact an encounter with himself from the far future. Himself so evolved, so changed, that he had become no longer a human being. Fat had remembered back to the stars, and had encountered a being ready to return to the stars, and several selves along the way, several points along the line. All of them are the same person.

"Fat," by the way, is short for "Horselover Fat," the main protagonist of the novel *VALIS,* who is himself an obvious and barely fictive stand-in for the historical Philip K. Dick, whose name means, literally, "Horselover Fat." In other words, another collapse of the fictive and the real.

This is another way to say it. The ancient history of religions, like the modern American history of science and technology recounted in these pages, is one big mistake, or, more compassionately, one long detour through the forest that circles back, loops back, to ourselves. As Lafitte has it in his own epigraph from Dick again, we are all amnesiac avatars of God, particles of light who have forgotten who and what we really are. Or, in his own frame now: "We are eternal beings imagining and simulating non-eternal beings," as in a virtual reality generator.

As an example of this realization, Lafitte cites Gene Roddenberry, the original creator of the oh-so-successful *Star Trek* franchise. In a classical gnostic manner (although the creator would probably have never put it that way), Roddenberry wrote that, "I believe I am God; certainly you are, I think we intelligent beings on this planet are all a piece of God, are becoming God. In some sort of cyclical non-time thing we have to become God, so that we can end up creating ourselves, so that we can be in the first place."

That is the point of Lafitte's book too. *That* is the spell to end our spell, the hex to end the hex. Mythically speaking, we are Luke Skywalker in the cave of Dagobah finally slaying Darth Vader, cutting his head off with his light saber, only to discover that that Vader's face is . . . his own.

I remember the same scene when the movie *The Empire Strikes Back* first came out in 1980. I was a teenager. The general public reaction was one of shock. So, I hope, will your own be to these pages and their scenes. Luke Lafitte is not just our Yoda. He is also our Luke Skywalker, and these cave-like pages, if you can enter them with the right intention, have the power, have the Force, to slay the widely held but fundamentally false dogma that the machines of technology are something external to us, something not human, that they will somehow overcome us. If you read him well, this real Luke, like the fictive one, will show you the true face of such machines and so remove the dystopian fear, which makes little sense when one realizes the truth of the things—that it is mind that has created the machine, not the other way around. The Darth Vader of our present techno-mythology is us.

I understand that it is sometimes difficult to think such thoughts with Lafitte and the authors and creators he details with such love and generosity. Darth Vader is all around us and seems all powerful. But that difficulty and darkness, I take it, is part of the point. These ideas are not easy to think precisely because they are not yet common ones. We find them odd or anomalous precisely because they are not yet established. Such ideas are forms of consciousness that have not yet become culture. That is why we have to imagine them first. So that we can someday think them, know them, become them. Entertainment serves a purpose here: it is the first step to awakening.

Something genuinely new, something fundamentally other, is trying to emerge here. This is what I have called the Super Story—an emerging mythology that is presently arising in and as us, and particularly through the new imaginaire of science and science fiction. This is why popular culture (or what my colleague Christopher Partridge so elegantly calls "occulture") is so important—this emerging mythology is coming up through the floorboards of American culture, as it were, mostly from California, it turns out. Think Silicon Valley and Hollywood. It certainly cannot emerge from the elite centers of American society (government, military, and higher education), although it also clearly depends on them as well. In truth, the Super Story is a moving back and forth between periphery and center, between fictive margin and real world. Because it still lacks any real cultural authority, however, its stories or mythmaking can only be safely imagined as "fictive" or as "fiction," that is, as "myth" in the simplistic and popular sense as "baseless entertainment."

Ah, but beware!

This is not what historians of religions mean by myth. Nor is it what Luke Lafitte means. They mean a story *so* fundamental, *so* powerful, *so* persuasive, and *so* intricately socially produced (through morality, law, education, commerce, religious belief, even child-rearing, emotion, belief, and, yes, entertainment) that those who live in it do not even know they are living in a story and what are really in fact a set of historically produced assumptions or beliefs. They are like fish swimming in the sea and not knowing what water is.

Do you want to know how such a myth reproduces itself? Just count the number of times you look at a screen each day, how many text messages you send, how many emails you read, how many streaming episodes or entire seasons you watch, how few real books you really read, how seldom you step out of your own beliefs and thoughts and question them, how you are shaped every moment of every day by the daily distractions and superficialities of social media—just twittered away. *That's* the mythology thinking and living you. *That's* the spell spelling you.

For a shorthand, Lafitte calls this entire mythical complex "mechanical-man." Such a mythical complex in turn depends on a myth lodged deeper still, the myth of materialism, that is, the assumed truth that is really a largely unconscious belief that there is only matter, that matter is insentient or "dead," and that this same dead matter somehow produces everything that is, including everything alive—like you.

Mechanical-man and its attending materialisms are *myths* in the deepest and most important sense of that much abused word: a fundamental cultural story that is *so* fundamental that people do not question it, do not realize that it is a story, that it is unnecessary, or, in this particular case, do not realize sufficiently enough that it is *always* human beings creating and guiding the technology that is supposedly going to overtake or overcome them. That is Darth Vader's decapitated head again, which turns out to be Luke's.

What might science and invention yet become?

Us.

Which is also to say: *you*. You might become your own screen and, even more so, your own projector (and were *we* not watching the *Star Wars* scene all along?). You might turn around in the movie theater called life and realize that nothing on the screen in front of you is as real, as stable, or as necessary as you thought. You might finally understand that it is all in fact produced, directed, edited, and performed over many generations, not by one or a few hundred individuals (in the credit lines on the screen), but by a few billion people, few of whom will ever be remembered, much less acknowledged. You might realize that the movie of culture is in fact constructed by historical processes too complex and too many to imagine, but that, finally, it is all projected by a light behind you.

What would happen then? This is the question or potential of this book. Reader beware! Beware of the cave of this book. If you choose to enter it, know that you are entering you. Know that only you can decapitate mechanical-man and realize that the Force is Mind, not machine. Only you can awake from this dark dream and help us all tell another brighter story, a super one.

JEFFREY J. KRIPAL
HOUSTON, TEXAS

Jeffrey J. Kripal is the associate dean of the Faculty and Graduate Programs in the School of the Humanities and the J. Newton Rayzor chair in philosophy and religious thought at Rice University. He is also the associate director of the Center for Theory and Research and the chair of the board at the Esalen Institute in Big Sur, California. Jeff is the author of eight books, including, most recently, *The Flip: Who You Really Are and Why It Matters* (2020), where he envisions the future centrality and urgency of the humanities in conversation with the history of science, the philosophy of mind, and our shared ethical, political, and ecological challenges. He is presently working on a three-volume study of paranormal currents in the sciences, modern esoteric literature, and the hidden history of science fiction for the University of Chicago Press collectively entitled *The Super Story: Science (Fiction) and Some Emergent Mythologies*. There he intuits and writes out a new emerging spectrum of superhumanities (in both senses of that expression). The website jeffreyjkripal.com contains his full body of work.

We are all sleeping avatars of God, with amnesia.

PHILIP K. DICK

Prologue

Man is twice-born—and by *man,* I mean *Homo sapiens, Homo habilis, Homo africanus,* and all the other hominids who were awakened by the Word. He—and I include *She* and any other genders—were awakened by the light, the archetype, the god within. But they were also awakened by the tool. Anthropologists surmise that the development of simple stone tools—pebble grinders and clubs of bone followed by hand-axes and cleavers—took selective pressure off the jaw and allowed the expansion of the cerebral cortex. The light brightened and took on texture, paradox, and personal identity. Tools included the wheel, the sling, the boomerang, and eventually the gear, spring, lever, axle, and axis. Tools gave birth to the weapons of war and the magic of stagecraft, from Vitruvius to William Shakespeare.

I pick up on the American frontier, where the light has become the mechanical-man, the conqueror of wilderness and agent of destiny, as well as the cowboy, sheriff, slaveowner, and snake-oil salesman. In part, my thesis is that the mechanical-man must go on a long-due journey to recover the light, the archetype, and the god within. During his journey, his two births—the birth of God into man and man back into God (material birth and spiritual birth)—confront each other again and again. This conflict appears in literature, in technology, in science, in computer circuits, in artificial intelligence, and in the themes and tropes of a culture that has long forgotten "him" and its origin, or the fact that its existential reality is held in trust by him alone.

This book demonstrates a connection between consciousness and the history of machines in the United States from the antebellum age to the present, specifically between the actual innovators and inventors of these machines and the power necessary to harness and activate our consciousness. The book's primary intention is to show that within the real and fictive history of American techno-supremacy, there lies a hidden quest to discover mankind's new consciousness and identify the powers to activate such an awakening. Accordingly, once Americans accept what our fictional thinking machines tell us, we will accept a new call of questing and individuation. We decide to avoid dualities no longer. Instead, we will lean into aspects such as paradox and return to society with the boon of spirit and service toward others.*

*The opening tropes are a blending of my own thoughts with those of Richard Grossinger, who was my final editor, sounding board, and co-parent of the mechanical-man.

INTRODUCTION

The Call to Remember

"This will begin to make things right."

Actor Max Von Sydow delivers this, the first line in *Star Wars: The Force Awakens.* He then passes something on to the resistance pilot Poe Dameron. It appears to be a computer chip of some sort that contains valuable information over which dueling forces may fight. Later, we discover it is a partial map leading to the whereabouts of the mythic hero, Luke Skywalker, the last Jedi, whom some believe never even existed—a mere fiction just like the Force itself.[1]

The above scene of *Star Wars: The Force Awakens* parallels what this mechanical-man quest on which you are about to embark will do for you and countless others. The mechanical-man, as mythic marker, will awaken you to the American techno-narrative ancillary to the history of harnessing the imagination to manifest objective reality. Once we unravel the materialist worldview that causes us to "project numinous value and meaning onto *things*" and onto our technological creations, we can defragment our subconscious to recognize the underlying nature of things.[2] Insight or, in a depth-psychological sense consciousness, must now expand if we are to break through the wall of reductionism and materialism. The mechanical-man, in the form of multiple archetypes, will imbue us with the tools of self-knowledge, integration, authenticity, and wholeness in order to expand and even change our consciousness to see the cybernetic magic that is the story of techno-manifestation. The history of American science and technology, when viewed through the dual prism of mythology and cybernetics, tells a story so powerful that it intimates a current transition in consciousness that Americans are undergoing at this very moment.*

Fear not, for you will have fail-safe(s) on this quest. We will utilize the term as both an adjective and a noun. Fail-Safe (adjective): causing a piece of machinery or other mechanism to revert to a safe condition in the event of a breakdown or malfunction. Fail Safe (noun): a system or plan that comes into operation in the event of something going wrong or that is there to prevent such an occurrence. If at any point you feel uneasy while reading—use the fail-safe of placing the book down for a while. Let us commence!

Never before have we been so wealthy and dominant as a species, but have our lives ever been as meaningless today? Materialism crushed most of the myths that lent significance to the lives of our ancestors. We've become orphans of meaning. We go on chasing one material goal after the other, as if there were a little bag of magic goodies at the end that would retroactively bestow meaning on the entire enterprise. This is akin to chasing ghosts. What do we live for? Life has turned into a mad scramble for the accumulation of things and status they confer, for the sole sake of leaving it all behind at death.[3]

For the United States and much of Western civilization, the twentieth century was an era of rapid technological advancement marked

---

*Mike Murphy also provides an analogy in *The Creation Frequency:*

> I also like to think of the conscious mind as the top of an iceberg, or what's visible above the surface of the ocean. The subconscious mind is the enormous bulk beneath the waves. The subconscious mind is permeable to the influence of others, society, and the life circumstances we find ourselves in. It's not just a personal set of beliefs, habits, and ideas. In fact, Jung believed we have a "collective unconscious," which might explain some of the deeper currents that drive us—patterns and archetypes that seem to go beyond the scope of our own life experiences. Imagine that the iceberg not only extends deep beneath the waves but connects to a vast subterranean ice continent. (22)

by material abundance. The era prompted novel psychological desires, fears, and truths for American culture, which sought to reflect on these changes through storytelling or myth. Joseph Campbell explains myth as metaphor and mystery: "Mythology is a system that endows the mind and the sentiments with a sense of participation in the field of meaning."* Myths teach us what we are unable to define in a logical and rational manner. To understand what one cannot otherwise comprehend, one must utilize symbolism, universal patterns, and archetypes drawn from the collective unconscious. In his book *The Inner Reaches of Outer Space,* Campbell says:

> For myths and dreams, in this view, are motivated from a single psychological source—namely, the human imagination moved by the conflicting urgencies of the organs of the human body, of which the anatomy has remained pretty much the same since c. 40,000 BC. Accordingly, as the imagery of the psychology of its dreamer, that of

---

*Early on during the writing of the mechanical-man, many argued that neither I nor anyone else for that matter could provide a definition of mythology. In an earlier manuscript, the following definition from Campbell's *Thou Art That* was included:

> Mythology may, in a real sense, be defined as other people's religion. And religion may, in a sense, be understood as a popular misunderstanding of mythology. Mythology is a system of images that endows the mind and the sentiments with a sense of participation in a field of meaning. The different mythologies define the possible meanings of a person's experience in terms of the knowledge of the historical period, as well as the psychological impact of this knowledge diffused through sociological structures on the complex and psychosomatic system known as the human being. . . . These images must point past themselves to the ultimate truth which must be told: that life does not have any one absolutely fixed meaning. The images must point past all meaning given, beyond all definitions and relationships, to that really ineffable mystery that is just the existence, the being of ourselves and of our world. *If we give the mystery an exact meaning we diminish the experience of its real depth.* But when a poet carries the mind into a context of meanings and then pitches it past those, one knows that marvelous rapture that comes from going past all categories of definition. Here we sense the function of metaphor [mythic-marker] that allows us to make a journey we could not otherwise make, past all categories of definition. (8–9)

> mythology is metaphorical of the psychology posture of the people to whom it pertains.[4]

For purposes of this study, we will utilize Campbell's definition of myth and his approach to consciousness of following a myth's development and progression.* Specifically, "myths are clues to the spiritual potentialities of the human life—they teach you that you can turn inward, and you begin to get the message of the symbols."[5] In addition, *myth* is "an organization of symbolic images and narratives, metaphorical of the possibilities of human experience and the fulfillment of a given culture at a given time."[6] With that said, one may argue that any story or mode of expression that communicates a message is a myth. However, to overcome this argument, we will specifically follow Campbell's consciousness formula for tracing the progression of myth. According to Campbell, myths involve a cyclical journey consisting of three main phases: a departure, an initiation, and a return. Utilizing the hero, trickster, and savior archetypes as mythic-markers, we will traverse these stages seeking clues and messages in America's industrial creation myth. This creation myth will center itself on phenomena blurring the lines between human and machine. Therefore, we will draw on these phenomena in analyzing issues of control, communication, and the exchange of information in fictive and real human-machine interactions.

Holding to the hypothesis that scientific revolutions and transitions can be placed within a mythic framework by utilizing the real and fictive, Eric Gould assists by saying,

---

*Whenever you stumble upon a definition you are having trouble with that is not defined in the primary text, look to the notes, and if the notes do not answer your question look for it in Jung's *Man and His Symbols.* When I started writing about the mechanical-man in 2005 I was mainly limited to definitions by Jung and Campbell. As my research progressed, Bernardo Kastrup was extremely helpful, and then finally, when numerous scientists read an earlier draft, they pointed my attention to Dr. Joe Dispenza and Gregg Braden.

> Myths apparently derive their universal significance from the way in which they try to reconstitute an original event or explain some fact about human nature and its worldly or cosmic context. But in doing so, they necessarily refer to some essential meaning which is absent until it appears as a function of interpretation. If there is one persistent belief in this study, it is that there can be no myth without an ontological gap between event and meaning. A myth intends to be an adequate symbolic representation by closing that gap, by aiming to be a tautology. The absent origin, the arbitrary meaning of our place in the world, determines the mythic, at least in the sense that we cannot come up with any definitive origin for our presence here. So what I continually turn to is the fact that myth is both hypothesis and compromise.[7]

Today, for the most part, we use scientific means to explain natural phenomena. Thus, stories reflecting scientific principles in fiction allow us to reflect on this mode of apprehension. In other words, fiction pertaining to scientific discovery allows us to grasp what the utilization of science and technology is doing to our human psyche and unconscious self. Stories of scientific discovery that "concern themselves with origins, with the gap between origins and present—with the human condition, and with humankind's relationship to the cosmos" allow us to reflect on and explain what takes place between events and meaning.[8] When Gould speaks of the ability of myths to explain what takes place in the gaps between events and meaning, he intimates that the real and fictive have reciprocal value in explaining such gaps—we must therefore analyze the specific events in scientific discovery along with the meaning of these events to which social scientists and fictive writers give voice. This analysis seeks to answer the following questions: How do myths we construct about technology and cybernetics make us see manifesting reality in a new or different manner; specifically, how do myths make us see issues of control, communication, and the utilization of information in human-machine interactions differently than merely studying these aspects of technology and cybernetics in a technical manner? In addition, how do these myths explain the gaps between the events and the meanings of such events when performing such analyses as these? The presumption for this analysis is therefore the following: The placement of the fictive and real story of control and communication between humans and machines—within a mythic framework—will make us see these interactions in a different manner than if we had just looked at the events and meaning of these events separately outside the context of a mythic framework.

The task of explaining what Gould calls "gaps" will lie in the reconstruction of the interactions between machines and humans in the fictive and real. In analyzing these gaps, our task is to show the interplay between American "society's practical technology and its intellectual and spiritual culture." This will elucidate the thought patterns and values of such a culture as affected by human-machine interactions. Further, by placing such an analysis in a mythic framework, "a historical process [of human-machine interactions] which is carried out by its agents largely without conscious awareness and hence without commentary [can] be documented and reduced to objective proof."[9]

Utilizing Campbell's definition and formula of myth, we will discover how and why myths affect the way we view and perceive cybernetics. In addition, we will come to understand that without myths to aid us in perceiving technology and cybernetics, we would dismiss this heroic and creative journey, thereby overlooking the symbolic representations of what takes place in the gaps between events and meaning.

Attempts to invent a mechanical-man in both story and reality will act as our mythic-marker in this analysis. This mythic-marker is a hero, trickster, and savior archetype. The term *mechanical-man* should not be taken literally; rather, when we speak of mechanical-man in the fictive or real sciences of technology and cybernetics, we are using it as a symbol or metaphor to analyze human and machine interactions in the context of control and communication. This marker is an industrial creation myth that shows how humans are becoming gods through technology. By following the evolution of the mechanical-man in American history, we see how these gods tell the story of their own creation. My purpose is not merely to demonstrate a historical relationship between the fictive and real via a cross-disciplinary study of the mechanical-man. Rather, it is to show that to understand the changes that take place in society during technological shifts, one must analyze the reality of technology and invention in conjunction with the fictive dreams of speculative writers in terms of technology, cybernetics, and mythology.

It consequently is not my purpose to present a history of the numerous tales of the mechanical-man in fiction or a history of the invention of machines imbued with human attributes. Although I will make statements about how tales of the mechanical-man became a regular literary feature during the late nineteenth century, this analysis is not a history of the mechanical-man in American literature. In addition, while I will show that tales of the mechanical-man coincide with advances in science, I am not attempting simply to prove that "science fiction operates (both prospectively and retrospectively) as the analytic component of scientific discovery and technological innovation."[10]

While the term *mechanical-man* is a catchall phrase for android, robot, machine, automaton, artificial intelligence, and living information, the term is also a meeting ground where extra-dimensional communications between different forms of matter occur. We cannot separate the machine from this middle ground of consciousness and unconsciousness, matter and spirit, dormant and manifest, and the original projection that invents a machine and the subsequent interactions that redefine it. While our human projections onto machines make humans more mechanical, these projections paradoxically allow us to recognize spiritual freedom and service hidden behind the wall of materialism. By becoming mechanical and seeing that our collective definition of perfection is flawed, we become spirits of service and sacrifice. The very term *robot* as "slave" is turned on its head during this transformation. The robot morphs into a spirit of service in teaching us to recognize our enslavement to materialism, thus liberating us to project mind into nature and receiving back the essence of service and agape.

DEFINITIONS

On the issue of control, James Beniger puts forth a multitude of definitions of the term in his book *The Control Revolution: Technological and Economic Origins of the Information Society*. For purposes of this study, the word *control* is utilized as a more general concept to "encompass the entire range from absolute control to the weakest and most probabilistic form, that is, any purposive influence on behavior, however slight."[11] It is not enough merely to say that control equates to the "purposive influence toward a predetermined goal." Dealing with the phenomena of human-machine interactions calls for a broader definition, namely, that there exists a "two-way interaction between the controller and controlled." Of course, in certain situations, this two-way interaction between human and machine, especially when looking at the real interactions rather than the fictive, will appear tenuous at best. Nevertheless, one of the purposes of this study is to analyze how these apparent weak connections lead to strong connections, or "strong control," in human-machine interactions.[12]

Beniger's definition of control also holds:

> Inseparable from the concept of control are the twin activities of information processing and reciprocal communication, complementary factors in any form of control. Information processing is essential to all purposive activity, which is by definition goal directed and must therefore involve the continual comparison of current states to future goals, a basic problem of information processing. So integral to control is the comparison of inputs to stored programs that the word control itself derives from the Medieval Latin verb *"contrarotulare,"* to compare something "against the rolls," the cylinders of paper that served as official records in ancient times.[13]

The idea that control equates to a purposive activity that is by definition goal directed is the fundamental theme of the first discussion in this analysis in the fictive works of Herman Melville, Edgar Allen Poe, and further in this quest, Nathaniel Hawthorne. These writers hold that control is of the utmost importance to know the goals of humans that endeavor to create a mechanical-man. In addition, to know the goals of the controllers, one must also be able to define the interactions between the controlled and controller, since this underlying interaction tells us the purpose of such activity. In addition, the subjectivity of both the controlled and controller is important to these authors and is a fundamental tenet of defining information that is exchanged between human and machine.

Implicit in this definition of control is the communication between controlled and controller. Norbert Wiener articulated this nexus when he coined the term *cybernetics* in 1948. Since Wiener's cardinal concern in his theorizing dealt with human and machine interactions, this study will utilize the definition of communication found in his book *The Human Use of Human Beings: Cybernetics and Society,* published in 1950, which corresponds to and overlaps with Beniger's definition of control. Indeed, on the issue of control, Wiener writes:

> The commands through which we exercise our control over environment are a kind of information which we impart to it. Like any form of information, these commands are subject to disorganization in transit. They generally come through in less coherent fashion and certainly not more coherently than they were sent. In control and communication we are always fighting nature's tendency to degrade the organized and to destroy the meaningful. . . .[14*]

Technology and information are the lynchpins of control and communication. The simplest definition of communication is the exchange of information. Wiener gives the following example: "When

*Later, we will delve deeper into Wiener's biography and his contribution to the quest.

I communicate with another person, I impart a message to him, and when he communicates back with me he returns a related message which contains information primarily accessible to him and not to me."[15]* Cybernetics is therefore the equivalent of what one can do with information and communication.

This analysis will present situations in which the returned message does not appear to be primarily information accessible to the return sender. Nevertheless, the message imparted by the apparent actions of the return sender fits within this definition of communication. For example, early in this analysis, we will look at the hoax of a chess-playing automaton, which is not an automaton or machine at all. It is a man underneath a desk that controls the movements of a mannequin. The import to this analysis is not whether this chess-playing hoax is a machine or automaton; rather, its necessity to this study is the meaning or message that the hoax is able to communicate to those that have

---

*When we speak of communicating archetypes to one another, we must properly define *archetype.* Stephen Hoeller says the following in *Jung and the Lost Gospels:*

> Jung defined an archetype as "a figure—be it as daemon, a human being, or a process—that constantly recurs in the course of history and appears wherever *creative fantasy* is freely expressed." When the human being encounters one of these images, the impact felt is one of intensity and novelty. As Jung expresses it, "It is as though chords in us were struck that had never resounded before, or as though forces whose existence we never suspected were unloosed." Human beings inherently know that the archetypes are autonomous, they obey their own sovereign laws, and that while they are in the nature of internal experiences they reflect themselves onto the screen of external human experience. Archetypes are thus present simultaneously in the internal structures of the human psyche and also in the arena of history. In the course of his life, Jung came to differentiate between the archetype as such and the archetypal image. The archetypes as such, he said, do not reach consciousness, for it is lodged in an inaccessible region of psychic reality, 'the invisible, ultraviolet end of the psychic spectrum.' Archetypal images, on the other hand, regularly manifest to the conscious mind in dreams, visions, imaginative experiences, and altered states of consciousness." (58)

As we will see much later in the quest The Gospel of Philip gives expression to archetypes holding: "Truth did not come into the world naked, but it came to them in types and images. It (the world) will not receive it in any other fashion" (Hoeller, 59).

either heard of such or saw it play. What messages does this hoax communicate to audiences that changes how they view human-machine interactions? The messages returned to the sender in this analysis may not always be patent. Nonetheless, the implicit messages received and procured by the sender by viewing the actions of the sendee is within the context of Wiener's definition of communication and control.

In fact, Wiener addresses this issue by pointing out, "There are detailed differences in messages and in problems of control, not only between a living organism and a machine, but within each narrower class of beings."[16] At the very core of cybernetics is the attempt to develop techniques that ameliorate these gaps and problems of control and communication and to discover the relevancy of these problems to the overall understanding of human-machine interactions. The following example illuminates how the communication process between machines and humans goes through nuanced and complicated stages of transference:

> When I give an order to a machine, the situation is not essentially different from that which arises when I give an order to a person. In other words, as far as my consciousness goes I am aware of the order that has gone out and of the signal of compliance that has come back. To me, personally, the fact that the signal in its intermediate stages has gone through a machine rather than through a person is irrelevant and does not in any case greatly change my relation to the signal.[17]

The reason why Wiener believes that the intermediate stages that a signal goes through are irrelevant has a great deal to do with his definition of information. Generally, he says that "information is a name for the content of what is exchanged with the outer world as we adjust to it, and make our adjustment felt upon it."[18] For purposes of this study, a more precise definition of information is needed in the application of myth to human and machine interactions.

N. Katherine Hayles, in *How We Became Post Human: Virtual Bodies in Cybernetics, Literature, and Informatics,* adds to Wiener's definition of information by showing that the messages exchanged must have a low degree of uncertainty. Hayles gives two examples that aid in this analysis:

> Information is identified with choices that reduce uncertainty, for example when I choose which book, out of eight on a reading list, my seminar will read for the first week of class. To get this information to students, I need some way to transmit it. Information theory treats the communication situation as a system in which a sender encodes a message and sends it as a signal through a channel. At the other end is a receiver, who decodes the signal and reconstitutes the message. Suppose I write my students an email. The computer encodes the message in binary digits and sends a signal corresponding to these digits to the server, which then reconstitutes the message in a form the students can read. At many points noise can intervene.
>
> Uncertainty enters in another sense as well. Although information is often defined as reducing uncertainty, it also depends on uncertainty. Suppose, for example, *Gravity's Rainbow* is the only text on the reading list. The probability that I would choose it is 1. If I send an email telling my students that the text for this week is *Gravity's Rainbow,* they will learn nothing they did not already know, and no information is communicated.[19]

In these examples, Hayles is expressing the axiom that the certainty and uncertainty of events defines information, which is dependent on predictability and unpredictability in the context of pattern and randomness.

This nuanced definition of information is evident in the actions of many of the fictive tales of the mechanical-man. Specifically, the actions of these agents will reveal that information from events is at its best when it fits into a spectrum of opposites, for example, absolute predictability or absolute unpredictability, or patterned or patternless, thus, utter randomness. Hayles incorporates fictional tales of human-machine interaction by looking at the "shift from presence and absence to pattern and randomness" in contemporary literature.[20] This definition delves into aspects related to the subjectivity of the reader and the author of the text. Knowing the subjectivity of both allows one to understand better the information exchanged. For our purposes, information exchanged between fiction and the reader will concern itself chiefly with patterns that coincide with the meaning that social historians have found during their study of the time periods during which such fiction was written. Once more, placing the phenomena of human and machine interaction within a mythic framework will allow us to fill in the gaps between events and meaning in analyzing issues of control, communication, and information. Extrapolating from the patent and latent problems in the interactions between machines and humans allowed cybernetics to enter not only into fields of mathematics, engineering, and electronics, but also into areas of psychiatry, psychotherapy, neurology, quantum physics, and the social sciences.

Finally, we will utilize David F. Channell's vital machine world view to aid us in this analysis of human-machine interactions. In understanding a society's values, the perceived tension between technology and organic life coalesces into "a new bionic view that transcends the earlier thinking"—trapped in either a purely mechanistic or organic view of life. Channell bases the vital machine worldview on a synthesis of the earlier two. On this view he holds:

> This world view is based on a new symbol that I call the vital machine, which emphasizes the role of interactive process or dualistic systems in understanding the world . . . the symbol of the vital machine provides new insights into the relationship between technology and life in light of recent developments in Artificial Intelligence, genetic engineering, and biomedical engineering.[21]

Being placed in this worldview allows our mechanical-man to traverse through liminal stages of evolving developments in areas of human and machine control and communication. Channell's vital machine worldview, later transformed into the vital machine holographic simulation, will allow the mechanical-man to illuminate America's cultural values and relationship to technology through these technological and cybernetic liminal changes. Since we are still stuck in the materialistic paradigm for the first half of the book, for chapters 1–7 you should think of "the organic and mechanical merging into one" as our base definition of *vital machine.* However, once we cross the threshold into idealism, chapter 8 onward, the definition we use is one of combining the spiritual and the physical—the word machine remains the same while the term *vital* now morphs into the very essence of one's consciousness or soul.

I will argue that there is a blurring between the fictive and real in terms of the mechanical-man myth. What this implies is that the exchange of information between humans and machines is so multidimensional that it is no longer possible to distinguish meaningfully between "the biological organism and the information circuits in which the organism is enmeshed."[22] Indeed, throughout the progression of mechanical-man as mythic-marker, this blurring or coupling of human and machine interactions will correspond to shifts and transitions in how the very idea of control, communication, information, and signification is understood and experienced. We will discover that the very idea of the fictive intimates a communication between the conscious and unconscious. In addition, a culture's view and definition of control is how it imagines the fictive into the real and subjectivity into objectivity.

MYTH IN POSTMODERN AMERICA

Cyborg mythologists Rushing and Frentz observe:

> If we are right that postmodern writing represses Spirit into the unconscious, then Spirit seems to be contained in the Other of post-

> modernism just as surely as it was in the sovereign rational subject of modernism. Whereas postmodernism has succeeded in bloodying the reputation of the ego, it is surprisingly similar to modernism in terms of the attempt of each to entrap Spirit within its own lauded domain and to co-opt its power in its struggle against its opposite.[23]*

In an era of technological perfectionism, society appears to lose its spiritual roots. Thus, such a society appears mythless when it replaces spirit with perfectionism, repressing vital forces of social life. Faced with the possibility of self-destruction, we try to reconnect to our roots that have lain dormant for centuries "in the hope that the nourishment from those depths may somehow counterbalance the sterility of the perfect machine."[24]

The archetypes utilized in this analysis evoke imagery of creation and of God as creator within man. The sterility of the perfect machine is counterbalanced by the imperfection of the creator. However, the mechanical-man myth, in both its imagined form through science fiction and its historical form through biography and social history, shows that our myths allow us to see God and spirit in both perfection and imperfection—in the fictive and the real or the unconscious and the conscious. As the machine progresses through transitional phases, so

---

*Rushing and Frentz discuss many of the themes presented in the mechanical-man, specifically, that of lost spirituality and new consciousness:

> In postmodernism the descent of Spirit fuses others into Other as a loose coalition of disparate parts. Moreover, the condensation of spiritual energy, while every bit as partial as that coopted by the modernist ego, functions to invest Other with an aura of its own. While in one sense it is positive that the Other is re[-]conceptualized as powerful because of its assumption of spiritual overtones, it is nonetheless dangerous for Spirit to operate from such a debased state. In myth, when any basic human instinct is repressed, it may condense into or "team up" with other ignored aspects of the human pantheon. The Great Mother, for example, when repressed as a vital force in social life, becomes shadow—something dark, loathsome, terrible, no-I, Other—and she tends to wreak revenge on her unsuspecting oppressor. (Rushing and Frentz *Projecting the Shadow,* 26–27)

too does the American public's collective unconscious. The development of machine into mechanical-man and the myths that accompany such development assists a postmodern society that appears fragmented to find stability.

Myth that crosses the boundary between the fictive and the real, and that can even blur such a distinction, illuminates that we live in a "hyper-real" environment only at the conscious level. Thus, when Jean Baudrillard speaks of societies that "no longer live in an age of production, but of reproduction, where there is no more original or natural or real thing from which copies are made," he is speaking of society only at its conscious level.[25] However, underneath this level is another plane of thought that can be explained through myth. Postmodernism tells us that we are "decentered—dispersed into the margins with no ego, no historically coherent sense of self at the nucleus."[26] However, I argue that underneath this surface, we are more centered now and have more of a sense of self than ever before. The fictive worlds embedded in our unconscious have been overflowing into reality. Thus, the American narrative illuminated in this analysis will show that we are not "condemned to live a perpetual present with which the various moments of a person's past have little connection and for which there is no conceivable future horizon."[27] This narrative shows that while materialism subtly pervades "our expectations, value systems, goals, and nearly every aspect of our lives," the process of individuation will evince that the wall challenging us now spiritually and philosophically is the very system that we pay allegiance to.

Breaking through this wall is part of the quest's initiation phase. If you want left-brain scientific proof and experiments of what the mechanical-man as archetype and mythic-marker tells us simply by means of scientific criteria, you should read Lynn McTaggart's book *Intention Experiment*. We will look at many authors who have experimented in the realm of the imaginal, creating new models of the self and nature.

The mechanical-man will recognize himself as a vital machine acting in a holographic simulation. And yes, there are numerous books that give voice to the current experiments that prove these models of objective and subjective imaginings. But that's not all the mechanical-man quest is about. His cardinal objective is to get us to see how the mythological formula of *quest* in a cyber-techno society elucidates our ability, including the process of imagining this ability, to create or invent more than just the light bulb or space rockets. We invent a new American Dream, no longer based on material riches; rather, one based on ideals. In order to balance out the mechanistic and materialist paradigm separating you from your higher self, we must morph our allegiance from the cowboy mythos to that of the mechanical-man.[28]

Our inventions and tools that we think we depend upon tell us a story of how wrong we are in our perceptions of what it means to be human. The following quest is a process for you to reverse the dead-end path of reductionism, where logic and materialism are the twin saviors of Western society. The end of this analysis will have witnessed a broadening of the reader's consciousness to see potentialities, synchronicities, and the act of creation in new ways. The very heart of intellectual-spiritual development has been under attack for the past fifty-plus years. It is time for modern depth psychology to replace physics as the ultimate purveyor of human truths. It is time that the American Dream equate to recognizing and perceiving the unseen. With the mechanical-man as our hero, trickster, and savior, we can accomplish all these feats and more.

SUMMARY OF CHAPTERS

Chapter 1 explores the relationship of myth, history, and science. It describes how the practice of invention cuts across demographics, creating a unifying story that projects a message to society and causes people to recognize and evaluate their values. This chapter is an introduction to the major myths of America's past: pastoral idealism, a promised land, dominion over nature, manifest destiny, capitalistic materialism, and romantic idealism. It describes how these myths must be transformed to fit the needs of an industrialized nation that now depends on automation, mechanization, and technology. While America's past myths still hold a place in creating a single, unifying story, the role that our dreams, archetypes, and the unconscious play in telling such a story must enter a new phase of adaptation—from the cowboy to the mechanical-man.

Chapter 2 concerns itself primarily with the interpretation of the messages, archetypes, and language of the unconscious in the first popular stories related to the mechanical-man, specifically Edward S. Ellis's *The Steam Man of the Prairies,* Frank Baum's *The Wonderful World of Oz,* and Ambrose Bierce's "Moxon's Master." In these stories, we find Americans coming to terms with the country's transformation from an agrarian household economy to one based on management, efficiency, technology, and industry. In order to understand these changes, our myths also morph. Since the nation now enters the departure stage of myth, which allows people to leave behind the myths of old and the life they once knew, society must imagine new heroes that represent technology and its promise.

Chapter 3 explores how creative forces at work in human reality become woven into myth to give it a coherent meaning for society on a conscious level. This chapter defines the role that dreams, archetypes, images, and the unconscious play within the framework of human reality and the practical sciences. The discussion answers such questions as the following: Why is Nikola Tesla called a sorcerer? Why is there a blending of the fictive with reality, and what does this tell us about our past myths and how we should interpret our new dreams and images? The chapter also introduces us to Lisa Nocks and her assertion that, "The boundary between speculative fiction and technological activity is porous." The fields of science and technology depend on myth and the fictive for inspiration. We will also examine social climate, history, politics, economics, and geography as contributing factors. Overall, the study portrays science and technology as shifting from a practice of independent inventors to a collective endeavor controlled by corporate and militaristic goals. The inventors themselves, not the mechanical-man as in Chapter 2, thus become our mythologized figures. The wizard and the sorcerer, Thomas Edison and Nikola Tesla, create new ambiguities about the conscious and the unconscious with their inventions.

Chapter 4 analyzes how myth changes and adapts to the times by examining the technological carnage of World War I and the technological sublimity of massive technological achievements during and after the Great Depression. In this chapter, a new myth, conveyed through science fiction, serves as a bridge to the old myths of America's founding. Here we recognize that America itself and its very creation constitute a myth of rebirth and redemption for the human race. This chapter shows how closely related science fiction is to the reality of science and technology, illuminating a symbiotic relationship between those who practice science and those who prognosticate about it. Here as well, we begin to recognize the importance of this study's arrangement, the format of which alternates between chapters on those who practice science and those who write speculative fiction thereon. However, the reader now begins to realize that this is merely a tactic to show how myth allows one to come to terms with and reconcile contradictions in human reality.

Chapters 5 and 6 return to the practical sciences by focusing on Grace Hopper and Richard Feynman; I present their contributions to advancements in robotics, computer technology, and artificial intelligence in detail. The narrative of myth in their biographies allows us to consider these pioneers as archetypes. For example, Hopper acts as our Queen Goddess, endowing the robot or computer with the tools it needs to learn and even teach. On the unconscious level she also acts as the Great Mother, by showing us that in the fields of technology and science the female form is as valuable as the male's. In the life of Feynman, we see a return to myths of the past that recognizes their importance in helping us understand our present narrative and ideologies. The practice of science and technology has now reached a point where it has equaled and, in certain areas, surpassed the fictive.

Chapter 7 addresses the research questions and clarifies why technology currently has surpassed our ability to process its functionality. The computer HAL in the film *2001: A Space Odyssey* is the embodiment and purest representation of the new American myth in terms of how we communicate our cultural values, beliefs, and ideals to our conscious selves. HAL allows us to reconcile the numerous dichotomies and ambiguities confronting society and imbues us with the foresight of a new worldview upon the horizon, namely, the vital machine in a holographic simulation. Thus, this chapter sets the stage for, by far, the most important chapters of this analysis—chapters 8, 9, 10, and 11. Only a fictive mechanical-man such as HAL has the gnosis to get the reader ready for what they are about to read at the end of, now, the reader's own questing from the seen to the unseen!

Chapter 8, which is prefaced by definitions of key terms (implicate and explicate order, fragmentation, holographic model, and cosmic consciousness), discusses the fragmentation stage of the mechanical-man's quest. It begins by investigating Philip K. Dick's and David Bohm's contributions to the "conceptual framework of control and communication" through consideration of their biographies, influences of the mechanical-man, and ideas. Before moving into a sustained close reading of Dick's "The Electric Ant," the chapter covers the quest's refusal stage and the trickster's role, integrating Bohm's theories with those of Joseph Campbell, David Channell, and Michael Talbot. These theories also contribute to the chapter's evaluation of "Electric Ant" protagonist Garson Poole as an exemplar of the questing mechanical-man/trickster. Poole is a unique mechanical-man because he can interpret and "program" fragmented reality; moreover, he is an example of "the vital machine [as. . .] the mechanical-man in fragmented form. . . put back together again as pure knowledge." Such an investigation of Poole paves the way for an interrogation of "illusion" and the holographic simulation model. The chapter concludes by showing, through the figure of Poole, how the mechanical-man's quest moves from fragmentation to defragmentation and results in the ability to distinguish between realities. Ultimately, the mechanical-man completes his quest through negotiating implicate and explicate reality while embodying and enabling inter-dimensional communication and extra-dimensional control.

Chapter 9 argues, through its interpretation of the Dagobah cave section in the film *The Empire Strikes Back,* that Yoda, Luke Skywalker, and Darth Vader (Anakin Skywalker)—as archetypes, human-technology hybrids, and mechanical-men—embody the concepts of information, communication, and control respectively. An initial comparison of the Dagobah cave to Plato's cave shows how *Star Wars* reveals the solution to conquering reality's limits, while also stressing how the quest undertaken by the mechanical-man is always accompanied by the dangers of traveling between implicate and explicate realities. The chapter then characterizes the experiences of the creator of *Star Wars,* George Lucas, to explore "how one returns from the implicate with the knowledge one has gained," contending that the filmmaker does "return" successfully from implicate-explicate travel. Utilizing Vader and Luke as examples, the argument expands the mechanical-man quest metaphor to all humans, claiming that "each quest is one of finding the author of oneself in the implicate." Then, remaining within the *Star Wars* context, the chapter discusses the quest's impact on information, communication, and control, exploring how Lucas's characters rely on the Force to navigate and distinguish the implicate from explicate within their holographic reality. It concludes by reasserting the significance of self-knowledge and the idea that "man must recognize his status as a vital machine acting within a holographic simulation" in order to "author his. . . life."

Chapter 10 examines the call to a quest that Americans have been hearing since the 1960s, when HAL awakened to find his fail-safe being activated by Dave. This call has been ignored, given lip service, and even mocked. Considering science fiction in the popular imagination since the time of *Logan's Run* through the time of the plague of 2020, in the twenty-first century, we find popular culture, philosophy, and psychology merging in a series of books that gather together articles, mainly for fans of individual programs. Regarding the mechanical-man, these books are as important as the stories discussed previously. We uncover the ways in which these individual authors dissect *Star Trek*'s Spock, Data, Q, and the Borg to find the mechanical-man awakening to his cosmic consciousness. We do the same with *The Matrix* and *Iron Man*. In this penultimate chapter, we discover our new model and paradigm of existence in changes to the implicate and explicate forms of Neo as well as in Tony Stark's alcoholism and the use of paradoxes in *Iron Man,* thereby allowing for the transformation of consciousness from the material realm to the idealistic realm. The aim of this transformation is to manifest objective reality through the emotions of gratitude, agape, diversity, compassion, and empathy. The result is a better understanding of fail-safes for thinking machines, including humans, and of what thinking machines have been telling us since HAL, namely, that we are not mechanical men or even vital machines but mechanical gods with amnesia regarding our cosmic and divine consciousness. In addition, we hear exactly what these sci-fi films and books have been telling us over and over for the past 50 years, namely, to wake up to our ideals.

Chapter 11 presents a formula of symbols to help explain a new model and paradigm of existence—the vital machine existing in a holographic simulation. This chapter explores the biographies of Norbert Wiener and Steve Jobs to characterize the liberation of the human spirit from the machine, claiming that our mythology and purpose in life has been fragmented and usurped by our technology and subsequent malaise. Utilizing Wiener's and Jobs's fears of technology usurping human liberalism, the metaphor of the mechanical-man becomes myth itself. It is now the job of the hero to liberate the mechanical-man so as to liberate our mythology and meaning in life. This New Age myth meets techno-creation genesis, bequeathing us three paradoxes: agape, solipsism, and holography. These paradoxes are what the mechanical-man quest leads to when drawn out to its final conclusions: the mechanical-man myth is one of selfish cooperation, selfish service, and shared autonomy, recognizing that one is merely a fragment of the whole and simultaneously God with amnesia. It concludes by asserting that Gnosticism, specifically, the lost gospels found at Nag Hammadi Egypt in 1945, explains the mechanical-man quest and imbues us with the ability to take back our myth and spirit from the machine by acting out these paradoxes. In addition, Gnosticism reasserts the importance of combining disparate elements of religion and science, the real and the fictive, organic and mechanical—into a religiosity of self-awareness and self-knowledge. It is a dormant precursor to everything that makes the American experiment great: rebellion, protest, and a self-awakening. Gnosticism is the lost blueprint for our future merger with machines leading us to form immortal vital machines or cyborgs. Indeed, the very discovery of the Nag Hammadi scriptures shows a divine synchronicity of which we are only now becoming conscious, recognizing our psyches' need for such a finding from our entry into the atomic age. The chapter ends with Elon Musk, perhaps our ultimate savior in this techno-quest. If Musk is, in fact, our deliverer, medicine man, and savior, the one to thrust us to Mars and simultaneously into a new consciousness, we may ask whether the mechanical-man foretold to us his coming and whether Musk and his inventions proclaim that the singularity is here.

In the end, our task will allow us to reconstruct a period of interaction "between a society's practical technology and its intellectual and spiritual culture."[29] Analyzing issues of control, communication, and the exchange of information involved in human-machine interactions within a mythic framework will illuminate the values and thought patterns in American society affected by this interaction. In addition, we will recognize the precise processes and channels through which this human-machine interaction takes place. While this may appear to be a historical process "carried out by its agents largely without conscious awareness," such awareness will be reduced to objective proof by utilizing the real and the fictive within the mythic framework of these interactions.[30]

We will discover that there is only consciousness. Every machine, android, robot, and cyborg will lead us to this conclusion. Every eye that opens, whether that of a human or a robot, does so only because of consciousness. As Richard Grossinger writes in his book *Bottoming Out the Universe: Why There Is Something Rather than Nothing,*

> Physics and neuroscience are the wrong tools for locating the source of either consciousness or matter. . . . Finally there is only consciousness—conscious consciousness and unconscious consciousness. The universe is not only a trillion trillion trillion atoms but also a trillion trillion trillion eyes opening like spiders, and they are the same eyes. You can no more shut it than you can extinguish a biblical burning bush.[31]

Interactions in the form of control, communication, and the exchange of information between humans and machines show us exactly what Richard is talking about: animate and inanimate matter emerge from the same source. And, since everything is consciousness, everything is conscious. The differing levels of consciousness and perceptions of subjectivity are what make one conscious source feel separate from another. Since we are all fragments of consciousness acting as a source that is both outside of and within time, time is harnessed variously depending on the conscious level that each of us inhabits. None of this would be possible without space and time. The surprise thing in all of this is that our interactions with machines are leading us to remember that we are all service and spirit. Moreover, no matter what conscious level of dimensionality one is experiencing, service and spirit remain the ultimate boon of the quest to live the archetypal lives of the hero, savior, and trickster by projecting mind into nature.

The Christian mystic Neville Goddard prepares us for this journey by saying:

> HOWEVER MUCH you seem to be living in a material world, you are actually living in a world of imagination. The outer, physical events of life are the fruit of forgotten blossom-times—results of previous and usually forgotten states of consciousness. They are ends running true to oft-times forgotten imaginative origins.[32]

1
Origins of the Mechanical-Man

The American myth of a new beginning amid rural areas insulated from the dire consequences of industrialization creates a "regenerative power which is located in the natural terrain: access to undefiled, bountiful, sublime Nature, which accounts for virtue and special good fortunes for Americans."[1] This promised land would allow humankind to reach its foreordained potential. In effect, America was the New Jerusalem where the ideals of the Enlightenment and Christianity could germinate, unencumbered by Old World corruptions and outdated scientific models of explanation. Justified by this mythos, Americans were free to harness natural resources for their advancement. Indeed, these chosen people were free to create whatever corresponded to their ideas of conquest and destiny.

D. H. Lawrence wrote, "The most idealistic nations invent most machines. America teems with mechanical inventions, because nobody in America ever wants to do anything. They are idealists. Let a machine do the thing."[2] When the machine entered the pastoral Garden of the American landscape, the populace viewed its advent as either an enhancement or a defilement of the virginal frontier. Ostensibly, it was a contravening force but, viewed more closely, a magnificent harmony of opposites: individual/collective, superstitious/rational, independent/

dependent, agrarian/industrial, sacred/secular, realist/idealist.

The machine brought the American myth of a promised land and chosen people to new heights. There was now a place, a magical garden, replete with rivers and streams, arable land, and conducive weather, where the machine could act as a catalyst to help Americans realize their destiny of becoming a shining beacon of hope. (Though it must be acknowledged that unfortunately this also resulted in forced resettlement and violence against Native Americans already living on the land.) Moreover, the machine allowed Americans to realize that the tools we invent enable us to move beyond our own humanness, accessing sources of power that for centuries were thought to be held only by the gods, forgotten by a sense of amnesia to the collective consciousness.

The machine acts as our first cultural symbol in terms of finding a specific narrative related to technology in America. Carl Gustav Jung says, "What we call a symbol is a term, a name, or even a picture that may be familiar in daily life, yet that possesses specific connotations in addition to its conventional and obvious meaning." Indeed, the symbolic meaning of machinery and industrialization itself "implies something vague, unknown, or hidden from us." Leo Marx in *The Machine in the Garden: Technology and the Pastoral Ideal in America* and David F. Channell in *The Vital Machine: A Study of Technology and Organic Life* both refer to the machine as a "cultural symbol" that conveys "a special meaning to a large number of those who share the culture."[3]

Whereas Marx and Channell discuss the importance of symbols' ability to transform our perceptions of the world, this analysis will show how symbols reinforce the mythic dimensions of America as the Chosen Nation, Christian Nation, Millennial Nation, Innocent Nation, Pragmatic Nation, and Idealist Nation, while simultaneously making Americans take inventory of their flaws and shortcomings. The machine, in its real and fictive form, quickened our conscious awareness of a collective destiny replete with optimism and anxiety. With the emergence of machinery and automation in America, our collective narrative had to be reevaluated and reinterpreted. Our very concepts of control, messages, and the flow of information—placed in Channell's vital machine worldview, where the mechanical and organic models are merged—necessitated new definitions, forms, and patterns of development for the American mythos.

A YANKEE AUTOMATON

America in 1827, though still an agrarian society, was nonetheless showing signs of mechanization and automation. When the Walker brothers, New England inventors, first saw a chess-playing automaton, which was operated by a man beneath a desk in New York City, their first thoughts pertained to the practical: How could they build a replica? In that ambition the Walkers were displaying the blind idealism of a country that believed it had a destiny in conquering its limitations. The two brothers saw no limits to what could be copied or built, no matter how complex it might appear. Therefore, the Walkers typify an American public yearning to create automatons and machines that seem to be operated by unknown and even mysterious powers—they sought to harness and control forces of nature in their pursuit of profiteering within the realm of human-machine interactions.

The Walker brothers copied the automaton-hoax from its earlier creator, Johann Maelzel, whose exhibition had appeared a year earlier in New York City at the National Hotel. New Yorkers were fascinated by the "celebrated and only automaton chess-player in the world."[4*] A

---

*Maelzel purchased the original automaton, the "Turk," from Wolfgang Von Kempelen, its original creator. Maelzel was the first to bring the automaton to America, and the Walker brothers were the first Americans to build a replica of the original. For a more thorough discussion of Kempelen, Maelzel, and the Walker brothers, see Tom Standage, *The Turk*. Interestingly, Standage ends his book by stating:

> In any case, it is clear that in 1769 Kempelen had conjectured that playing chess and holding conversations were the two activities that most readily indicated intelligence. Nearly 200 years later, the computer scientists of the twentieth century came to exactly the same conclusion. It is ironic that the Turing test relies on concealment and deception: on machines trying to act like people, and

writer at the time observed that "nothing of a similar nature . . . has ever been seen in this city, which will bear the smallest comparison with it."[5] What was so intriguing about a chess-playing hoax? This was the vexing question on the minds of many. We will find that we learn as much from the artificial as we do from the real, if not more.

Maelzel had displayed his automaton-hoax around the world for years prior to its arrival to America. Consequently, he surmised that American audiences would be similar to others in viewing the machine with astonishment. To his surprise, however, American audiences reacted quite differently from their European counterparts, who were amazed by the contraption but said nothing about its mechanics. Maelzel later recalled, "I had not been long in your country before a Yankee came to see me and said, 'Mr. Maelzel, would you like another thing like that? I can make you one for five hundred dollars.' I laughed at the proposition. A few months afterwards the same Yankee came to me again, and this time he said, 'Mr. Maelzel, would you like to buy another thing like that? I have one ready made for you.'" The Walker brothers represented a practical society, one that looked upon the automaton not as a spectacle but as a tool or instrument to achieve certain goals—the cardinal goal for them consisting of making money.[6]

The American audience did not regard the chess-playing automaton as a puzzle; instead, they "agreed that human agency played a role in its movements."[7] The Yankee reinventors saw it as an opportunity to build a replica of the machine. Like Oliver Evans (1755–1819) in developing an automated conveyer belt, the Walker brothers were opportunistic entrepreneurs. They would fashion a machine that made the same movements

---

people acting like machines. Kempelen would surely have been surprised to discover that, in a sense, little has changed since he unveiled his mechanical Turk. He regarded the automaton as an amusement and tried hard to prevent it from upstaging his other, more serious, accomplishments. But by embodying Kempelen's pioneering insight into the curious relationship between technology and trickery, intelligence and illusion, the Turk proved to be his greatest and most farsighted creation. [This is why] the automaton has had the last laugh after all. (246–47)

and projected the same façade as the original chess-player. This is the difference that Maelzel recognized in America as compared to the rest of the world. It reflected the adage, *If I can conceive it, I can build it.* So perhaps Lawrence was right when he wrote that America was a land of idealists who created machines to do their work. Journalist Joshua Foer believes the difference rests in how Americans and Europeans differ on the value of human memory. He says that "maybe, as one European soberly suggested to me, Americans have impoverished memories because we are preoccupied with the future, while folks on the other side of the Atlantic are more concerned with the past."[8] For Americans, anything was possible, even building a replica of a chess-playing automaton-hoax that the majority of Europeans thought was real.

For the Walkers, however, the problem with Maelzel's machine concerned the device's apparent "willful manner. . . . It had to be regulated by a person; humans could mind their own movements, but machines could not."[9] At least that was thought to be the difference between machines and humans prior to the Walkers' invention. Machines were not seen as willful creatures: they were neither spontaneous nor unpredictable; they could not make their own choices. Like those of a clock, their movements were precise and controlled. But America was becoming a land attuned to more complex processes like those of Oliver Evans's mills and Eli Whitney's interchangeable parts. Moreover, "these new machines and new productive arrangements were thought by many to blur the self-regulatory difference between humans and machines."[10]* The dynamics of these machines,

---

*On the steam engine, in *Mind the Machine,* Stephen Rice holds:

> The steam engine thus added to the notion of the machine a sense of power, vitality, and almost willful autonomy. Even textile machines, so captivating in their unwavering devotion to a set course, seemed small in comparison to the steam engine. . . . The steam engine also conveyed a new sense of power by displaying the human triumph over the forces of nature. Antebellum writers and lectures frequently described advances in machinery—for production, for transportation, and for communication—as akin to harnessing nature. (18, 20)

Harnessing the powers of nature is a theme threaded throughout this entire quest.

when combined in certain ways, produced a novel view of man's relationship to machines, and suggested to Americans that machines were similar to our own human makeup.

This blurring was articulated by John Howe in describing pin-making machines: "As one pair of forceps hands the pin along to its neighbor, it is difficult to believe the machine is not an intelligent being." Oliver Evans himself saw in the self-regulation of his conveyor-belt mill a sense of order not possessed even by some people. The machine was not a slave to passions but quintessentially rational in its operation. The great orator Daniel Webster (1782–1852) echoed this sentiment when he said, "Machinery is made to perform what has formerly been the toil of human hands with a degree of power to which no number of human arms is equal, and with such precision and exactness as almost to suggest the notion of reason and intelligence in the machines themselves." The machine, during a time of political uncertainty prior to the Civil War, gave Americans a sense of security. Thus, their relationship with the machine, at least in part, derived from insecurity during such chaotic and transformative years.[11]

The process of mechanization's appearing to be rational and the performance of the chess-playing automaton's appearing to be real created a "sublime transfer." Leo Marx identifies this appearance as the "technological sublime," which taps into our collective unconscious and causes an awe-inspired reaction to "wondrous" technological inventions. While it is easy to debunk Maelzel's chess-player as a hoax and prove that a machine is not capable of rational thought, the wonder that accompanies a machine's operation is not thereby diminished. Maelzel's chess-player hoax acted as a bridge between technological realism and technological sublimity. "That is, while [the chess-player] was exposed as a kind of mimetic fraud, the erstwhile authority of a fraudulent object was ultimately incorporated into . . . the authority of a text that has been reconceived as a technical device in its own right."[12] In the bourgeoning sense of human qualities in the actions of machines, we see a society "falling from innocence to experience." This motif in mythic narratives "focuses upon the change from a state of innocence to a state of sophistication in which the protagonist has learned from his experience that life is not the Edenic world he assumed it to be."[13] The machine, like the country it found itself in, was following the same mythic pattern from innocence to sophistication. Like a child, the machine as mythic-marker began its journey by questioning its environment and the verities of old.

The chess-player hoax is a cardinal example of technological inventions and hoaxes that were fueling the imaginations of artists and writers of the time—who now had to address the psychological changes that accompanied these worldview shifts. Following mythic patterns and using the symbol of mechanical-man as a marker, American writers would respond to these changes by crafting an industrial creation myth. Humans now had the ability to obtain heretofore godlike attributes through science and technology.

The chess-player hoax is of particular importance to this study in the messages the public receives from the actions of the automaton-hoax. It was a catalyst in getting the public to begin thinking about human-machine interactions on a new level, which is evidenced in the fictive tales of the popularized American writers of the time. This new level of understanding germinated ideas of uncertainty related to controlling the actions of machines. In addition, numerous questions were now, like weeds, taking shape in the garden. Specifically: If the chess-player was real and not a hoax, who is in control of the machine's actions? Why does an automaton-hoax possess value in the cultural messages it sends when it is in fact known to be a hoax? If it was possible for mechanical processes to appear to have human qualities of rationality and rhythm, the chess-playing hoax furthered this fantasy. The hoax prepared the public for the first stage in our myth of human-machine interactions: specifically, sending the message that humans and machines are not separate entities; rather, both are projections of mind.

POE'S MISSING LINK

Edgar Allan Poe is the first American author who pronounced Maelzel's automated chess-player simply a "spectacle, a complex visual trick, an elaborate casing with nearly nothing inside, save for space in which a human chess-player hides. No machine, it is just an entertainment whose key is the impressive fantasy it creates. . . ."[14]

In his short story titled "The Man That Was Used Up," published in 1838, he proposed that "in this new age our sense of [the] human is effectively a fiction fashioned by circumstance, a trick our world plays upon us." Thus "we too can be psychologically constructed by the fashions and enthusiasms of the day, and [by] how the wonders of the age hold in their imitative powers a certain capacity for deception." Nevertheless, this susceptibility to deception could be utilized to invent the imagined and seemingly impossible. Superstition, deception, and illusion are not science per se; nevertheless, they could be utilized to explain the world and how it is ordered similar to myth.[15]

Poe begins "The Man That Was Used Up" by recounting a mystery of appearance: "I cannot remember when or where I first made the acquaintance of that truly fine-looking fellow, Brevet Brigadier-General John A. B. C. Smith. Someone did introduce me to the gentleman, I am sure, at some public meeting, I know very well, held about something of greater importance, no doubt, at some place or other, I feel convinced, whose name I have unaccountably forgotten." Even in the story's beginning there is an element of mystery about how the first-person narrator came to meet the General. Nevertheless, the person remembers that the General was "remarkable," recording that "this is but a feeble term to express my full meaning about the entire individuality of the personage in question. . . . There was an air distingué pervading the whole man, which spoke of high breeding, and hinted at high birth." In short, the question one asks when reading Poe's description of the General is whether this man is really human.[16]

When Poe's narrator attempts to ferret out this man's origin and identity, the General says:

> There is nothing at all like it[;] we are a wonderful people and live in a wonderful age. Parachutes and railroads, man-traps and spring guns. Our steam boats are upon every sea, and the Nassau balloons packet is about to run regular trips between London and Timbuktu. And who shall calculate the immense influence upon social life, upon arts, upon commerce, upon literature, which will be the immediate result of the great principles of electromagnetism! Nor, is this all, let me assure you! There is really no end to the march of invention. . . . I say the most useful, the most truly useful mechanical contrivances are daily springing up like mushrooms.[17]

The narrator then speaks to those who once knew the General, but they all acclaim him in almost the exact same words. Behind this seemingly scripted litany of attestations the attentive reader begins to suspect an automaton. Here we recognize the importance of Poe's story, wherein we glimpse an apparent merger of machine and human, or at least some mythically anthropoid variant of the latter category.

When Poe's narrator meets again with the General at the story's end, he discovers that John A. B. C. Smith is not what he appears to be:

> It was early when I called, and the General was dressing, but I pleaded urgent business, and was shown at once into his bedroom by an old [black] valet, who remained in attendance during my visit. . . . There was a large bundle of something which lay close by my feet on the floor, and, as I was not in the best humor in the world, I gave it a kick out of the way.[18]

The bundle that the narrator kicks is the General, a mechanical automaton that emits "the funniest little voices, between a squeak and a whistle."[19] As it begins to morph back into the General, Poe leaves the reader with several unanswered questions. Was the General once a man, or is he merely an automaton? Are the people whom Poe's narrator meets during his attempt to discover the truth about the General's origins merely automatons repeating the same scripted *récit?*

According to Joan Tyler Mead's analysis of "The Man That Was Used Up," Poe demonstrates that the myths of old once utilized to incite American society to their predestined role in the affairs of humankind were now seen as artificial or "used up."[20] Similarly, David Ketterer in *The Rationale of Deception in Poe* acknowledges that Poe's tale is about the efficacy of deception, but he also says that our myths must have new heroes who deal with technological and scientific changes during the Industrial Revolution. Such is Poe's fictional exploration of a mechanical-man. Jonathan Elmer finds a pattern of repetition "pointing to the ideological construction of the very idea of man" in a post-Jacksonian America.[21] For Poe and the Walker brothers, this myth begins with a new hero—a "self-made" man—who crosses the threshold between reality and deception. On aspects of control and communication between humans and machines, appearances and deceptions are of value in the messages humans perceive from such pertaining to the potentialities of human-machine interactions and creativity.

Poe wants one to question the appearance of all experiences and discover its origins. By unlocking the origins of deceptive experience, one can unlock the simple laws that go into creating the actual thing itself. This is how the Walker brothers approached creating a replica of the chess-player. Further, Poe never explicitly says the General is an automaton; rather, the human aspects of the General were "used up" causing him to utilize invention/prosthetic limbs to create or bring life to the lifeless. In this respect, the General and those that invented his limbs were tricking others into believing that nature had created a perfect specimen.[22]

Nevertheless, it was the simple act of tricking others into believing that the General was a perfect specimen that caused those that were tricked to invent the actual thing itself. Poe is setting up the American public to invent the very things that they are tricked into believing are true or capable of invention. In this way, superstition, illusion, deception, and escapism can be utilized as tools to incite one to build what may appear impossible or frightening. These ideas are important to the evolution of the mechanical-man in that they cause those that practice science to not discard the ideas of imaginaries. They set up the American psyche in the realm of science to include what others may consider unscientific in their theorizing. In addition, these tools make us look at the value of information and communication exchanged between machines and humans in new ways. Since the messages communicated between machines and humans are not always patent, we must look at how the fictive and real interact and push each other forward in coming to terms with how human-machine interactions are affecting our psychological desires and fears. In the actions of the Walker brothers and the writings of Poe we begin to understand how the fictive and real interact, which is at the very heart of Joseph Campbell's formula of myth—taking one on a ritualized journey of inner and outer discovery.

MELVILLE'S SLAVE

It was not a chess-playing hoax that caused Herman Melville to imagine a mechanical-man; rather, it was the need for "some decent and orderly relationship between him and his God." Charles Fenton in "'The Bell-Tower': Melville and Technology" discusses why Melville's mechanical-man represents some of the same hidden fears and excitement that Poe saw in the American unconscious. Not simply one invention, in Fenton's view, impacts Melville's imagination regarding technological change. What influenced the author instead are the underlying processes and methods that were accepted as bringing about such inventions.[23]

The assembly-line process and the monotonous rhythms of the factory led Melville to opine that technology causes man to discard myth and storytelling, placing an emphasis on science as the sole mode of understanding. Fenton is correct when he argues that Melville viewed advances in science as impoverishing what it means to be human. Nineteenth-century faith in science caused Americans to hide behind an illusion of truth. In the *Piazza Tales,* written in 1856, Melville looks upon illusion as a venal practice that is akin to religious superstition. In his eyes, illusion causes man to become blinded by his own vanity and ignorance. Thus, by looking at the world through the prism of illusion and superstition, man becomes a slave to illusion, a slave to the forces of nature and the status quo.[24]

In his short story "The Bell-Tower" Bannadonna is building a tower to challenge the heavens, as in the biblical account of the Tower of Babel. Bannadonna wanted "to solve nature, to steal into her, to intrigue beyond her, to procure someone else to bind her to his hand, . . . to rival her, outstrip her, and rule her."[25] In short, Bannadonna sought to usurp God, to utilize nature for his own bidding, allowing his human apparatus to be guided by the hands of a new slave—a fully functional automaton.

In Melville's story, the public soon realizes, as they watch the tower's construction, that the inventor appears to be engaged in the creation of something beyond the mere casting of bells. Indeed, "those who thought they had further insight would shake their heads, with hints, that not for nothing did the 'mechanician' keep so secret. Meantime, his seclusion failed not to invest his work with more or less of that sort of mystery pertaining to the forbidden."[26] Melville thus foreshadows the coming age of independent inventors such as Nikola Tesla and Thomas Edison who harnessed the powers of nature in the isolation of their laboratories. In the same way that Bannadonna's solitude caused him to experiment with nature and attempt to control natural forces for his own creative purposes, so too would self-made American inventors like Edison and Tesla—utilizing the solitude of the laboratory—distance themselves from the whimsical desires and fears of the public who may have looked upon their work as acts of sorcery.

In Melville's work, Marvin Fisher identifies two dilemmas

plaguing nineteenth-century America: "unquestioning faith in technological progress and man's domination of the processes of life and enslavement of lesser beings."[27] However, Melville's vision extends beyond his own time, for Bannadonna's automaton, which acts as a surrogate to ring the bell, is not merely a slave or replica of his master but has its own identity. In its emphasis on pragmatic instrumentalism, the Civil War provided a context for complex scientific thought, a necessary element for establishing the idea of the mechanical-man in Americans' collective imagination. And war itself was now seen as a way to advance the creation of machines. Historian Robert Bruce writes:

> The national pride that supported wars to win or keep territory also spurred scientists and gave them a claim to public support. Their ambitions for American Science had overtones of Manifest Destiny and the War for the Union. Territorial expansion drew them back for a time toward descriptive natural history and gave them government employment. The strengthening of democratic ideals and representative government forced American scientists to cultivate the general public.[28]

In analyzing Melville's view of technology, specifically how technology can be utilized by humans to bring about dreadful ends, Bruce notes, "American editions of both *Frankenstein* and *Faust* came out in the 1840s, and those motifs haunted Melville. The forces released by technology evoked in him a sense of the mystical or supernatural, an instinctive resort to satanic imagery, an echoing of folklore and classical mythology in their dread of the knowledge that outruns wisdom and challenges divinity."[29] Here, though, we glimpse a contradiction in Melville's philosophy. How can Melville say that scientific technology has the ability to create the same dread as religious superstition? Bannadonna is much more than a mere "mechanician" or conflation of a mechanical engineer and magician. Of his actions Melville writes:

> [H]e had not concluded, with the visionaries among the metaphysicians, that between the finer mechanic forces and the ruder animal vitality some germ of correspondence might prove discoverable. As little did his scheme partake of the enthusiasm of some natural philosophers, who hoped, by physiological and chemical inductions, to arrive at knowledge of the source of life, and so qualify themselves to manufacture and improve upon it. Much less had he aught in common with the tribe of alchemists, who sought, by a species of incantations, to evoke some surprising vitality from the laboratory. Neither had he imagined with certain sanguine theosophists, that, by faithful adoration of the Highest, heard-of powers would be vouchsafed by man.[30]

Melville thus is distinguishing true science from pseudoscience. He says that Bannadonna is a "practical materialist" for whom "common-sense was theurgy; machinery, miracle."[31] This statement places Bannadonna in the realm of both pure and instrumental pragmatic science. How, then, do we assess Melville's overall view of technology? Was he averse to all the sciences? In technology Melville saw an extension of humanity. It was important for him to distinguish between true science and pseudoscience to illuminate the utility and value in both explanatory spectrums. In addition, since technology is an extension of humanity, it too has characteristics that are plagued by subjectivity. Science and invention, like all human endeavors, are not bereft of bias and prejudice. Melville's task for society is that they recognize these biases and prejudices in their modes of explanation. In short, it does not matter to Melville whether Bannadonna was a pragmatic materialist or a metaphysician. By knowing Bannadonna's goals, we are able to determine the worthiness or unworthiness of his creation. The sciences or methods utilized in inventing such an automaton are superfluous. Melville consequently foreshadowed an American technology predicated on the usurpation of power, which anticipates corporate and militaristic goals dictated by the rhythms of capitalism. Melville's concern is the issue of control and the motivations behind one's yearning to control. As the country begins to show signs of emphasizing military and economic strength abroad in the years that follow Melville's writing, it follows that those that control technology will have goals predicated on capitalism and militaristic superiority, where the ends justify the means. In the inventions of Thomas Edison and Nikola Tesla we will see how these issues play out—with Tesla, in the end, refusing this techno-supremacy controlled by elitists.

At the end of "The Bell-Tower," its narrator points out that the "creator was killed by the creature" and that the "blind slave obeyed its blinder lord; but in obedience, slew him. . . . So the bell's main weakness was where man's blood had flawed it. And so pride went before the fall." What type of pride is Melville speaking of when he writes these final lines? In the short story we find him coming to terms with reality versus illusion, superstition versus science, and the rational versus the imagined. If "technology is at bottom a mere instrument" or pragmatic instrumentalism governed by utilitarian laws, then "the inquiry [into] what guides [it] becomes a task in its own right." Such is Melville's contribution to this proleptic nineteenth-century debate.[32]

We must conclude that Melville did not detest science per se but rather regarded it in terms of how it guides one. Melville shows how we as a society go about creating a collective imagination as a key ingredient in advancing common action and scientific progress. In this way "The Bell-Tower" is an important contribution to the fictive and factual history of mechanical-man. Melville is not averse to technological advancement, nor does he oppose the invention of an automaton. His ideas do transgress, however, the status quo of his time. Thus, he warns us that we cannot defer intelligent control for self-delusions produced by "common goals" and "common action" that ignore issues of control.[33] The idea that automatons can have consciousness and make rational decisions may alter from one simulation to another:

> "[S]ituations change, undergoing alteration and maturation over time. A situation that seems insoluble at one moment may in the next display clear avenues of resolution." Melville explores the inter-

> play between imagination and scientific advancement that will produce the mechanization of men. And within this interplay we must know the motives of the actual inventors. This problem arises within the context of human-machine interactions. The question arises: How is it possible to know the motives of inventors or the motives among the controllers of the controlled? Conversely, how is it possible to know all possible motives of both the controller and controlled in their relations to external stimuli?[34]

CONCLUSION

With this in mind, we come to realize that Poe wants us to question reality and be critical of self-delusion by analyzing these ambiguities of control and communication within technological creation. For Melville, Americans have a hard time distinguishing delusion from illusion—the former is tricking yourself and the latter tricking others. For his part, Melville wants us to be aware of where such delusion comes from and understand why we defer to the status quo. Both writers, finally, want us to take the final step in our recognition of self-delusion and the status quo as variables in science. Poe and Melville recognize the inherent problems of control and communication in human-machine interactions. In order to understand these interactions, our definitions of control, communication, and information must be broadened to understand the effects such interactions have on a culture. In their writings, Poe and Melville are creating the fundamental questions that others will be compelled to attempt to answer when human-machine relationships become more complicated and nuanced in real-world situations. We are still early on in the mechanical-man narrative. Our first-person subjective experience of machine intelligence is limited, to say the least. On this point, Richard Grossinger asserts that

> machines—even ones that talk back and play chess—are not free agents; they are projections of our existence. They simulate habitual

> behavioral patterns and data outputs, rearranging generic incidents into familiar syntaxes; they ape the speech algorithms of the masses, thus sound like personalities. Yet they do not replace their own molecules or keep remaking themselves. They are not illuminated from within by awareness.[35]

How do we go about *bottoming out* beliefs that are based entirely on the individual and collective experience of human-machine interactions? We do exactly what Grossinger does in his meditations and musings on why there is something, rather than nothing, in the universe: analyze consciousness, unconsciousness, and imagination as the holy trinity when it comes to living a particular subjective experience. Whether our bodies are encased in metal or carbon, we seek to determine how each form of consciousness finds its essence by projecting mind into nature so as to receive back service and spirit.

As we travel through the Newtonian, linear world, we are also traveling in a "deeper world that lies inside and behind it."[36] This is a realm where we "no longer focus on one thing but suddenly are perceiving everything."[37] Melville and Poe have been helping us to leave our human-oriented frame. Their writings code the imaginal dimension where we "slide completely out of our frame and totally into the others . . . and into ours"[38] and new forms of communication, control, and information allow mechanical men to communicate with us. The task for us in this new dimensionality of the imaginal is to discover commonalities regarding what it means to be conscious in a seemingly amnesiac world.

2
Accepting the Challenge
The Departure

Abraham Lincoln's second inaugural address articulated a transition in American life. He reminded the victorious Unionists and defeated Confederates that during the war they both prayed to the same God, but that neither side knew whether God approved of their actions. Lincoln thereby placed the power to create, build, and control the forces of nature back into the hands of the American people, a people that had witnessed the deaths of over half a million brothers, fathers, husbands, and sons on battlefields like Gettysburg and Antietam. The next year this American Prometheus was assassinated. Three years later, in 1868, Edward S. Ellis, in his popular dime novels, created the first mechanical-man in American literature—the Steam Man. Thirty-two years later, L. Frank Baum, in *The Wonderful Wizard of Oz,* presented the Tin Woodman. And between these two figures a mechanical-man resembling Maelzel's automaton-hoax, the subject of Ambrose Bierce's short story "Moxon's Master," would act as the centerpiece of America's imaginary history of the mechanical-man.

In these first literary versions of identifiably American mechanical-men, we see the genesis of a mythic rite of passage for ourselves in coming to terms with the technology we create and in recognizing our human relationship to the mechanical-man working alongside us in the garden. In these fictive tales the very notions that we once held pertaining to control and communication between machines and humans take new form. In addition, humans are edified through fiction concerning both how machines and technology reflect the departure stage of our mythic narrative as well as how ambiguities within the scope of human-machine interactions will not be easy to resolve, and in so doing we are recognizing our new dependence upon technology and each other, and attempting to understand novel definitions of what information and control mean in an emerging techno-society. Nevertheless, the optimism brought about by new inventions during this era, between the Civil War and World War I, propels American society to enter this departure stage of human-machine interactions.

THE RAILWAY SYSTEM AS MECHANICAL-MAN

Ralph Waldo Emerson likened the railroad to a magician's rod—it had transformed America into a great industrial nation with merely one wave. In 1820 the United States was a disjointed nation composed of lonely and self-sufficient farmers. But the railroad would make us something different. It welded the nation together, creating an American outlook, an American point of view.[1]*

---

*One should have a book of Emerson sayings or writings when reading these first chapters, just like one should have a book by Dewey or William James's *Varieties of Religious Experience* available in the later chapters. It will not be clear how important the Transcendentalists and Pragmatists (Metaphysical Club) are to the mechanical-man mythos until we cross the threshold from materialism to idealism. This self-individuation will be different for all. Personally, for myself, I crossed the threshold once the implicate and explicate modes of existence were defined by David Bohm.

By the Metaphysical Club, I mean the pragmatists of the late nineteenth century, which include Oliver Wendell Holmes Jr., William James, John Dewey, and Charles S. Peirce. (*See* Menand, *Metaphysical Club.*)

This description of the Transcendentalists by Perry Miller, on the jacket of his book *The Transcendentalists,* provides a helpful definition of that group:

> Before the Civil War, in Boston and its environs, a group of writers and thinkers belonging primarily to the Unitarian Church and its cultural milieu, profoundly affected by the Romantic movement in literature that was spreading

Both philosophies (Transcendentalism and Pragmatism) lay down the principles of utilizing the imagination as a tool for creating one's own individuation or rebirth—the former based on the Universal Mind, and the later based on the skeptical mind—the fusion forming the foundation for *pragmatic manifesting.*

The cowboy archetype showed an American collective consciousness changing to one of competition, taming nature, and finding one's rugged individualism. The mechanical-man is one of changing our consciousness to a skepticism marked by love and compassion, where the selfish gene is not in competition but looking out for the entire race of genes to survive. It's similar to the *feelings* of Isaac Asimov's *Law of Robotics.* Social Darwinism's concept of competition is replaced by the sacrifice made in following the Laws of Robotics or, in our case as humans, the *Law of Assumption.*

During the Civil War everyone witnessed the transformative power of the railroad and how it changed military history forever. In the North steam engines were seen as saviors, delivering supplies efficiently and predictably. Previously disdainful of railways, intellectuals now looked upon them as glue that would connect the states, never allowing

---

across the Atlantic from Europe, began to think and write in a new manner—a manner that its style and subject later became seen, especially in the examples of Ralph Waldo Emerson and Henry David Thoreau, as "distinctly American."

For our quest, I see the Transcendentalists, especially Emerson, making the soul of America aware of its infinite powers of the mind while the Metaphysical Club in post–Civil War America consists of pragmatic transcendentalists. They are skeptical about the technological advancement of the late 1800s to the early and mid-1900s. Metaphysics is the branch of philosophy dealing with abstract concepts. When the mechanical-man recognizes his vitality, he is abstracting from reality. So, when I say that the Metaphysical Club pragmatized the ideas of the Emersonians, it simply means that they were turning what we would consider pseudoscience into science, always skeptical as to what we take as truth and what we take as real, but nevertheless, accepting such belief as truth if it works in that given culture's collective consciousness. This enhances the cybernetic ideal, allowing for more flexibility in how we view and experiment with control, communication, and information and even how we define these terms from one collective consciousness morphing into a new collective consciousness.

another succession from the Union. To others—Indigenous Americans, plantation owners in the South, and those who feared that the railroad would destroy the western frontier and the yeoman farmer lifestyle—steam engines were monsters, destroying all of nature and humanity in their path. Nevertheless, the victorious Union states now envisioned a country connected by an interlocking network of railways. With such conflicting views of the railway system and the steam engine, it was only logical that the first mechanical-man in American literature would reflect ambivalence.

The best-known literary genre that draws on myth's foundational power in America has always been the Western, with its emphasis on the individualistic cowboy. "Progress associated with the western frontier," writes Eugene Rosow, "meant a geographic movement away from the decaying civilization of Europe toward the fresh possibilities offered by the wilderness, an optimistic march away from the past toward the future of America, home of the brave and free."[2] If America were to shift from this established narrative into the technological future by way of myth it needed to do so in a gradual manner. Therefore, the hero archetype in the form of a mechanical-man necessitated an environment that fascinated the emerging technological fantasies of Americans but one with which they were also familiar, namely, the Wild West, now replete with inventors and machines.

The American imagination needed to make a pivotal transition from the "iron horse to an iron human."[3] Sam Moskowitz, in *Explorers of the Infinite: Shapers of Science Fiction,* discusses the first tales that personify steam power as a transcendent force behind westward expansion and a burgeoning industrial economy. Among such tales, he notes, was *The Steam Man of the Prairies,* published in August of 1868, as number 45 of Beadle's Dime Novels. At the time, Edward S. Ellis, the Steam Man's creator, was a superintendent of schools in New Jersey and had authored fifty books on American history. In that same year, Zadoc P. Dederick, a machinist from Newark, New Jersey, was granted a patent on March 24 for a steam carriage constructed to resemble the human body. According to a newspaper report, Dederick's invention was "seven foot tall, the boiler was disguised by dressing the machine in a suit. When it was time to refuel, the driver was supposed to have unbuttoned the vest that hid the burner door." The article even stated that the Steam Man had a "cheerful countenance" and a stovepipe hat.[4] Ellis describes the imaginary Steam Man in a similar manner:

> It was about ten feet in height, measuring to the top of the stovepipe hat, which was fashioned after the common order of felt coverings with a broad brim, all painted a shiny black. The face was made of iron, painted a black color, with a pair of fearful eyes, and a tremendous grinning mouth. A whistle-like contrivance was traded to answer for the nose. The steam chest proper and boiler were where the chest in a human being is generally supposed to be, extending also into a large knapsack arrangement over the shoulders and back. A pair of arm-like projections held the shafts, and the broad flat feet were covered with sharp spikes, as though he were the monarch of base-ball players. The legs were quite long, and the step was natural, except when running, at which time the bolt uprightness in the figure showed different [*sic*] from a human being.[5]

During this time periodicals such as *Scientific American* sought to exploit the public's curiosity about the potential of inventions in a technological society by discussing the practical purpose of building a Steam Man as early as 1849: "Our own opinion is that there is nothing wonderful about it. There are hundreds of mechanics in our own land who could make steam men, if they received orders to do so and good pay for their labor." When Dederick and Isaac Grass were granted a patent for a steam carriage constructed to "resemble that of a human body," the event became news in local as well as national publications.[6]

Ellis's anthropomorphizing of the railway system illuminates a new optimism regarding technological innovation. Northern industrialists, entrepreneurs, and businessmen could harness the power of the railway system they had witnessed during the Civil War and leverage it for profit, while immigrants could be exploited for manual labor, Indigenous Americans forcefully removed, and yeoman farmers bought out. The Steam Man reflects this nexus of conflicts and contradictions in American society leading up to and persisting throughout the Gilded Age. This mechanical emblem became a symbol of how harnessing nature allowed mankind to achieve its larger goals.

Ellis's android transformed the railway system into a quasi-human form. Like a train, it was powered by steam and made almost entirely of iron, but the Steam Man was not hindered by tracks that it must follow. Moreover, Ellis gave the apparatus "a pair of arm-like projections," "legs," and a "face" with "fearful eyes" and a "tremendous grinning mouth."[7] These details about the mechanical-man's face reflect traits of American society at the time. His fearful eyes are symbolic of how Indigenous peoples, slaves, farmers, and immigrants viewed this emerging technology and the capitalistic avarice of robber barons who monopolized and controlled all aspects of the railway system. Conversely, the figure's grinning mouth is symbolic of how middle- and upper-class whites viewed these same corporate moguls as captains of industry who were shifting the country away from an agrarian household economy and toward an industrial empire. Ellis's Steam Man thus captures the country's divided vision of Andrew Carnegie, John D. Rockefeller, J. Pierpont Morgan, Cornelius Vanderbilt, and John Jacob Astor as being both captains of industry and robber barons.

Although the Steam Man could not think, speak, or converse, it nevertheless was imbued with both human and machinelike qualities. The figure was programmed to serve the needs of the one who controlled the apparatus. Thus, the first mechanical-man in the American popular imagination is a conflation of the human brain and a machine, since the former controls the latter. The mechanical-man's dependency on his inventor is indicative of the numerous dependent factions in the United States before and during the Civil War: African slaves and plantation owners, Indigenous Americans and federal or state governments, immigrants and employers, farmers and the federal government. This curious relationship between groups and institutions had been noted earlier by French visitor Alexis de Tocqueville, who lauded the American white male for his independence and individuality but who also found it troubling that so many others were very much dependent on others for their well-being and rights. These dependents, Tocqueville noted, seemed incapable of finding their individuality without inevitable conflict.[8]

THE STEAM MAN AS HUNTER AND HUNTED

> *The wagon dragged behind was an ordinary four-wheeled vehicle, with springs, and very strong wheels, a framework being arranged, so that when necessary it could be securely covered. To guard against the danger of upsetting it was very broad, with low wheels, which it may be safely said were made to hum when the gentleman got fairly under way.*
>
> DOUGLAS, *ALL ABOARD*

The function of the Steam Man becomes more complex as Ellis's tale progresses, but his major function is still one of transportation. He replaces horses in dragging a wagon behind him at a rate of speed far surpassing that of any animal. In this age between the Civil War and World War I, the Steam Man thus violates the laws of natural selection. It is now the Steam Man that stampedes through the wilderness and transports farmers to new land; it is the Steam Man that accomplishes the work of miners more efficiently than humans. Turning the complex into the simple by human ingenuity shows how even the laws of nature can be manipulated and changed by the creative will of man.

In an era of Social Darwinism, Ellis suggests that in order for Americans to survive the challenges of new frontiers, whether the West with its "savage Indians," mining towns plagued by lawlessness, or urban cities with overcrowded tenements and gang warfare, they had to adjust to these environments and use their ingenuity to create mechanisms still capable of being controlled by the human mind. People had to become more dependent on technology and each other if they wished to fulfill the American Dream in terms of conquest and destiny. For Ellis, in this post–Civil War period, man was no longer capable of surviving without the help of technology. It was no longer an age of romantic transcendentalism but rather one of realism and pragmatism. Even war itself was no longer looked upon as a means to achieve heroic stature in terms of myth and romance; rather, war now became a means to marshal into the social psyche technological awareness along with the necessity of technological advancement and production. The carnage produced by war, now based more on technology and machinery rather than tactics, necessitated a need to spur technological growth. It also led to thoughts of technology as a dubious force, which can foster both life and death.

The biggest threats to all the characters in Ellis's tale are Indians and "recalcitrant" Mexican settlers. *Manifest Destiny,* a term that came into circulation in the 1840s, declared that "the United States had a *Manifest Destiny* to overspread the continent and claim the desert wastes," thereby giving Americans justification for conquering new territories and expelling supposedly inferior races that were deemed unfit to assimilate into American culture.[9] To the United States government, rapacious land speculators, and European entrepreneurs, so-called Indians were a hindrance to development and profit. Furthermore, the symbolic power of the railway and prodigious laying of track showed that the United States would not default on its destiny. One scholar writes,

> The story of the building of the transcontinental railroad has been told in many books, with special reference to the financial buccaneers who made the undertaking infamous to several generations of Americans. Sometimes neglected is the all-too-human side of the story, the saga of the thousands upon thousands of individuals who toiled for years to lay the rails, of the human habitations they left behind and the storm and stress involved in their Herculean effort.[10]

In its inauguration into American literature, the mechanical-man embodied America's hopes, dreams, aspirations, and concerns. Oftentimes, a society's dreams and concerns are hidden, only to be uncovered by the imagination in our collective endeavor of storytelling. Every technological advancement changes and reshapes society in unforeseen ways; the mechanical-man helps us see and affirm real-world enactments of these changes through the sharing and reading of fiction. The mechanical-man acts as an external expression, symbol, and model of mechanizing the world and human interactions. The external expression, symbol, and model that the mechanical-man offers through the fictive means of communication does not transgress society's verities of human and machine interaction; rather, it bolsters the idea that humans are still in control of technology since it is a human imagination that creates the mechanical-man to help us understand these rites of passage.

LINCOLN AND THE STEAM MAN EVOLVE?

> *He knew the Indians had horses somewhere at command, while neither he nor his comrades had a single one. The Steam Man would be unable to pass that formidable wall, as it was not to be supposed that he had been taught the art of leaping.*
>
> *Whatever the plan of escape was determined upon, it was evident that the steamer would have to be abandoned; and this necessitated, as an inevitable consequence, that the whites would have to depend upon their legs. The Missouri river was at no great distance, and if left undisturbed they could make it without difficulty, but there was a prospect of anything sooner than that they would be allowed to depart in peace, after leaving the Steam Man behind.*
>
> Ellis, *Huge Hunter*

At the end of Ellis's novel, the Steam Man is destroyed. Instead of being told to help the "beleaguered miners," a small group of Irish and Chinese immigrants, the mechanical-man is instructed to depart from Wolf Ravine. His creator obviously surmised that informing the miners of the gold and the abilities of the Steam Man to unearth such riches would cause chaos among the different factions and invite an attack to capture the invention. As the miners plan their assault, the Indians have simultaneously lost their superstitious fear of the android, realizing that a mere man controls the machine, and they too plan to capture it. Rather than allowing the Steam Man to fall into the hands of these mobs, Ellis scripts its destruction: "The shock of the explosion was terrible. It was like the bursting of an immense bomb-shell, the [S]team [M]an being blown into thousands of fragments that scattered death and destruction in every direction." To those who befriended its Yankee inventor on his journey, the Steam Man is almost a Christlike figure who sacrifices himself for the sins of others, intimating that the advent of technology in the garden will usher in a new era of hope and optimism. The Steam Man echoes the deep structure tropes of Judeo-Christianity that lay beneath the American psyche. It connects American society with its older mythic patterns based on religious text and a pilgrimage for a new beginning wrought with sacrifice. Moreover, the Steam Man frees them from the redundant toil once necessary simply to survive.[11]

When just before his death Joseph Campbell was interviewed by Bill Moyers, they had the following exchange:

> MOYERS: So the new myths will serve the old stories. When I saw *Star Wars*, I remembered the phrase from the apostle Paul, "I wrestle against principalities and powers." That was two thousand years ago. And in caves of the early Stone Age hunters, there are scenes of wrestling against principalities and powers. Here in our modern technological myths we are still wrestling.
>
> CAMPBELL: Man should not submit to the powers from outside but command them. How to do this is the problem.[12]

Campbell's response is indicative of how Ellis and Dederick were attempting to understand and command "powers from outside," specifically steam power. Dederick was harnessing the powers of nature by a patented invention, while Ellis was dealing with the implications of such a technological contrivance by means of his imagination. The American ethos of tackling the seemingly impossible is evident in both of their enterprises. Contemporaneous newspaper reports of mechanical-men reflect an awakening of mythic patterns set in a familiar setting but with new heroes such as the inventor and his mechanical-man.[13]

Campbell's response is also evident of the problems that arise in human-machine interactions. The question arises: Are machines natural powers from the outside that humans are capable of having absolute control over? Ambiguities in society are now becoming evident concerning whether man controls technology or technology controls man. For now, Ellis indicates that man must control machines or face dire consequences. Thus, machines cannot be left on their own—control must then always be the focus of any inventor that creates something that is impervious to human weakness. He also harks back to Melville's and Poe's concerns about the motivations and goals of the controller. Ellis shows how if the mechanical-man is placed in the wrong hands, for example a controller with maniacal goals, the weak will suffer.

THE TIN WOODMAN

The general outline of the plot of the Tin Woodman of Oz resembles that of many tales involving quests full of trials undertaken for the sake of a beloved, ranging from *The Odyssey* to the part of the German tale "Rapunzel" in which the blinded prince wanders through the woods.[14]

One of the most popular children's stories of all time, *The Wonderful Wizard of Oz,* published in 1900, brought to the American public's imagination a mechanical-man almost thirty-two years after

Ellis's Steam Man, but this mechanical-man was different. Unlike the Steam Man, the Tin Woodman was once a human being. While the Tin Woodman represents many of the same conflicts and ambiguities as his predecessor, he also displays a new optimism about technology in America. How things had changed in less than four decades! "By 1900 America had reached the promised land of the technological world, the world as artifact. In so doing they had acquired traits that have become characteristically American. A nation of machine makers and system builders, they became imbued with a drive for order, system, and control."[15] Concomitantly, there had arisen a need for differentiation given the emergence of a pluralistic population. "This differentiation led to a coherent system of inventions, which was put on display at the Centennial Exhibition in 1876 and marked a technology moving from the simple to the complicated, all the while maintaining a sense of differentiation and coherency for purposes of mass production."[16] The Tin Woodman is an example of how one's individual differentiation in a technological society leads to creativity, making diversity a key element in an increasingly pragmatic and pluralistic society.

Advancements in technology did not come without a cost. Factories in the North had replaced plantations in the South; the industrial factory worker had replaced the African slave. There was a new system of servitude in America, but inventions were proliferating and, for most people at the time, technology was regarded as a way to improve one's quality of life. Electricity, the telegraph, telephonic communication, the incandescent lamp, new farming techniques, the airplane, skyscrapers, and trolleys inclined Americans overwhelmingly to embrace technology, which now was taking America down a path of its own choosing. Harnessing the power of the natural environment was merely the first step in this process. At the turn of the century Americans were turning their attention toward harnessing the power of social diversity, profiting from the ideas of ethnic minorities that were new to the garden. Technological inventions would unify this plurality into a singular national identity and renewed sense of shared purpose.

TRANSFORMATION AND TRANSHUMANISM

There was one of the Munchkin girls who was so beautiful that I soon grew to love her with all my heart. She, on her part, promised to marry me as soon as I could earn enough money to build a better house; so I set to work harder than ever. But the girl lived with an old woman who did not want her to marry anyone. . . . Thereupon the Wicked Witch enchanted my axe, and when I was chopping away at my best one day, for I was anxious to get the new house . . . the axe slipped all at once and cut off my leg.

BAUM, *THE WONDERFUL WIZARD OF OZ*

Before Baum's Woodman was transformed into a tin android, he fell deeply in love with a Munchkin girl. Can a mechanical-man love? This is a much more complex question than Ellis posed with his Steam Man.

The Woodman's transformation into a mechanical-man with emotions illuminates a changing culture in urban America. By the century's turn, the United States had accepted technology as a tool to bring about change and a better quality of life. People no longer feared technology but looked forward to Edison's next invention in Menlo Park, New Jersey. The idea that such a mechanical-man can love is an indication of the humanization and acceptance of technology, a shift in the human consciousness. The idea of a machine able to love has profound implications; specifically, the very idea of love as emotive response indicates that technology was affecting the human brain in ways we were unaware of. Technology itself was asking us for acceptance and attempting to foster a relationship with us. It was imbuing us with feelings. The emotive responses of the Tin Woodman to his mechanization lead to what has previously been defined as the technological sublime—a reaction of awe toward the wonders of technology. The Tin Woodman's machine-like body allows him to enter into this "sublime" state when he recognizes his ability to better control the natural environment with his new powers—he is stronger now, does not age, and is impervious to physical injury. For America at the time of Baum's novel, awe and wonderment were felt when imagining that bridges, skyscrapers, and electricity were technological spectacles that came out of the human imagination: "As electrical lighting transformed the appearance of streets, bridges, skyscrapers, public monuments, the National Bridge, and Niagara Falls, it became not only the double of technology but also a powerful medium of cultural expression that could highlight both natural and technological objects and heighten their sublimity."[17]*

After accidentally chopping off all of his limbs and even his head the Woodman, with the help of the Tinsmith, morphs into a mechanical-man. His entire body becomes one of tin. He has no heart, brain, or human organs, yet he still possesses some human qualities. He displays pride in his new investiture by saying, "My body shone . . . brightly in the sun." He also shows a bit of stoicism in saying, "But, alas! I had now no heart, so that I lost all my love for the Munchkin girl, and did not care whether I married her or not."[18] Although the Tin Man is bereft of a human garment (body), he still has numerous human qualities such as pride, stoicism, pessimism, fear, and the ability to think—something the Steam Man was never able to do. When juxtaposed with Ellis's automaton, Baum's Tin Man typifies an evolution of the American psyche toward a harmonious rapprochement with technology.

As the American population became more pluralistic and complex, so too did technology. While this complexity caused difficulty for many

---

*What exactly is the *sublime?* Is it the same as a *peak experience?* At this point, let us merely take Nye's definition in *The American Technological Sublime* and settle on it as we move forward:

> The experience, when it occurs, has a basic structure. An object, natural or man-made, disrupts ordinary perception and astonishes the senses, forcing the observer to grapple mentally with its immensity and power. This amazement occurs most easily when the observer is not prepared for it; however, like religious conversion at a camp meeting, it can also occur over a perilous of days as internal resistance melts away. (15–16)

Americans, they soon would see that on the other side of that apparent chaos lay efficiency, control, precision, and individuality amid diversity. And they would realize that diversity was a key ingredient for realizing technological advancement and the mythos of American conquest.

It appeared that new inventions were the work of a singular people with a singular destiny and purpose, as evidenced by the emergence of a mass consumer culture. Even within this emerging singular destiny, though, there was room to manifest personal identity, as the Tin Man and the innovators of the time showed. Grossinger points out that:

> Personal identity is the turnkey; it differs from consciousness in that it recognizes itself *as itself*. It is how consciousness inserts itself into a universe that does not otherwise express agency. When the self experiences *its own* existence, things seem to happen to it as an individual, divorced from all other individuals. That is a radical situation, even in as simple an entity as a worm.[19] (emphasis in original)

The American mythos of personal identity becomes the centerpiece of a pluralistic society manifesting itself in one out of many. This paradox was a reflection of the collective consciousness growing alongside individual consciousness. Pluralism did not contradict individual freedom and autonomy but rather enhanced it. We will later use this paradox to reanimate our own material bodies with spiritual energy, recognizing that we too are working to activate and enact service and spirit derived from nature by harnessing the power of the imagination.

DEPENDENCY

While Ellis's Steam Man had a human creator, the same cannot be said of Baum's Tin Woodman. The former human, after asking the Tinsmith to replace his limbs and torso, muses: "I had time to think that the greatest loss I had known was the loss of my heart. While I was in love I was the happiest man on earth; but no one can love who has not a heart, and so I am resolved to ask Oz to give me one. If he does, I will go back to the Munchkin maiden and marry her."[20] In an age of massive immigration, the Tin Man was another nationality, adding yet another valuable variable to a pluralistic society. In his pilgrimage, this mechanical-man represents the formula for creativity and invention, namely, diversity. The mythos of American conquest and destiny is predicated on how a society deals with issues of diversity while maintaining a singular vision.

For Baum there was no escaping technology at the turn of the twentieth century. The seeds in the mythical garden had now germinated, and there was no way of stopping their growth. However, Baum's mechanical-man is not a servant to others, nor is he involved in mundane acts; instead, he is a slave to his mechanical apparatus, which he cannot escape. Incapable of acting out his human emotions, of going back to a time when relationships were simpler—with few to no ambiguities seen in human-machine interactions—he is a servant to his yearning desire to love again, to have a heart. His imprisonment in a mechanical body is an example of a population now locked within an industrialized society, where reason and logic reign supreme. Nonetheless, we learn that the Tin Man is "kinder than ordinary men," which conceivably suggests that beneath such technological transformations humanity itself still lies, simple creatures looking to imagine, love, and create.[21] Also, there is the aspect of self-creation; the Tin Man did not require a human to control his actions but rather was self-defined and realized, echoing the American mythos of self-reliance and transferring that to a machine. The aspect of self-creation and self-reliance takes one back to the times of Oliver Evans, Eli Whitney, and Robert Fulton—when the country witnessed the pragmatic ingenuity transferred from the minds of these inventors into the mechanics of novel means of mass production, which came to be called the American system of manufacturing. The mechanical-man now acts as a symbol for this mythos of self-reliance, foreshadowing an American technological system predicated on sustaining itself.

Ellis and Baum view man's dependence on technology differently. Whereas Ellis was able to draw a clear distinction between his mechanical-man and its creator, Baum's vision is one not of humans controlling technology but rather of humans becoming technology. Ellis never realized a fusion of technology and humanity in his tale of the Steam Man. He made it clear that it is better for the Steam Man to maintain the status of mechanical-man as machine. In the American mind, the trade-off of humanity for technology and independence for dependence was worth it in order to achieve the ends of national destiny by means of technological manifestation.

Ellis was never able to witness the inventive genius of Thomas Edison, Nikola Tesla, Alexander Graham Bell, and many others. He was unable to witness these men's numerous inventions that brought an augmented quality of life—each in its own way providing an aspect of human enhancement via technology; to hear better, see better, fly, function, and produce more efficiently. And while these inventors and their creations were not mechanical-men per se, their inventions were fragmentary aspects of the mythic/fictive mechanical-man, since such inventions are technological enhancements of human capabilities. Nevertheless, Ellis did anticipate how people would react to such inventions, and he allowed us to stand on the threshold of a shift in consciousness. Baum, however, lived in the midst of this technological revolution and witnessed the potential of such inventive genius. Though Ellis was apprehensive about where technology was taking the country, underneath this anxiety and fear there was a fascinating journey that titillated the imagination.

FREE WILL AND AMERICAN INDIVIDUALISM

Other historians have taken a different view of the Tin Man's symbolism. Hugh Rockoff, for example, remarks:

> The tale is a powerful representation of the populist and socialist idea that industrialization had alienated the working man, turning an

> independent artisan into a mere cog in a giant machine. The joints of the Tin Woodman have rusted, and he can no longer work. He has joined the ranks of those unemployed in the depression of the 1890s, a victim of the unwillingness of the eastern gold[-]bugs to countenance an increase in the stock of money through the addition of silver. After the joints are oiled, the Tin Woodman wants to join the group to see if the wizard can give him a heart. He, too, will learn that the answer is not to be found at the end of the yellow brick road.[22]

While such scholarly analysis gives insight into social and political problems of the time, it fails to recognize the underlying essence of the Tin Man. He may represent the alienated working man, but he also represents the working man's acceptance of technology. He may represent a cog in a giant machine, but he is not devoid of human qualities or self-knowledge. Finally, his yearning to see whether the wizard can give him a heart is not indicative of a mindless cog; rather, it illuminates the cognitive ability of a mechanical-man to make decisions, thus imbuing him with subjective free will. This "will" or *Geist* does not cause him to self-destruct like the Steam Man; rather, it enables him to understand how his technological apparatus can be used to his advantage, for "even the cowardly lion cannot hurt the Tin Man with his long claws."[23] Like the American pioneers of technology who came before him—men such as Eli Whitney, Oliver Evans, Samuel Slater, Robert Fulton, and Cyrus McCormick—the Tin Man recognizes the importance of utilizing his mechanical apparatus as an instrument, reflecting the principles of American pragmatism.

The Tin Man's possession of free will is a reflection of the American populace's gain of sociopolitical independence by means of technological innovation. For the Tin Man, technology does not supplant free will; it may even increase the options one possesses. One choice he makes, for example, is to avoid killing other living creatures. Thus, when he

> stepped upon a beetle that was crawling along the road and killed the poor little thing[,] this made the Tin Man very unhappy, for he

> was always careful not to hurt any living creature; and as he walked along he wept several tears of sorrow and regret. . . . "This will serve me a lesson," said he, "to look where I step. For if I should kill another bug or beetle I should surely cry again, and crying rusts my jaws so that I cannot speak."[24]

Baum's use of tears in this experience indicates that the Tin Man, even though devoid of a heart and brain, has a capacity within him that guides his actions and awakens sympathy. "You people with hearts," he says, "have something to guide you, and need never do wrong; but I have no heart, and so I must be very careful. When Oz gives me a heart, of course, I needn't mind much."[25]

Surely Baum did not realize it at the time but he suggested in these simple words is that the mechanical-man has the ability to evolve as it interacts with its environment. Nevertheless, the question arises: Where do these innate emotions and ideas come from? In the case of the Steam Man, one can argue that his creator assigned some of his own human qualities to the android. Perhaps the Tin Man's innate emotions derive from his past experiences, yet one must ask how these traits are passed on and why they are necessary for the human-machine culture that was then emerging. In subsequent chapters, we will see the importance of harnessing traits and understanding how they are passed on within the framework of a human-machine culture.

Rockoff writes, "The little party heading toward the Emerald City reminds me of 'Coxley's army' of unemployed workers that marched on Washington, D.C., in 1894."[26] Such commentary does not illuminate the unique traits of the characters, who are set in a transformative environment and experiencing a momentous technological change. In addition, such analysis overlooks the unity of the characters, who while different are still very much bound together by a common cause. The Tin Man is different from Dorothy but shares equal status with her. Theirs is a cooperative endeavor that draws on all aspects of their differences for the good of the whole. Their sojourn to and within the Emerald City is reflective of the American mythos of the city on the hill. Now the city on the hill was a place where industry and the future were being manufactured. The Tin Man's move from the forest and rural setting to the city is a metaphor for the shift that was occurring socially in the country at the time. No longer would a mere garden mirroring the likes of the agrarian myth bear the fruits society needed in an emerging global economy. The yellow brick road is indicative of this bridge of not only human and machine but rural and urban. These tropes are all metaphors reflecting a changing American mythos of the city on the hill while still displaying the patterns and imagery of the original myth.

In the fiction of Ellis and Baum, we see a transformative moment when the simple mechanical-man becomes a complex system of thoughts, emotions, and consciousness. This transformative moment is rife with potential failure. We know that the Tin Woodman and the Steam Man are neither Odysseus nor Aeneas in terms of mythic patterns. The quests they undertake are extremely "sterile" and "renew nothing" in themselves.[27] Richard Tuerk proposes that the Tin Woodman is an antihero, quoting Victor Brombert as saying, "The anti[-]heroic mode involves the negative presence of a subverted or absent model."[28] Tuerk would argue that both the Tin Man and the Steam Man are picaresque figures embarked on a pointless wandering. However, there is no denying the fact that the Tin Man "achieved a stage of love that few mortal saints achieve." The Tin Man then is a fusion of the countless ambiguities that technology entailed for Americans. He is both heroic and anti-heroic: "[H]is true love is so selfless that he inflicts self-damage when he harms another creature."[29] Similarly, the Steam Man self-destructs, although some may argue it is destroyed, so as not to cause friction between competing classes. Technology in terms of myth attempts to balance such contradictions and opposites through these cybernetic imaginings.

These contradictions and ambiguities are most evident in the fields of control and communication in terms of human and machine interactions. The Tin Man shows how, in complex systems. a pluralistic approach is best able to solve problems. In a society moving toward techno-systems, one must sacrifice one's independence in order to enjoy the benefits of systems technology, which is a web of industries acting together. Both the Tin Man and Steam Man are dependent on others—the Tin Man for his oil and the Steam Man for his coal. James Beniger calls this type of cooperation, "Reciprocity Selection." He holds that:

> Programming for altruistic behavior that benefits the recipient more than it costs the donor will be selected, if the same behavior is likely to be reciprocated, because the average fitness of all individuals with the program is thereby increased. Because it will be to each individual's advantage to "cheat" by accepting benefits without incurring the costs of reciprocity, however, this form of cooperation requires much greater information-processing. . . .[30]

In effect, the Tin Man and Steam Man display the need for human and machine interactions to be attuned to the dynamic behavior between programmer and programmed, creator and created, and controller and controlled. The cooperation and messages shared within these relationships, in the context of human and machine interactions, require a kind of information-processing that takes into consideration the goals of both.

MOXON'S MASTER: THE RHYTHMS OF TIME, LABOR, AND CONSCIOUSNESS

Ambrose Bierce, in "Moxon's Master," attempts to answer important questions haunting the American psyche at the turn of the century: What is intelligence? What is the relationship between mechanistic and living systems? Can a machine ever attain consciousness? To address these questions, Bierce, like Ellis and Baum, writes of a mechanical-man in the midst of a far-reaching technological shift. Bierce is the first

American writer to offer concrete definitions of machine intelligence. In addition, he imbues his automaton with human emotions and carnal desires, not unlike those of the Tin Man, but with a different type of bodily apparatus, one that appears animalistic rather than mechanical.

Bierce begins his strange tale by having his narrator discuss the definition of machines with his inventor friend Moxon: "The word has been variously defined. Here is one definition from a popular dictionary: 'Any instrument or organization by which power is applied and made effective, or a desired effect produced.' Well, then, is not man a machine? And you will admit that he thinks or thinks he thinks." Moxon's own definition of machines centers on applied power to bring about a "desired effect."[31] This story was written in 1894 when the United States was only years away from defeating Spain, colonizing the Philippines and Guam, and becoming a power-broker on the world stage. Indeed, in the realms of business, politics, and social relations, the country was on the verge of fulfilling its ascendancy in terms of technological superiority.

Moxon, as an independent inventor similar to the likes of Thomas Edison and Nikola Tesla, illuminates the scientist's relationship with machines. This relationship involved looking for a "practical outcome, which interests the inventor; the search for causes is but the search for the means which would produce the result."[32] The independent inventor's devotion to the study of man as automaton opened the floodgates to forging such relationships with machines. The machine was following an evolutionary path displaying a reflection or coevolution of a simple external machine to a more nuanced and humanized fetish. The mechanical-man was now serving as a mirror to its human counterpart, informing the trends of social evolution in an industrializing age. Searching out first causes and means that would produce results in the human sphere led inventors to look for the same in machines. In this way the inventor takes the simple "cause from effect" or "means from end" and creates a complex idea of the whole "because of its superior practical importance."[33] In Bierce's short story, Moxon thus comments:

> As Mill points out, we know nothing of cause except as an antecedent, nothing of effect except as a consequent. Of certain phenomena, one never occurs without another, which is dissimilar: the first in point of time we call cause, the second, effect. One who had many times seen a rabbit pursued by a dog, and had never seen rabbits and dogs otherwise, would think the rabbit the cause of the dog.[34]

Moxon's friend and inquisitor, the narrator, believes that Moxon's "devotion to study and work in his machine shop had not been good for him. . . . Had it affected his mind?" When we look at the lives of Tesla and Edison, we will come to understand how seclusion in the laboratory affected their outlook on the nature of the universe and the relationship between man and machine. Such seclusion led Moxon to believe that "a machine thinks about the work that it is doing."[35] Thus the machine, like a human being, has value and is consciously aware of itself if it is accomplishing an act that yields a product. The actions of a machine/mechanical-man define its intelligence and perhaps its very existence. Writes one scholar:

> There are no intermediate appliances, no adjustment of means to remote ends, no postponements of satisfaction, no transfer of interest and attention over to a complex system of acts and objects. Want, effort, skill, and satisfaction stand in the closest relations to one another. The ultimate aim and the urgent concern of the moment are identical; memory of the past and hope for the future meet and are lost in the stress of the present problem; tools, implements, weapons are not mechanical and objective means, but are part of the present activity, organic parts of personal skill and effort.[36]

If "tools, implements, weapons are not mechanical and objective means," they are an extension of human consciousness; therefore, we are organically tied to them. The mechanical-man in Bierce's story is not merely a mechanical tool or objective means; instead, he is a part of our "activity" of adjusting to a transformative environment when the "memory of the past and hope for the future meet and are lost in the stress of the present problem."[37]

The human-machine is worthy of its functioning, for it is an integral cog in the organizing principles of society. Moxon's friend capitulates and agrees that "you may be able to infer machines' convictions from their acts." Moxon sets a high value on action, which is embedded in the very definition of machines. What then of American technology at the turn of the century? Moxon gives us an example of American technology's maturation by referring to Herbert Spencer's definition of life. "Life, he [Spencer] says, is a definite combination of heterogeneous changes, both simultaneous and successive, in correspondence with external coexistences and sequences."[38] America was in a moment of technological transformation, and with this transformation came new complexities and systems. "Late-nineteenth and early-twentieth-century Americans in all walks of life acted as if their world was composed of systems, and worked to design or arrange these systems' diverse components so as to achieve peak efficiency. Such was the case in technical education and organization."[39]

In terms of Moxon's reference to Spencer's definition of life, America was doing just that—coming alive. Through heterogeneous changes in all technological fields, the simple garden was maturing into a complex environment full of growth and potential. In Bierce's story, the discussion ends with Moxon espousing that Spencer's definition of life can be applied to machines, for they too have systems that correspond with external coexistences and sequences. Bierce is now setting the stage for Moxon's Master, the automaton, to enter the story. He is suggesting that the activity of disputes between actively involved elements of society does not impede technological progress; instead, it opens up new dimensions of possibility in society, including its economic and political spheres. Historian Howard Segal writes: "These spheres of activity, however, by no means exhausted the possibilities. Although Americans frequently disputed the elements or order of systems, the notion of system itself pervaded virtually every aspect of American life in the half-century after 1870. In sum, the country's inhabitants *reinvented and remanufactured the nation*."[40] But there is also an anxiety that underlies these changes accompanying Spencer's definition of machines. Spencer's definition is echoed later by Norbert Wiener when he says:

> A complex action is one in which the data introduced, which we call the input, to obtain an effect on the outer world, which we call the output, may involve a large number of combinations. These combinations, both of the data put in at the moment and of the records taken from the past stored data, which we call memory. . . . For any machine subject to a varied external environment to act effectively it is necessary that information concerning the results of its own action be furnished to it as part of the information on which it must continue to act.[41]

While Wiener recapitulates Spencer's definition of machine intelligence, he extrapolates on it by explaining the problems inherent in communication and control between humans and machines; specifically, the impossibility of controlling or knowing the infinite combinations of data input and variable output. This will become more evident when we discuss issues related to predictability in the actions of machines and the messages we exchange with them. But for now, it is important to note that even our early definitions of machine intelligence intimated problems in areas of control and communication due to unknown variables in the combinations of sequences a machine must utilize to interact with a multiplicity of environmental variables.

The very concept of machine intelligence creates a "complexity of perceptual feeling" that consciously opens the doors to perception of the unseen.[42] We sit with a machine or a mechanical-man and ask ourselves not only about its mechanics or intelligence but also, and perhaps more importantly, *How does it feel?* By asking this question, we seek to experience the inherent meanings that flow through the world in order to initiate communication between the so-called animate and inanimate. As Stephen Harrod Buhner writes:

> Something new enters [our] experience when we reach out with that nonphysical part of us and touch the wildness of the world. . . . Importantly, hidden within the secret kinesis of the world, within the feeling of any particular wild place or thing, are the mythic dimensions that belong to the world in general and that thing or place in particular.[43]

You soon notice that that which you are touching through your own awareness has its own aliveness, its own awareness, and even its own capacity to communicate—albeit in ways different from other things about which we have been taught. You find yourself in a different state of mind in which one can encounter the living reality of machine intelligence. No longer are you concerned with scientific concepts of measurement and efficiency; rather, your chief concern in this "state of analogical, feeling perception and thinking," is the information perceived by means of novel modes of control and communication. Information from this "secret kinesis" is communicated when we perceive something in a "different state from the common,"[44] as Thoreau says,

> It is only when we forget all our learning that we begin to know. I do not get nearer by a hair's breadth to any natural object so long as I presume that I have an introduction of it from some learned man. To conceive of it with a total apprehension I must for the thousandth time approach it as something totally strange. If you would make acquaintance with the ferns you must forget your botany. You must get rid of what is commonly called knowledge of them. Not a single scientific term or distinction is the least to the purpose, for you would fain perceive something, and you must approach the objective totally unprejudiced. You must be aware that nothing is

> what you have taken it to be. [To perceive something truly] you have got to be in a different state from the common.[45]

The mechanical-man will gradually take us from consideration of human and machine relationships by means of the common states of sensory perception to uncommon states of paradox, cosmic consciousness, and spiritual freedom as we move forward in our collective as well as individual questing.

RHYTHMS

After Moxon finishes discussing the definition of life and consciousness, it is discovered that he is hiding something, perhaps something sinister. His friend asks, "Moxon, whom have you in there?"[46] At this point, Bierce delves into the question of consciousness: "Do you happen to know that Consciousness is the creature of Rhythm?"[47] The narrator leaves Moxon's laboratory and muses about what has just occurred. The reader at this point becomes aware that Bierce is speaking of an American public now becoming aware of its "possibilities" given a heterogeneous population.[48] Indeed, there are rhythms, both psychological and social, in systematizing. America was not only coming alive during this shift in technological complexity, but was also becoming conscious of its prowess in global economies and militaristic operations. Bierce's narrator remarks:

> That faith was then new to me, and all Moxon's expounding had failed to make me a convert; but now it seemed as if a great light shone about me, like that which fell upon Saul of Tarsus; and out there in the storm and darkness and solitude I experienced what Lewes calls "the endless variety of excitement of philosophical thought." I exulted in a new sense of knowledge, a new pride of reason. My feet seemed hardly to touch the earth; it was as if I were uplifted and borne through the air by invisible wings.[49]

Moxon is now the Master, his inquisitor having shifted in how he views the relationship between machine and man. And by the very use of the word *inquisitor,* Bierce implies a magical element that harkens back to Melville's automaton. That there is still something mysterious or pertaining to alchemy when discussing the mechanical-man through the utilization of human logic and reason implies that America is still an orphan, searching for its origins of consciousness. Through the utilization of myth and storytelling, Americans were unlocking these unconscious origins. The narrator admits that "Consciousness is the creature of rhythm. . . . All things are conscious, for all have motion, and all motion is rhythmic."[50] Once he realizes this and returns to Moxon's laboratory, there is a turn of events. He opens the door and, to his dismay, sees Moxon playing chess with an automaton. He describes the mechanical-man as follows:

> He was apparently not more than five feet in height, with proportions suggesting that of a gorilla, a tremendous breadth of shoulders; thick, short neck and broad, squat head, which had a tangled growth of black hair and was topped with a crimson fez. A tunic of the same color, belted tightly at the waist, reached the seat, apparently a box, upon which he sat; his legs and feet were not seen. His left forearm appeared to rest in his lap; he moved the pieces with his right hand, which seemed disproportionately long.[51]

Bierce's mechanical-man is quite different from its predecessors in being more animal than human, suggesting that technology and machinery at the turn of century were imbuing us with a better glimpse into our animal instincts. As Moxon's friend then watches the chess game, an allusion to the hoax of Maelzel's famous automaton, the "gorilla" becomes angered when it hears Moxon say "checkmate," whereupon the automaton leaps across the table and kills its creator. Immediately after this event Bierce's narrator records these details:

> I now became conscious of a low humming or buzzing which, like the thunder, grew momentarily louder and more distinct. It seemed to come from the body of the automaton and was unmistakably a whirring of wheels. It gave me the impression of a disordered mechanism which had escaped the repressive and regulating action of some controlling part.[52]

CONCLUSION

All of these mechanical-men reflect an emerging anxiety embedded within the American Industrial Revolution. As James Beniger points out: "Increasing the speed of an entire societal processing system, from extraction and production to distribution and consumption, was not achieved without cost."[53] Harnessing the power of steam and applying it to transportation led to the development of more complicated machines "to sustain the high processing speeds and volumes" of America's Industrial Revolution.[54] Man was now placed in the precarious situation of having to figure out how to control processes and movements of material goods, people, messages, and information "at speeds faster than those of wind, water, and animal power."[55] The anxiety that these mechanical-men reflect is evident in the very interactions Americans were experiencing with machines and technology at the turn of century. There was an emerging Control Revolution occurring, "particular to the material aspects of information processing, communication, and control."[56] The principles of control and communication were now uncertain and unpredictable in terms of human-machine interactions. The task for Americans was to figure out how to control a growing interconnection of extraction, production, distribution, and consumption. The interaction with and understanding of machines would aid us in this task. However, with machines being asked to do more tasks at faster speeds, Americans needed to reevaluate how to control and communicate with them. Needless to say, with steam and soon electricity driving mechanical processes and speeding them up, these issues of control and communication would only grow more complicated and nuanced.

In retrospect Ellis, Baum, and Bierce explore prophetic issues. Today medical technology can provide that which the Tin Man sought—a heart. We now have the capability both to replace an entire heart and to create an artificial replacement that accomplishes the same function. In addition, prosthetics replace lost limbs. The Tin Man elucidates many of the same issues we now tackle in terms of medical ethics, specifically whether fabricated organs or prosthetics make one less human. Baum for one realizes that what matters is how an apparatus interacts with the environment around it to bring about or implement change, which is now a cardinal variable in unlocking verities in artificial intelligence research and development. Americans were beginning to see their natural environment and even the social and psychological constructs of the human mind as an elaborate laboratory. Human beings, their ideas, and their machines were now seen as tools in this pluralistic society.

Ambrose Bierce was the only one of these writers that fought in the Civil War. He left his boyhood farm for the battlefield, enlisting in the Union army at the age of eighteen. Historian Drew Faust says he is "the most significant and prolific American writer to actually fight in the Civil War."[57] The war haunted him for the rest of his life. While his writings about war "are often cited as the beginnings of modern war literature and as a major influence upon both Stephen Crane and Ernest Hemingway," his use of a mechanical-man as mythic-marker reflects his approach to writing and life: "Cultivate a taste for distasteful truths. And . . . most important of all, endeavor to see things as they are, not as they ought to be."[58] At the heart of Bierce's writing rests an inquiry, a paradoxical riddle that Melville and Poe also wrestled with. The riddle of death, brought to the attention of all during and after the Civil War, mandated a reevaluation of American values, truths, and myths. The mechanical-man was now the mythic-marker that would guide Americans out of an Edenic garden and through the battlefields and urban jungles of uncertainty. In *The Republic of Suffering,* Faust sums up the riddle of death:

> Death itself becomes war's end, the product of its industrialized machinery; there is no more transcendent or glorious purpose; northerners and southerners lie mingled together, "fame or country least their care." But they now understand what in their youthful zeal for battle they did not . . . for the pieties and pomposities of war have dissolved. The dead have discovered as well the answer to the riddle that Melville cannot know, the riddle "which the slain / Sole solver are." Beginning in such innocence, they are brought by war to an ultimate knowledge that even their survivors lack. The living remain captured in uncertainty.[59]

There remained uncertainty after the Civil War, but now on an ever expanding scale in terms of control, communication and the exchange of information in human and machine interactions. Katherine Hayles cogently points out that: "The ultimate horror is for the rigid machine to absorb the human being co-opting the flexibility that is the human birthright."[60] Would human beings work alongside machines or would they merely become "cogs and levers and rods"?[61] One thing was certain, this new threshold of uncertainty was based on the thesis that "control involves programming, that programs require inputs of information, that information does not exist independent of matter and energy."[62] How then does one go about bringing certainty and predictability to such programs and programming? New issues and problems with human and machine interactions were now appearing and within the context of both—the fictive and real—scientists and inventors in America were now having to deal with such issues head on.

Jeffrey Kripal in *Authors of the Impossible: The Paranormal and the Sacred,* aids us in our analysis of human and machine interactions within the context of the real and the fictive by saying:

> Reality itself—or so I am assuming—does not change, but what is generally possible and impossible to experience as real does appear to change from culture to culture, as each culture actualizes different

> potentials of human consciousness and energy. Such a dialectical model, I should stress, is both universalistic and relativistic at the same time. There is radical Sameness. And there is radical Difference. And neither can be sacrificed to the other.[63]

This begs the following question: When does what is generally possible and impossible to experience as real appear to change from culture to culture? The fictive's appearance in reality in the form of myth occurs when the conscious communicates with the unconscious—this is the first step in a culture's discovery and actualizing different potentials of human consciousness and energy. The next step is when science and invention changes what is generally possible and impossible to experience as real—this occurs when the conscious finds ways to control the information that the fictive brings out in reality. "Consciousness both occults or hides itself in material and symbolic forms and allows itself to be seen, as if in a mirror, so that it can be cultivated and shaped into definite, but always relativistic forms."[64] The mechanical-man is a symbolic vehicle of communication between conscious and unconscious systems. The vehicle is a connector and when this connector is broken or unrealized a society feels rootless. When we look at the definitions of cosmic consciousness and super consciousness during the return of the quest, we will look back at how the narrator in Bierce's "Moxon's Master" reaches the level of super consciousness upon learning from the "Master" his definition of "machine consciousness." He says, "[A]s if a great light shone about me, like that which fell upon Saul of Tarsus; and out there in the storm and darkness and solitude I experienced what Lewes calls 'the endless variety of excitement of philosophical thought.'"[65]

This endless variety of excitement is consonant with the variety of consciousness—the possibilities of manifesting whatever reality one wishes to experience. You see, the act of creating, or technology in general, is the boon of consciousness on the material plane. When we later ask why some things are possible in one time period and not possible in another we see that manifesting technologies is solely dependent on the consciousness activated within that space and time. Saul of Tarsus (Paul) did not only see the varieties of consciousness but he also saw cosmic consciousness, which showed him the point of consciousness at its zenith in terms of projecting the imagination into nature and receiving back information on service and spirit, allowing for any and all possibilities—including any imagined and yet unimagined technologies—there is no limit.

Grossinger appropriately ends this chapter for us when he writes:

> Technology is not just the result of incremental applications of thermodynamics and properties of material to the landscape in which creatures find themselves. It is the precise meeting ground of mind and matter evolving through Stone, Bronze, Machine, and Atomic Ages. . . . Cars, planes, radios, and computers materialize almost magically out of consciousness even as they are simultaneously constructed of minerals and petrochemicals. They were on this planet somewhere at its genesis—dormant during the Ice Ages, latent in the pumps and war wagons of ancient Assyria and Persia. As we rushed (and rush) pell-mell toward mechanism, the shadow of the machine precedes the machine.[66]

3

The Sorcerer and the Wizard

On July 1, 1898, one month before the armistice ending the Spanish-American War, a well-known inventor, Nikola Tesla, applied for a patent to create prototypes of remote-controlled machines. His application stated,

> The problem for which the invention forming the subject of the present application affords a complete and practicable solution is that of controlling from a given point the operation of the propelling engines, the steering apparatus and other mechanisms carried by a moving object, such as a boat, or any floating vessel, or carriage, such as an automobile, whereby the movements and course of such body or vessel may be directed and controlled from a distance, and any device carried by the same, brought into action at any desired time.[1]

Earlier that summer, American warships had destroyed the Spanish fleet in Santiago Bay. Although only one American life was lost in this battle, reasonable minds foresaw the destruction of imperial powers with the aid of modern technology. Tesla's invention, as he conceived it, would both save lives and prevent war:

> The plan . . . of my invention . . . employ[s] any means of propulsion to impart to the moving body or vessel the highest possible speed, to control the operation of its machinery[,] and to direct its movements from whether a fixed point or from a body moving and changing its direction however rapidly, and to maintain this control over great distances, without any artificial connections between the vessel and the apparatus governing its movements, and without such restrictions as these must necessarily impose.[2]

The idea of American warships being controlled by humans on the mainland was a practical necessity in the mind of Tesla. Wireless communication would permit human beings to control automatons and machines over vast distances, but such ideas were thought to be the stuff of science fiction, the ramblings of a modern-day sorcerer.

However, this idea and its patent, awarded later that year, was neither fiction nor the fantasy of a magician. Instead, Tesla's "invention not only established all of the essential principles of what came to be known a few years later as the radio; it also lay as the basis of such other creations as the wireless telephone, garage-door opener, the car radio, the facsimile machine, television, the cable-TV scrambler, and remote-controlled robotics."[3] So, while America was making its mark on the world in terms of its military prowess and economic wealth, Tesla was making an entirely different mark in the realm of ideas, imagination, and invention. Tesla's automaton in the field of wireless communication was necessary in a more complex environment than the mythical garden. The United States' entrance into world economics, markets, and politics acted as a catalyst for utilizing wireless inventions in new ways. The nation was no longer a secure republic that could hide behind the vastness of two coastal oceans. Tesla's automated invention is both a product of this shift and also a casualty of such transformation.

CHAOS FOMENTS INVENTION

The ravages of war were unavoidable for Tesla early in his adult life. It was mandatory in his country of birth, Croatia, for Tesla to serve three years in the military. However, with the onset of war against the Turks, Tesla's father urged his son to seek safe haven in the surrounding hills. "For most of the term," he recounted, "I roamed the mountains, loaded with a hunter's outfit and a bundle of books, and this contact with nature made me stronger in body as well as in mind . . . , but my knowledge of principles was very limited."[4] In the mountains, Tesla recognized natural forces that could unleash untold amounts of trapped power. There he also experienced "a separation from the world, a penetration into some source of power and a life-enhancing return."[5] Coupled with his mechanistic view of the universe, this background prompted Tesla to create inventions that would usher in a time of peace, with war becoming nothing more than a remote-controlled game.

A total of 385 Americans died in the Spanish-American War. Although then Secretary of State John Hay called it a "splendid little war," Tesla saw these deaths as avoidable. However, another aspect of modern warfare haunted him, the idea that weapons of mass destruction were becoming more and more powerful.[6] When he unveiled his remote-controlled boat that he called the "telautomaton" at Madison Square Garden's Electrical Exposition in May of 1898, the Spanish-American War was at a crisis point. "Beating the Spanish with modern instruments of destruction became the overriding theme of the [E]xposition. Tesla would have far and away the most sophisticated construction."[7] In a demonstration that uncannily paralleled the telautomaton's cousin, Maelzel's chess-player, Tesla intentionally created an aura of mystique about his creation. "In demonstrating my invention before audiences," the inventor admitted, "the visitors were requested to ask any question, however involved, and the automaton would answer them by signs. This was considered magic at the time, but was extremely simple, for it was myself who gave the replies by means of the device."[8]

The telautomaton was a by-product of the wireless revolution that took hold of the United States and the world at the turn of the century. Wireless communication was a means of conquering the oceans. The ship-building spree in America and other countries in the 1880s through the 1890s necessitated a practical means to communicate over large distances. Tesla's telautomaton was one of these inventions, one indicative of countries' becoming more dependent upon each other and less dependent upon nature.[9]

> When a society passes from pre-industrial to industrial conditions, which is what happened in the United States in the years between 1870 and 1920, people become less dependent on nature and more dependent on each other. . . . Put another way, this means that in the process of industrialization individuals become more dependent on one another because they are linked together in large, complex networks that are, at one and the same time, both physical and social: technological systems.[10]

Tesla's invention was merely the next step in the revolution of wireless communication. Here was a mechanical automaton that could receive and exchange messages and perform complex tasks. The telautomaton went beyond ideas of humanoid automatons; it was an extension of the human brain. At the same time, its dependency on those who communicate with it over vast distances is a cardinal example of a more complex relationship between society and technology than we have seen heretofore.

TESLA IN AMERICA

When Tesla landed in the New World, America was in a state of transition. He found the country uncivilized and "a hundred years behind the lifestyle of the great European cities." However, the country was in the process of building itself anew. New York City was the epicenter of this new growth and expansion: "Tesla's ship dropped anchor in New York in late spring of 1884, just as the monumental decade-long project, the Brooklyn Bridge, was being completed and the last components of the Statue of Liberty were being hoisted into position."[11] The New World was a perfect laboratory for Tesla.

Urban America in the 1880s relied on immigrants from Ireland, Slovakia, Germany, Poland, and Bohemia. Tesla saw this environment as a golden opportunity to change the world. If the cities of America were in their incubation state, attempting to become civilized, immigrants brought with them their experience, labor, and tools with which to capture the mythos of the American Dream. Tesla brought to America what would eventually lead to the creation of his telautomaton, specifically, his AC invention and accompanying machinery. In New York he entered a competitive world of inventors not unlike that of the era's corporate robber barons. Tesla yearned to become larger than life, a wizard of creation, and one of the figures he sought to emulate was none other than the famed Wizard of Menlo Park. "In an entirely different realm of invention, besides being a better technician than anyone else, Edison was a creator; his most original work was a machine that talked: the phonograph. With this device, Edison entered the realm of the immortal; he was the Wizard of Menlo Park."[12]

When Tesla met Edison, "possibly aware of the proximity of Transylvania to Tesla's birthplace and a resurgence of interest in the tales of Vlad Dracula, the fifteenth-century alleged vampire who lived in the region, Edison inquired whether or not the neophyte had ever tasted human flesh." Tesla replied in the negative and asked what Edison's diet consisted of. After this discussion, Tesla began following Edison's dietary regimen. Tesla writes:

> The meeting with Edison was a memorable event in my life. I was amazed at this wonderful man who, without early advantages and scientific training, had accomplished so much. I had studied a dozen languages, delved into literature and art, and had spent my best years in libraries . . . [but] felt that most of my life had been squandered.[13]

Tesla, in his description of Edison, illuminates the difference between American inventors and foreign mechanics, model-makers, and engineers who helped men like Edison create a system-oriented science for American technological growth. Tesla thus makes the following observation about Edison:

> If he had a needle to find in a haystack he would not stop to reason where it was most likely to be, but would proceed at once with the feverish diligence of a bee to examine straw after straw until he found the object of his search. . . . I was almost a sorry whine to his doings, knowing that just a little theory and calculation would have saved him 90 percent of his labor.[14]

Tesla saw a new kind of science emerging in America, one based on pragmatic trial and error in manifesting results. His impressions of Edison elucidate a deep chasm between those who knew science and those who merely practiced it. Tesla would fuse the two, which is why Tesla has been called the only pure inventor of his age, and he started to make connections between areas of knowledge and study that heretofore had appeared insoluble.

For Tesla, American science and technology were too result oriented; there was no time for theory and calculation—no time to allow your unconscious to breathe. For Americans, the telautomaton in its early stage of development was merely a model that promised little or no practical benefits. Tesla recognized in the American public a desire for practical results that they could witness with their own eyes rather than a device that could deliver dividends in the future. The telautomaton's demise, in other words, was owing to a culture that valued instrumental immediacy. The American public, therefore, did not herald Tesla's invention at the Electrical Exhibition of 1898 because it was regarded as a child's toy. He nonetheless believed that it was only a matter of time before Americans began to harness their childlike imaginations, utilizing such as tools to bring about adult inventions.

While working for Edison, Tesla also learned that the Wizard of Menlo Park was fallible. In Edison he glimpsed a deep-seated hatred for his competitors, especially George Westinghouse in their struggle over AC versus DC electricity. The project of science in America at the time was a personal battle in which fights over first place in the patent-race signified alliances. It also was a game of espionage. Tesla too would meet the same fate as Edison's competitors, leaving the stubborn wizard in order to create his own legacy, but he did not part company with Edison until he had learned the uniquely American system of science and systems-building that would give life to the telautomaton.

With this knowledge, Tesla was able to contribute and add an important and necessary outside element to the mechanical-man myth. Like the Tin Man and characters in *Oz,* he represents the American mythos of self-reliance. He invigorated aspects of the electrical revolution with the foreign eyes of an outsider, illuminating the importance of diversity in a society striving for hegemony. As seen in the dime novels of this era, Tesla resembled the actions of those fictional inventors of mechanical-men in his practical and creative endeavors. He sought to utilize the power source of electricity to bring life and progress to the masses similar to those writers that were now moving from steam to electricity to bring life to their fictive mechanical-men. Only an immigrant or outsider could reinvigorate the electricity discussion on such a large scale as Tesla did. While tales of mechanical-men were attempting to explain the American mythos of American prowess and conquest by means of invention and technological progress, Tesla was displaying such conquest by means of curiosity and invention on the stage of reality with his most cherished tool, human imagination.

TESLA ENTERS THE DEPARTURE STAGE

In February 1895, after leaving his partnership with Edison, Tesla created the Nikola Tesla Company. The directors were a hodgepodge of engineers, inventors, and corporate financiers. "This endeavor was made possible by Tesla's assiduous work on the Niagara Project a year earlier in which it was said that Tesla played the pivotal role in this, the largest electrical enterprise in the world." Of Tesla's achievement the *New York Times* wrote: "To Tesla belongs the undisputed honor of being the man whose work made the Niagara enterprise possible. . . . There could be no better evidence of the practical qualities of his inventive genius."[15] Working with Edison had taught Tesla two lessons: first, he now saw the importance of a team that included model-makers, engineers, mechanics, and financial wizards who knew the ins and outs of an emerging science based on cooperation; second, he recognized that, in order for his inventions to enter the public domain, they needed to demonstrate practical utility. Implementing these ideas, he soon became the master of electrical oscillators and a revolutionary new AC system of lighting, both of which systems are the crux of remote-controlled robotics. Although Tesla was never able to fit into the corporate world as his competitors did, he still made the largest strides in the field of electricity and electromagnetism. He was not a showman, but with this independence also came constraints.

Biographer Mark J. Seifer notes of Tesla's relationship with John Jacob Astor, financier of many independent inventors during the boom of the 1880s and 1890s, that Astor was reluctant to fund Tesla's project for remote-controlled devices, as a letter dated January 18, 1896, reveals:

> Dear Mr. Tesla,
>
> [Regarding your] letter offering me some of your oscillator stock received . . . 95 seems rather a high price; for though the inventions covered by the stock will doubtless bring about great changes, they may not pay for some time as yet, and, of course, there are always a good many risks.
>
> Wishing the oscillator as much success as I could if financially interested, and hoping soon to be able to use one myself. . . . [16]

The oscillators for Tesla, of course, were "never ends in and of themselves." Since a fire had destroyed his laboratory in New York, now was the opportunity to build anew. "Tesla took a train to Colorado Springs in late February 1896," recounts Seifer, "to look over a prospective site for a new laboratory and also to conduct the kinds of wireless experiments he had wanted to undertake before his laboratory burned to the ground." It was here that his telautomaton came to life, and when Astor heard the news he rushed to assist Tesla.[17]

The change in Astor's response to Tesla was dramatic. "Come to Cuba with me," invited the financier, "where you can demonstrate your work upon the insufferable scoundrels." Astor, or the Colonel as he now was called, "donated $75,000 to the U.S. Army to equip an artillery division for use in the Philippines and lent the *Nourmahal* to the Navy for use in battle. The tall [steam-driven] ship was nearly a hundred yards in length, was equipped with a corps of military seamen [. . . and] made a formidable warship." Colonel John Jacob Astor, later honorary inspector general, was one of the first robber baron moguls to attempt the fusion of scientific technology and the military. Thus, even as far back as 1898, the military-industrial complex was well on its way to achieving amalgamation. Beating the Spanish with modern instruments of destruction was on the minds of all the heavyweight tycoons who had previously been financing inventors for profit. Now they were turning their attention to war, imperialism, and American hegemony abroad, which would further increase their economic dominance. This time, however, it was Tesla who turned down the Colonel, saying that he had been called "for a higher duty."[18]*

*While Tesla discarded spirituality and religion as merely an intellectual process on the brain involving imprints one has already experienced, his comments on the imagination are intriguing and perhaps tell a different story. Tesla writes in *My Inventions,*

> For a while I gave myself up entirely to the intense enjoyment of picturing machines and devising new forms. It was a mental state of happiness about as complete as I have ever known in life. Ideas came in an uninterrupted stream and the only difficulty I had was to hold them fast. The pieces of [the apparatus] I conceived were to me absolutely real and tangible in every detail, even to the minutest marks and signs of wear. I delighted in imagining the motors constantly running, for in this way they presented to the mind's eye a fascinating sight. *When natural inclination develops into a passionate desire, one advances*

What exactly this "higher duty" was no biographer to date knows. One can only speculate that Tesla's idea of profits did not pertain to ego, nationalism, or market resources; rather, the profits he sought reflect a different aspect of the American mythos, specifically, the importance of ideas bereft of hindrances in the public sphere. Tesla was displaying the same spirit the Founding Fathers displayed in the political arena during the country's mythic past. He sought an environment that was conducive to the free exchange of ideas. Nevertheless, Tesla's telautomaton was not commercialized or accepted by either the public or the military. Like many of his inventions, it was not a commercial success "for reasons difficult to understand."[19] In short, it was a complete failure financially. The invention that the public flocked to instead was Marconi's wireless detonation system. The public wanted what it could understand and did not fear—this space and time that separated human from machine equated to lack of control and predictability.

Seifer writes, "Tesla's telautomaton remains one of the single most important technological triumphs of the modern age. In its final form, it was conceived as a new mechanical species capable of thinking as humans do, capable of carrying out complex assignments and even capable of reproduction. The invention also comprised all the essential features of wireless transmission and selective tuning. Here was a true work of genius."[20] The imagined mechanical-man was now jumping from the pages of fiction into reality. The American mythos of creationism ancillary to the mythos of the Christian Nation and Millennial Nation was now being backed up by more than the writings of speculative entertainers of mechanical-men as mythic-markers.

> *towards his goal in seven-league boots.* In less than two months I evolved virtually all the types of motors and modifications of the system which are used under many other names all over the world. (49, emphasis added)

Here we glimpse the feeling of spiritual freedom in a technocracy, a freedom based on the imaginings of the Transcendentalists and the Metaphysical Club. If this autobiographical statement does not place one in what Colin Wilson calls a "peak experience," perhaps nothing will.

The mythic-markers were now taking the form of tangible inventions.

Meanwhile, an experience had occurred that would change Tesla's outlook on life and human nature. At a dinner on February 13, 1896, he met Swami Vivekananda, who preached, among many things, chastity as a path to enlightenment. Tesla had always been intrigued by such individuals since his time in Budapest. He "had first heard of the Swami during the summer of 1893 when the Hindu gained overnight prominence after speaking at the Congress of World Religions, which had been held at the Chicago World's Fair." At the Fair, Tesla was giving speeches on the mechanics involved in creating remote-controlled automatons, while the Swami was speaking of a universal mind and the idea that all historical events are stored in some over-soul of thought. In the words of Joseph Campbell, this "supernatural principle of guardianship and direction unites in itself all the ambiguities of the unconscious, thus signifying the support of our conscious personality by that other, larger system, but also the inscrutability of the guide that we are following, to the peril of all our rational ends."[21] In effect, the Swami was providing Tesla with a different viewpoint in how one should view the processing of knowledge and how the cosmos is a spiritual memory bank of feedback control, which possesses the hidden laws of nature. In terms of American mythos, the Swami was giving the immigrant a crash course on American Transcendentalism, causing Tesla to study the writings of Ralph Waldo Emerson, Henry David Thoreau, Margaret Fuller, James Freeman Clarke, Frederic Henry Hedge, Orestes A. Brownson, and George Ripley.*

*In the *Cambridge Dictionary of Philosophy* Robert Audi defines *Transcendentalism* as:

> [A] religious-philosophical movement held by a group of New England intellectuals whom Emerson, Thoreau, and Theodore Parker were the most important. . . . The Transcenden[talists] insisted that philosophical truth could be reached only by reason, a capacity common to all people unless destroyed by living a life of externals and accepting as true only secondhand traditional beliefs. On almost every other point there were disagreements. Emerson was an idealist, while Parker was a natural realist—they simply had conflicting a priori intuitions. Emerson, Thoreau, and Parker rejected the

Seifer offers the following observation: "To Swami Vivekananda, creation was a process of combining existing elements into a new synthesis. This idea of the eternal nature of existence with no beginning and no ending was appealing to Tesla."[22] It was at this time that Tesla formulated his ideas on the nature of humankind and automatons. "Though we may never be able to comprehend human life, we know certainly that it is a movement, of whatever nature it be . . . a mass urged by force."[23] The Swami was the catalyst Tesla needed to crystallize his concept of an automaton. Drawing on Newton's laws of force, action, and equal/opposite reaction, Tesla envisaged humankind as one physical mass. "Human energy will then be given by the product ½ MV, in which M is the total mass of man in the ordinary interpretation of the term mass, and V is a certain hypothetical velocity which, in the present state of science, we are unable to exactly determine."[24] This equation led Tesla to write, "I have demonstrated, to my absolute satisfaction that I am an automaton endowed with power of movement, which merely responds to external stimuli beating on its sense-organs, and think and act accordingly?"[25]

What then of the mind of such an automaton? Tesla claims that "I could easily embody in it by conceiving to it my own intelligence."[26] Thus, Tesla was of the belief that the conjuring up of an idea comes from something outside the brain. This dynamic was similar to the input and output of a computer. Once something is programmed into a computer, there can be an equal and opposite output. This realization is the first step in Tesla's formula; the next, in Newtonian terms, is the process within the machine; and the last is the output or creation, since invention does not come from nothing. In simple terms,

---

supernatural aspects of Christianity, pointing out its unmistakable parochial [and patriarchal] nature and sociological development. . . . Emerson, Thoreau, and Parker were also bonded by negative beliefs. They not only rejected Calvinism but Unitarianism as well; they rejected the ordinary concept of material success and put in its place an Aristotelian type of self-realization that *admires what is unique in people as constituting their real value.* (emphasis added)

Tesla believed that if he could imagine or conceive of something, it was possible. Consciousness and free will become obsolete in Tesla's universe. Consciousness is merely the process of reaction time, or the time it takes one's knowledge to go from input to output. This is Tesla's contribution to the interplay of the creative and practical. Tesla evolves the arguments earlier voiced by Poe and Melville on how the creative and practical interplay by merging the two. Thus, Tesla's philosophy on the mechanics of the internal and external or the conscious and unconscious creates the very foundations for a reciprocal relationship between the imaginers of science and those that practice it.

Scientist and historian T. G. White is critical of Tesla's argument and deductions. Regarding the idea that Tesla could transfer his mind to that of an automaton, he says:

> With all these slight differences from the Garden of Eden, Mr. Tesla has turned out a kind of electrical boat or animated bath tub whose movements can be controlled from a distance. This machine behaved just like a blindfolded person obeying directions received through the ear. It had a borrowed mind. In fact, it had no mind at all, except in Mr. Tesla's confused terminology. It had the same kind of mind as . . . a steam-engine has its own mind, and distinguishes what it ought from what it ought not to do. So did Maelzel's [chess-player]. So did Babbage's calculating device. So does every device.[27]

However, there was something different about Tesla's device. The telautomaton was not merely a blindfolded person obeying directions; instead, it was an extension of the autonomous human being. It possessed the ability to communicate. And while electronic signals are far from the complexities involved in human language or mathematical symbols, it unleashed new insights into how human-machine interactions of communication could potentially evolve.

With complexity comes sacrifice. The telautomaton's complexity is indicative of trade-offs, specifically independence for dependency and nature for technology. Indeed, the mechanical boat operated like a "blindfolded person obeying directions received through the ear."[28] Nevertheless, it was a starting point that allowed the mythical garden to transform itself. Tesla's automaton differed from Maelzel's chess-player in being an extension of the wireless revolution and of man's dependence upon himself and others rather than on nature. Controlling and communicating with machines was now separated by space and time, which brought to light more inherent issues in human-machine interactions—space and time between humans and machines equated to less direct control and predictability between the goals directed by humans and the subsequent actions accomplished by machines. What critics like White and others failed to grasp was that Tesla's "confused" terminology is at the very crux of cybernetics. More specifically, Tesla's ability to differentiate the telautomaton from Maelzel's chess-player hoax showed the need for a whole new set of different terminology and a novel worldview when discussing human-machine interaction.

Cybernetics is then not merely the study of communication and control but also a tool utilized to chisel the sculpture of a culture's collective unconscious out of the mystery that is the coevolution of the real and fictive. Campbell holds that the collective unconscious is:

> Used in recognition of the fact that there is a common humanity built into our central nervous system out of which our imagination works. The appeal of these constants is very deep. . . . Why does this group see themselves as the special people, though a group over there doesn't think that way—their deities have to do with the world of nature? What is it that gives those different pitches to the different culture systems—not special history and biography—but what about the general humanity? You can recognize a human being no matter where you see him. He must have the same kind of basic nervous system; therefore, his imagination must work out of a comparable base. What's so mystical about all that? That seems to me to be obvious. And that's what the term "collective unconscious" covers.[29]

The mechanical-man in the form of the chess-playing hoax and Tesla's telautomaton show us an emerging techno-culture with new life possibilities and requirements. With such enormous transformation in the forms of American society, the previous models and heroes were meeting their eventual death. Even God himself was said to be dead. The mechanical-man was a new hero that was his own hero and followed the path "that's no path."[30] A path of new ambiguities, anomalies, and synchronicities that are hidden in the present circumstances of experience now emerged. The facile magic of control and communication needed to reorient us with our environment and ourselves. But in order to do this society needed to change how it viewed the orthodoxy of defining control and communication between ourselves and our machines.

A COMPLEX DESIGN: THE MIND OF THE TELAUTOMATON

Tesla's automaton caught America off guard and was truly far ahead of its time. It integrated novel systems and networks in the fields of electromagnetism and wireless telecommunications. Also, in Tesla's blending of numerous systems, the age of automatons was now open to the military-industrial complex. The garden had been transformed from a pastoral ideal of autonomy and independence to a corporate and militaristic battlefield. Perhaps this is why Tesla shied away from imposing his automaton upon the public and why Americans preferred Marconi's detonation device over the mechanical boat.

Besides Seifer, another of Tesla's recognized biographers, Margaret Cheney, writes that the age of remote-controlled machines in America owes its inception to Tesla's sorcery, without which "the new technology of remotely piloted vehicles" would never have been created.[31]

Other models of automatons on the battlefield would emerge, but it was Tesla's telautomaton that fused simple ideas in telecommunication and electromagnetism to create an automaton that completed complex tasks. Tesla thus paved the way for others to introduce robotics in a technological environment dependent on networks, systems, and cooperation. What is still perplexing, however, and what no biographer has been able to discover, is why Tesla felt that the world was "manifestly unready" for his automaton at the turn of the century. Perhaps he realized that delving into some areas of curiosity is forbidden; perhaps he foresaw in the invention of automatons the possibility of a new race that could lead to the eventual destruction of humans. In this foresight, he too was collapsing reality and prophecy similar to the speculative writers of mechanical-men. Like those writers of speculative fiction, he foresaw the consequences of invention placed in human reality. Whereas the difference between those that speculate on the consequences of invention in their writings and those that speculate on the consequences of their inventions to human reality is self-evident, the bridge between the two is that both are imbued with the human attribute of foresight.[32]

In terms of American cultural mythology, Tesla had separated himself from the public's capricious desires. In this separation phase, like mythical heroes before him—"Prometheus, who ascended to the heavens; Jason, who sailed through the Clashing Rocks into a sea of marvels; and Aeneas, who went down into the underworld and crossed the dreadful river of the dead"[33]—Tesla had come down from the mountains of his homeland and landed in the urban jungles of New York City. From the chaos of this new environment, he too was able to steal fire from the gods. Tesla's telautomaton, however unpopular its reception at the time, had crossed the first threshold of the automaton as myth. In its invention, built on childhood fantasies, Tesla foreshadowed the emergence of American cyber-technology. He prefigured, in other words, the extension of the human to robotics and artificial life/intelligence where the imagination is the cardinal tool for creation.

Specifically, on issues of control and communication, the telautomaton functioned at a level far surpassing that of a mere closed clockwork basis. Norbert Wiener explains that receptors for messages of any kind must be looked upon as sense organs:

> These may be simple as photoelectric cells which change electrically when a light falls on them, and which can tell light from dark, or as complicated as a television set. . . . Every instrument in the repertory of the scientific-instrument maker is a possible sense organ, and may be made to record its reading remotely through the intervention of appropriate electrical apparatus.[34]

The remote-controlled activities between the controller and the boat contained inputs and outputs of complexity. Due to its relay coils and magnets, the boat is similar to later machines that Wiener says, "frequently consist of punched cards or tapes or of magnetized wires, and which determines the way in which the machine is going to act in one operation, as distinct from the way in which it might have acted in another."[35] The boat's communication with the controller was showing parallels to a human's ability to control entropy through feedback. Both of them have sensory receptors as one stage in their cycle of operation: that is, in both of them there exists a special apparatus for collecting information from the outer world at low energy levels, and for making it available in the operation of the individual or of the machine.[36]

Neurologist Donald Hebb, in 1948, showed that connections between neurons in the brain become stronger the more often one of them excites the other. *Connectionism* is the term used to describe the way neurons in the brain are interconnected. Hebb theorized that neurons become increasingly sensitive to and aware of one another the more often they are used—when they are exchanging information. As a result, pathways form for signals to travel across. As the pathways develop and establish themselves, learning and memory take place. The principle itself is called *Hebb's Law:* "nerve cells that fire together, wire together." Hebb's fundamental thesis holds:

> Any frequently repeated, particular stimulation will lead to the slow development of a 'cell assembly,' and diffuse structure comprising cells in the cortex and diencephalon, capable of acting briefly in a

> closed system, delivering facilitation to other such systems and usually having a specific motor function.[37]*

So imagine for a moment that the controller of the boat is neuron A, the boat itself is neuron B. As the boat sends signals back to the controller and vice-versa, the signals become stronger, and as these pathways develop and establish themselves, learning and memory takes place. For Tesla, the frequently repeated stimulation of signals facilitates a closed system displaying specific motor functions.

The "relay magnet" of the boat and the "relay coils" are the parts of the apparatus that send messages back to the terminal, messages pertaining to location, weight, etc. . . .[38] As certain magnets and coils are utilized more than others, the boat begins to react to its external environment, evolving with the environment. In addition, the more these magnets and coils are utilized, the better adapted they become in picking up signals or remembering how the external environment affects its relays. Ulrich Neisser harks back to the telautomaton holding that:

> We must begin by admitting that machines can behave intelligently and purposively. Such a direct use of these terms may be surprising. Is not a machine made of simple and unintelligent pieces? Indeed it is, but here as elsewhere the whole is more than the sum of its parts. Out of many thousands of simple operations by relays or transistors can come something unpredictable and

---

*Joe Dispenza writes in *Breaking the Habit of Being Yourself:*

> There is another possible consequence that I should mention, if you keep firing the same neural patterns by living your life the same way each day. Every time you respond to your familiar reality by re-creating the same mind (that is, turning on the same nerve cells to make the brain work in the same way), you "hard-wire" your brain to match the customary conditions in your personal reality, be they good or bad. There is a principle in neuroscience called *Hebb's Law.* It basically states that "nerve cells that fire together, wire together"; Hebb's credo demonstrates that if you repeatedly activate the same nerve cells, then each time they turn on, it will be easier for them to fire in unison again. Eventually those neurons will develop a long-term relationship. (45)

> adaptive, perhaps a new proof for a logical theorem or persistent search for an elusive target.[39]

On memory, Tesla stated, "It is but increased responsiveness to repeated stimuli." Thus, consciousness itself is merely "secondary reverberations of these initial external stimuli." There is no arguing that Tesla's telautomaton was "responsive to repeated stimuli," in that it exhibited the power to react to directions and make decisions based on the input it received.[40] Similar to the human brain that makes decisions based on the input of repeated environmental stimuli, the automaton was also capable of experiencing secondary reverberations of these initial external stimuli, and thus, in its reactions, displays the same qualities of memory that the human brain exhibits. David F. Channell alludes to Wiener's second edition of *Cybernetics* (1961), when he discusses the inherent evolutionary characteristics of machines. "Wiener discussed how an electronic device might be designed that could be capable of causing another electronic device to reproduce the specific pattern of the original device. His model was taken from genetics."[41] This intimates a convergence of mechanical and organic terms with worldviews.

The question then arises: How does Tesla's telautomaton fill in the gaps between the procurement of the invention and the meaning of the boat in terms of human-machine interaction? Channell answers this question by evidencing a change and convergence in society, transforming itself from viewing the world in solely mechanistic or organic terms into a fusion of the two. In the boat, Tesla saw human attributes of reason and the ability to learn. Instead of viewing his own body in mechanistic terms he viewed the boat as a living organism. In his application for the boat's patent he assiduously makes claims of the boat's autonomy in making its own decisions and its ability to learn and retain knowledge. Even today, when we think of the control issues involved in remote-controlled activities, we place little autonomy in the actions of the device that is controlled. However, Tesla's patent for the boat shows that he views the controlled device to be in the possession of the same attributes as the human is in possession of the signal device. In the boat, Tesla saw the ability to transform and change. As Channell points out, in the vital machine worldview: "Change did not have to be explained as some special situation; growth, development, transformation, and evolution were inherent characteristics of all phenomena."[42] The boat changed depending on the repetition and utilization of certain relay magnets and coils over others. This process allowed the boat to make choices based on inputs of the natural environment. The boat was no more the controlled than the controller itself in the eyes of Tesla. On constructing an automaton that would mechanically represent himself Tesla says,

> Long ago I conceived the idea of constructing an automaton which would mechanically represent me, and which would respond, as I do myself, but of course, in a much more primitive manner to external influences. Such an automaton evidently had to have motive power, organs for locomotion, directive organs and one or more sensitive organs so adapted as to be excited by external stimuli. . . . Whether the automaton is of flesh and bone, or of wood and steel, it mattered little, provided it could provide all the duties required of it like an intelligent being.[43]

Imbuing his replica with "one or more sensitive organs so adapted as to be excited by external stimuli" shows the importance that Tesla places on viewing mechanical creations as living organisms.[44] The automaton, like all living things, must have the ability of adaptation and change. Tesla's ideation of creating a replica of himself in addition to his views on the boat's purposeful attributes creates a concordance with Channell's view that there exists a contradiction between the mechanical and organic worldviews. It so follows that: "In the sense that it is a machine it must be treated with an attitude of control and responsibility while in the sense that it is alive it must be treated with an attitude of autonomy and freedom."[45] The boat was neither fully autonomous nor free; however, in the messages the boat sent back to the controller, Tesla saw freedom and autonomy; in the way the boat was free to interact with its environment, Tesla saw autonomy. And while there was still a degree of control and responsibility between the controller and the remote device, in Tesla's mind the relay magnets and coils were no different from "directive organs." Indeed, society now began to realize that within the nebulous of human-machine interactions, perception was seen to be as important as reality itself—and if our creations "are treated as organic beings with intrinsic values, it becomes impossible for their human creators to take [full] control of their creations."[46]

THE LEGEND OF THE WIZARD

In 1886 the French writer Villiers de l'Isle-Adam published a novel titled *L'Ève future* (*Eve of the Future*) that centered on the life and times of Thomas Edison. In it, Edison as protagonist manufactures the ideal woman but laments "Dead voices, lost sounds, forgotten noises, vibrations rippling in the abyss, too distant now to be picked up."[47] Edison had already tapped into the forbidden, even before limning his ideal woman. His invention of the phonograph, the ability to record spoken words and sounds, changed our relationship to both space and time. Moreover, his invention allowed one to create plausible replicas of every human sound. For Villiers, perception was the key to reality. Thus, it is one's perception of sound that matters and not its source. The primacy of perception for human cognition transforms man's relation to technology, causing him to question the constructs he has created. This parallels Baum's story of the Tin Man and the Wizard of Oz, for it is the Tin Man's perception of the wizard, rather than the reality of a small gray-haired man behind a curtain, that sustains his hope of acquiring a heart.

For Villiers's virtual Edison, the phonograph is the device that will either disprove or prove the existence of God. In the novel Edison says, "As for the mystics, I submit to them an observation, naïve, paradoxical, superficial, if you wish, but singular: isn't it disheartening to realize that if God, the Most High, the Good Lord, call Him what you will, the Omnipotent, to realize, as I was saying, that if He deigned to let us . . . record a simple phonographic proof of His true Voice . . . ?"[48] Villiers here recognizes the full implications of Edison's invention. Specifically, Edison was defining communication and control in new ways through his inventions. This would have far-reaching effects on human-machine interactions. Once we have a thorough understanding of the importance of the phonograph, Villiers directs our attention to Edison's fictional creation of the ideal woman. When a young English nobleman, Lord Celian Ewald, visits Edison, it appears as if Edison owes Ewald a favor. Ewald is in a desperate state, even voicing thoughts of suicide. Edison comes to his friend's rescue by inquiring into what ails him. It turns out that Ewald is in love with a woman who is "imprisoned, by some sort of occult punishment, in the perpetual denial of her ideal body." Thus, there is discordance between her body and soul; the woman's outer beauty is exquisite, yet her soul is corrupt and disdainful. "'To summarize,' states Ewald, 'this woman manifests a cynical candor and lack of awareness which I cannot ignore even from a distance. I am not, as I have said, one of those men who can accept the body while rejecting the soul.'"[49]

In this respect, the definition of *soul* is how others perceive a person's behavior. The reader never fully understands what the woman is really like; we just know her through Ewald's eyes. Nevertheless, that is Villiers's point. When it comes to androids, "soul" is nothing more than the interactions we perceive of them in the virtual environment in which they are placed. "Soul" is not some tangible mechanism installed within an android; instead, it is the emotions that we as humans come to associate with the robot based on our experiences with it. Thus, we need not know what is going on in the "minds" of robots so long as our perception or mythos of the robots' imagined "minds" adds to our understanding of what it means to be human, especially when we come to recognize that we too are experiencing a holographic simulation. This progression in how humans view the soul places an emphasis on the imagined. In Edison's creation of the idealized woman in fiction we see the connection between the mythos of Edison as wizard and the reality of Edison the inventor humanizing the sciences for society. In the same way the wizard in *Oz* utilized the human attribute of perception to procure hope in the hearts of his people, so too did Edison in fiction and reality. Indeed, the majority of his inventions were based on experimenting with human perception. Believing that an android is an idealized woman is no different from losing oneself in the ritualized cave of celluloid images projected onto a screen in a theater.[50]

Villiers's novel has Edison say the following about his virtual creation:

> I shall make a double of this woman with the sublime aid of light. And projecting her upon its radiant matter, I shall illuminate with your melancholy the imaginary soul of this new creature, who will astonish angels. I shall bring illusions down to earth! I shall make it our prisoner. In this vision I shall force the Ideal itself to show for the first time to your sense, palpable, audible, and materialized. I shall arrant in the expanse of flight the first hour of this enchanted mirage that your fancy pursues in fain . . . that I positively to make emerge from the clay of present-day human science a creature made in our image and how, consequently, will be to us what we are to God.[51]

Today we have the capacity to record sounds and images of human beings, which can make them seem alive even after death. It is as if we have the power to stop time's arrow, or at least reverse its trajectory temporarily. Similarly, the first step in creating a robot's "soul" is a simulacrum that fools us into believing in its empirical reality. The mechanical-man then becomes our personified self. Mechanical-men in fiction and automatons in reality were now in a sense becoming human. At the very least, they were displaying such actions as to fool us into believing in their empirical reality.

David Deutsch defines virtual imaging as "any situation in which a person is artificially given the experience of being in a specified environment." He goes on to explain that "we experience our environment through our senses[. A]ny virtual-reality generator must be able to manipulate our senses, overriding their normal functioning so that we can experience the specified environment instead of the actual one."[52]* For Villiers, Edison's phonograph was the first step in realizing virtual simulation. Similar to a child's toy that tricks the senses into believing the hyper-real, the phonograph and later the movie camera provide simulacra of the authentically human sphere.

For Deutsch, the brain itself is a "virtual-reality generator":

> We realists take the view that reality is out there: objective, physical and independent of what we believe about it. But we never

---

*Winner of the Paul Dirac Medal and Prize, David Deutsch, on the shoulders of Richard Dawkins, presents the ultimate theory for materialists. And with all the unifying theories presented since Einstein's own endeavor, by the preponderance of the evidence, it comes the closest to getting us to the end. But, as the *mechanical-man* shows us by the end of this analysis, materialism, like reductionism itself, has theorized itself down to nothing. There is no need for me to point out the "gaps" in this theory. Nevertheless, for our purposes, chapter 5 in Deutsch's *Fabric of Reality,* which discusses virtual reality, is significant, even today. He writes:

> One of the most important concepts of the theory of computation is universality. A universal computer is usually defined as an abstract machine that can mimic the computations of any other abstract machine in a certain well-defined class. However, the significance of universality lies in the fact that universal computers, or at least good approximations to them, can actually be built, and can be used to compute not just each other's behavior but the behavior of interesting physical and abstract entities. The fact that this is possible is part of the self-similarity of physical reality[.] (98)

He then delves into a discussion about virtual reality that is well worth one's time reading.

> experience that reality directly. Every last scrap of our external experience is of virtual reality. And every last scrap of our knowledge, including our knowledge of the non-physical worlds of logic, mathematics and philosophy, and of imagination, fiction, art and fantasy, is encoded in the form of programs for the rendering of those worlds on our brain's own virtual-reality generator.[53]

The brain is the organ by which we survive. Thus, "biologically speaking, the virtual-reality rendering of their environment is the characteristic means by which human beings exist."[54] What Edison created in Menlo Park, then, is a universal image generator that "can be programmed to generate any sensation that the user is capable of experiencing."[55] The android's environment is one of illusion, but so too is our own environment, since it could be argued that our brains create the illusion of consciousness and free will, which coincidentally is exactly what Tesla and Edison both believed.

Villiers touches on a problem that has plagued artificial intelligence research and robotics for years—namely, how does one create a "virtual-reality generator" for all possible environments? For Deutsch, "the heart of a virtual-reality generator is its computer, and the question of what environments can be rendered in virtual reality must eventually come down to the question of what computations can be performed."[56] Like Poe and Melville, Villiers places great emphasis on one's goals in utilizing instruments and inventions in order to fool the senses. Inventions such as Edison's phonograph, in effect, were supplanting their human originals. Villiers, in fiction, was doing the same thing to Edison that Edison was doing to the world with his recording devices, namely capturing and taking possession of him to do what he will and disembodying him from reality.

In Villiers's novel, we should note, Ewald begins to question the reality of Edison's fictional android by saying, "Just a moment. Whatever it is, it's my freedom of thought, my love itself, I would be deprived of if I submitted my spirit to its encounter."[57] At this moment,

Ewald begins to recognize that the creation is a function of his own brain's program. Thus, he too is in bondage, capable only of reacting within the confines of his own programmed "virtual-reality generator." Edison responds: "'What does it matter, if it guarantees the reality of your dream? And who is really free? The angels of the old legends, perhaps. And they alone have won the right to freedom, in fact! Because they are effectually delivered from temptation, for having seen the abyss wherein have fallen those who willed to think.'" "Without illusion," concludes Villiers, "everything would perish. No one can avoid it. *Illusion is illumination*."[58]

Villiers then, like Deutsch, believes that Darwin solved the problem of consciousness, and in the area of robotics this problem is now moot. Deutsch says that "the problem was solved by Darwin. The essence of the solution was the idea that the intricate and apparently purposeful design that is apparent in living organisms is not built into reality *ab initio,* but is an emergent consequence of the operation of the laws of physics."[59] Thus, an android's supposed consciousness is merely an extrapolation of the laws of physics in conjunction with the Neo-Darwinian theory of evolution; it emerges as a consequence of neural activity and the transferring of information in the brain. But reading Villiers more closely intimates there may be a two-person dance going on within all of us—our right brain dancing with our left brain—creating a first experience of subjectivity marked by feelings and a second person experience of objectivity marked by perception and measurement.

What Edison offers Ewald in Villiers's novel is "the possibility of preferring a positive, prestigious, eternally faithful illusion to deceptive, mediocre, eternally changing reality."[60] The creation of the android is then an act of free will. Moreover, free will involves the ability to create an illusion that is eternally faithful and positive. It involves the ability to create something that is unchanging. Herein lies the truth of Edison's inventions: he was engaged in creating machines that would give one a feeling of unchanging reality, a predictability that the world had not seen before. He was on the verge of preserving the voice and image of man for all time—and he succeeded. He discovered, in the end, that controlling your perception is the key if you believe it is the key. Even as far back as Edison and Tesla we begin to see how the art and science of manifesting on the individual and collective levels of questing. And again and again, we keep finding in the gaps patterns projecting that a "controlled imagination and a steadied, concentrated attention results in inventing one's ideal invention. And so it makes sense that ignoble thoughts and actions inevitably result in unhappy consequences."[61]

SIMULACRUM OR REALITY?

Once Ewald falls in love with his phantom android, he recognizes how serious the implications are. He says, "'What kind of man had imagined that this sinister automaton could move me by the aid of some kind of paradox embossed on her metal leaves? Since when did God permit machines to be spokesmen? And what derisory pride these electrified phantoms take on when they put on female form and meddle in our lives.'"[62] This paradox allows Ewald to regard the machine as a person, perhaps even a spokeswoman for other androids. Indeed, the android is elevated to a status above that of a human being. Neither machine nor human, it is able to communicate with both. The android ironically illuminates humanness, allowing its creator to reach a higher level of consciousness in terms of agape (unconditional love to self and selfless love to others).

In the end, Edison's android in Villiers's novel becomes more "flesh and blood" than Ewald's original love, Alicia. The android placates his emotions by telling him that it is his love that brought her to life and made her human; thus, it is his perception of her that imbues the android with human attributes, while also intimating that *love itself may be a natural force* that brings vitality to inanimate and lifeless objects. Ewald then leaves with the android in an ebony coffin, and they depart aboard a steamer named *Wonderful*. On the voyage he is accompanied by Alicia, whom he now can barely tolerate. Although she is human, Ewald recognizes in her the shortcomings of flux and imperfection: noticing her wan smile and fading beauty, he yearns for the eternal embrace of the android. His experience has edified him about human nature, space, and time. If conscious awareness of self was once elusive, it is now very tangible. Indeed, the android has held up a mirror to his face, showing him the inherent weaknesses of human evolution and transience. Our dependence on technology is furthered by this realization. Harnessing the power of the imagination allows one to prevent change, decay, and the process of aging itself. The implication of this is that humans were attempting to become more godlike, asserting control over the imagination and thus, reality itself. The American mythos of conquest and destiny was now taking on a different meaning with these new implications; specifically real people like Tesla and Edison were now seen as wizards and sorcerers due to their ability to change human perceptions of reality through their inventive genius.

Meanwhile, back in the novel, some days later, while reading a newspaper, Edison comes across a report that the *Wonderful* caught fire:

> A fire broke out in the back hold around two-o'clock in the morning in the merchandise compartments where barrels of mineral gas and spirits, heated to combustion by an undetermined cause, burst into flame.
>
> The sea was rough, and as the steamer was pitching considerably, a sheet of flame penetrated the baggage compartment almost at once. A wind from the west fed the blaze so fast that fire appeared at the same time as the smoke. . . . When the captain announced that the ship would go under in five minutes, all rushed toward the lifeboats which were set afloat in a few seconds. . . . During this horrible scene, a strange incident occurred in the middle bridge. A young Englishmen, Lord Ewald, having seized a bar of the hatchway, wanted to penetrate into the trunk and baggage section despite the

> flames. He knocked down the lieutenant and one of the first mates, who tried to stop him. No fewer than six sailors were required to restrain his insane plunge into the fire. While struggling, he cried that he wanted to retrieve at any price a trunk enclosing an object so precious that he would offer a reward of one hundred thousand guineas to anyone who would help him seize it. . . . It was necessary to bind him hand and foot and carry him unconscious into the last boat, whose passengers were picked up. The number of victims is estimated at seventy-two. A list of some of these unfortunates follows. (An unofficial list followed. One of the first names was Miss Emma Alicia Clary, lyric soprano.)[63]

Villiers gives us a glimpse of the bridge connecting the nineteenth century with the modern age, an age based on relativity, simulacra, and quantum physics. In effect, *Eve of the Future* is a treatise on American technological growth in the latter half of the nineteenth century. Like Alexis de Tocqueville in his analysis of Jacksonian America in the early part of the century, Villiers captures the mythic centrality of independence by describing one of the culture's most important inventors. And by doing so he allows his readers to look in awe at what awaits them on the other side.

Edison's simulacrum in Villiers's story anticipates the thinking of philosopher Jean Baudrillard. Villiers's virtual Edison becomes hyper-real—that is, the actions of the simulacrum, reflecting Edison's historical life, become more important than the man himself.

Reality could go beyond fiction: that was the surest sign of the possibility of an ever-increasing imaginary. But the real cannot surpass the model—it is nothing but its alibi. The imaginary was the alibi of the real, in a world dominated by the reality principle of simulation. And, paradoxically, it is the real that has become our true utopia—but a utopia that is no longer within the realm of the possible, that can only be dreamt of as one would dream of a lost object.

On why the United States is set up similar to the qualities of a hologram, I recommend Jean Baudrillard's *America,* where he writes:

> America is neither a dream nor reality. It is a hyper(-)reality. Is it a hyper(-)reality because it is a utopia which has behaved from the very beginning as though it were already achieved? Everything here is real and pragmatic, and yet it is the stuff of dreams. It may be that the truth of America can only be seen by a European, since he alone will discover here the perfect simulacrum—that of the immanence and material transcription of all values. The Americans, for their part, have no sense of simulation. They are themselves a simulation in [a] most developed state, but they have no language in which to describe it, since they themselves are the model. As a result, they are the ideal material for an analysis of all the possible variants of the modern world. No more and no less in fact than were primitive societies in their day. The same mythical and analytic excitements that made us look towards those earlier societies today impels us to look in the direction of American. . . . America is a giant *hologram* in the sense that information concerning the whole is contained in each of its elements. Take the tiniest little place in the desert, any old street in a Midwestern town, a parking lot, a Californian house, a Burger King or a Studebaker, and you have the whole of the US—South, North, East, or West. Holographic also that it has the coherent light of the laser, the homogeneity of the single elements scanned by the same beams. From the visual and plastic viewpoints too: things seem to be made of a more unreal substance; they seem to turn and move in a void as if by a special lighting effect, a fine membrane you pass through without noticing.[64]

Both the imagined Edison and the real Edison are equally important. This idea sheds light on the importance of analyzing the history of the mechanical-man in terms of their imagined history (literature/film) and real history, for both are equally hyper-real and create perceived models for learning and simulation. Thus, reality's importance is the appearance of the simulation. In short, appearance and simulation of reality correlate to myth in its ability to bring consciousness to what may be hidden or forgotten. Whether such simulation is contrived or said to be real is irrelevant. The word *reality* no longer has any meaning or priority in this hyper-real sphere. The myth of Edison adds to the overall American narrative of conquering and harnessing nature for conquest. The technological transformations he wrought illuminate the instrumental value of diversity and competition in the arena of scientific ideas. In addition, his life demonstrates how dependency on diversity and competition to further technological progress does not go against the grain of the American narrative of conquest and destiny. Instead, it magnifies the myth's importance, especially for immigrants who embrace the rags-to-riches dream, for which machine culture and industrialization were central to their achievement of this dream.

THE REALITY OF THE WIZARD

At the end of World War I, the United States "refused to recognize the various treaties by which co-belligerents had already divided the spoils of victory."[65] It was as if the refusal indicated that we had already embarked upon a pilgrimage of virtue, the war itself being America's way of sanctifying itself, "the necessary prelude to mankind's salvation."[66] World War I changed people's conscious awareness of how they viewed the mechanics of humanity on a larger scale. The machine was no longer an intruder into the pastoral setting of early America; instead, people now envisioned the machine in the urban inner cities and on the war-torn battlefields of Europe. Of the Great War, historian Robert H. Wiebe writes,

> The nation threw itself into the conflict and with surprising efficiency made its resources available to the Allies. Its navy helped break the grip of submarines on transatlantic commerce. [I]ts raw

> materials, food and petroleum in particular, fueled the Allies' sputtering war machine, and by the summer of 1918, its troops, arriving far earlier than most had anticipated, gave essential movement to the counterattack that unnerved Germany's weary leaders.[67]

For Wiebe it was not blind hubris that propelled America to victory but rather observable facts. "The society that so many in the 1890s had thought would either disintegrate or polarize had emerged tough and plural."[68] America, by the time of the war, had found its identity.

The nation had spent the last fifty years since the Civil War and the tumult of Reconstruction reinventing itself out of chaos. Lightning had struck in America during this epoch, and from the country's industrialization emerged a titan whose politicians glimpsed the emergence of a new world order. Other men, however, were influential in this national awakening, scientists like Thomas Edison who sought to change nature, perhaps even altering time through the use of celluloid images and voice recordings. All in all, it was a crusade to venture into the forbidden source of power that is life, defying the orthodoxy of old and turning the universe upside down.

For Edison, however, the course of public events was full of surprises that could come unexpectedly. Germany had just sunk the RMS *Lusitania* in May of 1915, and Edison, who had vowed neutrality, looked upon the imbroglio with new eyes. He now sought to defend against the German threat. Nonetheless, he asked, "How can we help by going to war? We haven't any troops, we haven't any ammunition, and we are an unorganized mob."[69]* Organization arising from chaos was an

---

*Edison held the pragmatic view that the idea of organization was a tool, but so too was the idea of chaos. The task of the inventor was to create a productive union of the two. In the "American system" of practicing science, we see Edison defining incorporeal ideas into tools. The foundation is being set for the mechanical-man to look at processes and words like *organization* and *chaos,* extrapolate on these ideas within an idealistic paradigm and see the imagination as the ultimate tool for innovation and manifesting. In Edison's inventions we see each invention as a part (fragment) of the whole that is the American simulation as seen by Baudrillard. See Hughes, *American Genesis.*

axiom Edison often used in describing his scientific experiments, perhaps not realizing that the same principle operated in the political and social arenas of America during the years leading up to the Great War.

A pacifist for many years, Edison was moving toward calling America to action. The military was all too willing to take advantage of the wizard's changed outlook. In 1915 Edison accepted an invitation by the government to head a board of scientists and inventors that would advise the military on the construction of weapons to defeat the enemy. The machine was poised to leave the pastoral garden once and for all. Edison left behind his belief that science and technology should be as separate from the military as church and state. He thereby showed how different human emotion is from the cold and calculating logic we build into our replicants. With World War I acting as the catalyst, the Wizard of Menlo Park was thrust into having to conceive anew how he viewed the relation between humans and machines. Utilizing his imagination as a tool he would juxtapose the imagined to that of practical reality and recognize early on the need for a different kind of machine on the battlefields of war-torn Europe.

THE WISE MAN

We must remember that twenty years earlier the military had laughed at Tesla's remote-controlled automaton. In sharp contrast, when President Woodrow Wilson approved the proposal for an advisory board of inventors, Edison was deemed "the man above all others who can turn dreams into realities."[70] The motive for appointing Edison as head of this first scientific consultancy in the United States was that he symbolized a pluralistic community of technological inventors. Unlike Edison, Tesla was never able to realize that pluralism in a democratic environment could act as the foundation for a new scientific age. On Edison as a symbol, Secretary of the Navy Josephus Daniels, who was in charge of selecting the right man for the job, wrote: "I feel that our chances of getting the public interested ... [in] this project will be enormously increased if we can have some man, whose inventive genius is recognized by the whole world, to assist us."[71] The fusion of technology and the military needed a figure whom the public trusted as a wise man of counsel. Edison "now believed that he was helping in the defense of his government and people against the danger of an attack."[72] Little did he know at the time that his persona was being exploited, and that his independent inventor's autonomy was being replaced with militaristic goals and corporate financiers. Nevertheless, the fusion of the military and technology was also a fusion of the convention of war and the mythos of nationalism. Similar to how the interactions of machine and man propel each other forward, the fictive mythos of nationalism and the reality of war were now propelling American society to push forward in technological endeavors that were once laughed at.

Edison promptly advised Secretary Daniels that such a board should be composed of men from leading scientific and engineering societies. Edison realized at an early stage that the board needed to consist of gentlemen who liked one another. Nevertheless, Daniels proposed nominating only famous inventors, failing to recognize the dynamics of rivalry within the scientific community. Even Edison despised many of his famous counterparts, but he had emerged as a larger-than-life myth.

CONCLUSION

> *Plants, animals, and humans don't come about mysteriously and existentially any longer, getting born—they are "made," created by information. We should be able to read their plan, follow it, and back-engineer them too. We should be able to manufacture and patent autonomous entities, robots with agendas, humanoids as real as humans. . . . Trapped in a labyrinth of Artificial Intelligence; we accept the appearance, the imitation of consciousness, as consciousness.*
>
> Grossinger, *Embryos, Galaxies, and Sentient Beings*

We recognize that this imitation of consciousness meets us on a different dimensional plane of awareness. We start by mythologizing matters we can't understand to fill in the missing gaps. Within these gaps we quickly find previously hidden powers for the imagination to harness. If we are to harness the powers of the mind, there has to be a story of ideas first, in order to manifest the reality of such ideas; an evolution of ideas in the realm of the unseen, where ideas slowly densify and materialize into waking actualities of cause and effect—pragmatic manifesting. As Goddard says, "[Y]ou win by assumption what you can never win by force. An assumption is a certain motion of consciousness. This motion, like all motion, exercises an influence on the surrounding substance causing it to take the shape of, echo, and reflect the assumption. A change of fortune is [not] a new direction and outlook, [it is] merely a change in arrangement of the same mind *substance—consciousness*."[73] We see the American collective consciousness evolve no different from a child growing into adulthood.

Tesla and Edison's mechanical-man myth displays an evolution on many different levels. First, the phenomenon demonstrates a new source of power. Unlike steam, electricity is mysterious and was once thought to be the source of animating life. In addition, the electric-man is linked to a new genre in literature called "Edisonade." John Clute explains that these stories are usually about uneducated young inventors who achieve the American Dream through their own wits and determination. Thus, the mythography of Tesla and Edison gave the nation popular heroes. In addition, it displayed to a nation of builders and dreamers that "fictional dreams [can be] transformed into reality."[74]

In the "Edisonade" stories, the true heroes are the inventors. The evolution of mechanical-man from the time of Poe, Melville, and Hawthorne to that of Tesla and Edison is one of morphing the power source of life from steam to electricity. It is an industrial myth in that it tells us how we view our relationship with machines and each other in an ever-changing technological society not beholden to European metanarratives. In the stories written by Poe, Melville, and Hawthorne—and in the lives of Edison and Tesla—we see that the mechanical-man myth is co-created by the real and the imagined. In this myth is uncovered the unconscious awareness of a unique American technological spirit living alongside the mythos of a collective American destiny. And in this ongoing and evolving myth was now a reflexive reality that was formed by the union of human-machine interaction and evolution. As we will see in other biographies, the mechanical-man was now bringing about a metaphysical "philosophy for the people." Emerson tells us to look for guides during the quest. These biographical accounts are guiding us through the conscious and unconscious ethos of spirit and selfless service in America. Americans were quickly learning that "*thoughts are causative.*"[75]

4
Knowing Thyself
The Initiation

After the technological destruction wrought by World War I that threatened economic instability globally, the manner in which Americans viewed technology was dubious at best. World War I marked the beginning of the mechanical era and the end of the nineteenth century. Soldiers became parts of metal organized within the sphere of technology. Technology was no longer solely synonymous with inventive progress; rather, the sphere of technology now encompassed a means of organized and efficient death during times of war. The mechanical-man was now at the epicenter for debating the values and challenges of technology. No longer was the mechanical-man regarded merely as an instrument of service; rather, he now became a symbol of the practical choices involved in engineering and scientific decisions. The most popular writers of science fiction immediately after World War I were mathematicians, chemists, astronomers, and engineers, many of whom participated in wartime intelligence and weapons initiatives. That fact notwithstanding, the overall theme of how the mechanical-man interacts with society in terms of control and communication within these stories implies that problems within a technological society can be solved by responsible engineers and scientists.

During this period, science fiction and scientific development merge. As L. Sprague de Camp said, "The man who has neglected to keep himself informed concerning the frontiers of science, or, even having managed that, fails to be reasonably knowledgeable about any field of human activity affecting his story, or who lacks a fair knowledge of . . . current events—failing in any of these things, he has no business writing speculative fiction."[1] Indeed, with a public becoming more knowledgeable about scientific developments after World War I, the imaginers of mechanical-men had to embrace the technology of which they wrote. In addition, they needed to have the acumen to discern the new methods of control and communication in human and machine interactions. John Diebold, president and founder of the Diebold Group, Inc., a management consulting firm pioneering in the field of automation, held the following at a cybernetics symposium it held in Washington, D.C., in November 1964:

> The [I]ndustrial [R]evolution is finally over. At some time in the last twenty or thirty years, the seeds of the next era were sown. We have but to look about us to see these seeds growing. This shape of this new age is still unknowable. . . . To begin with we should be absolutely clear about the meaning of the machines and the technology that constitute the applications of cybernetics. They are agents for social change. One has but to examine the phrase "the industrial revolution" to realize this.[2]

Stories and myths prevalent during this period reflect a culture moving from steam and electricity toward atomic energy—leaving behind the Industrial Revolution and entering the Control Revolution. With this evolution came a mechanical-man that appeared more godlike than human, reflecting a growing understanding of ourselves in terms of our relationship with technology. In addition, these stories utilize the mechanical-man to project a double or *shadow*. This motif represents the two faces of good and evil as objectified by the mechanical-man at war with himself in terms of how to utilize technology. There is also a quest motif in these tales that relates to the fall into experience. The mechanical-man is no longer a mere tool; instead, he is a reflection of what has befallen man by dint of his dependence on technology. These tales thus illuminate American culture's descent in terms of economics, national apprehensions, and social iniquities. Since the goal of any quest is "the lost treasure of innocence, which may be symbolized in various tangible and intangible ways,"[3] the mythos of the mechanical-man shows an awakening of America's rendezvous with destiny in terms of science and technology—technology being an extension of man was now seen as a tool for purposes of control and communication—ameliorating problems of dependency upon nature.*

By the 1920s radios and automobiles had helped create a consumer society increasingly focused on leisure, pleasure, and intimacy. Advances

---

*Many scholars who write about mythology and mythic-markers cite to secondary sources that are closer in publication time to their own work, but which utilize *Myths and Motifs in Literature* as their primary source. The fact that it was published in 1973 does not mean the ideas captured are obsolete; indeed, the Jungian bent of this quasi-"textbook" is the fulcrum we work from when discussing mechanical-man literature. We must always keep in mind that at the very heart of this study is the mechanical-man as the new American archetype, replacing the cowboy. Burrows helps us by defining, "[a]n archetype [as] a model, or pattern, from which all other things of a similar nature are made." I like to look upon archetypes as the dark matter that makes up most of the universe that our imagination takes from and molds into objective experience and material reality. He goes on to describe the archetype critics as inherently literary critics by saying, "[a]rchetype critics are literary critics, or myth critics, strongly influenced by the belief that primitive man is yet within culture—have the power to make us aware of the collective experiences of the race. The archetype critic is concerned with the enduring patterns and motifs and how these are reflected in literature."

When Burrows holds that "society, institutions and literature change; the human condition remains the same," he's speaking in a similar vein to what we have already seen in the writings by Joseph Campbell: that, biologically speaking, the biology of man has been the same for thousands of years; thus these archetypes and motifs don't change—they merely transform to fit with our collective consciousness at a given time. The beauty of the mechanical-man is that it shows that our biology may be the same on the level of the collective consciousness, but our imagination/consciousness is different on the individual subconscious level and the collective level. Some mechanical-men, like humans, still have the imagination of automatons, thereby not utilizing the tool of the imagination to bring about the sublime or transcendence. (Burrows, xiii)

in technology and medicine had improved the quality of life for many Americans. Indeed, even the Scopes trial could not stop the deluge of benefits that science brought to the daily lives of the masses. Historian Ruth Schwartz Cowan points out that "scientific management" sparked an "efficiency craze." The production techniques implemented by entrepreneurs like Henry Ford and Frederick Winslow Taylor created a workforce that believed "they would become artisans of times past, mini-entrepreneurs acting in their own interests, managing their own time to suit their own interest." It was a means for them to control their own labor and value as a commodity in society. On the issues of control and communication, the expression of individuality was augmented by the work of machines to free people from the toil of redundancy, which was now a controlling factor in an industrialized society of factories.[4]

Greater efficiency, scientific managers also thought, would make industrial work easier and pleasanter for the workers, who would then be able to realize the Romantic goal of expressing their individuality: being creative on the job, cutting loose from the mass of passive workers, increasing their wages and their status.[5]

In the late 1920s, however, the economic predictability of the country crashed. The never-ending search for the most pragmatic principles upon which to build society ground to a halt and would have to be resurrected again by a new generation of Americans in the midst of the Great Depression. Harl Vincent, a mechanical engineer and freelance writer who became popular in such early pulp magazines as *Argosy* and *Amazing Stories,* typifies views of science at the time. His short story "Rex," published in 1934 and set in the twenty-third century, depicts a society served by robots devoid of reason. Where one such robot named Rex experiences a change that confers on him the ability to reason, he sets out to bring order to the chaos that he sees humans as having created. The aftermath of World War I and the Great Depression's impact thus set the stage for a new kind of imagined mechanical-man.

Vincent's short story differs from many utopian novels of the late nineteenth century. In works such as Edward Bellamy's *Looking*

Backward and King Camp Gillette's *The Human Drift,* technology solved all of humanity's problems. Thus, "all the utopias were similar in equating industrial technology with prosperity and prosperity both with happiness and with democracy."[6] Vincent, unlike his predecessors, writes at a time when the idea that science could cure all social ills was dubious at best. Vincent, himself an engineer, saw two sides to science in a democracy, and via his fictive mechanical-man he reflected on what role, if any, technology would have in a troubled garden.

Rex's reasoning, we are told, "was that of a logician: coldly analytical, swift, and precise, uninfluenced by sentiment. No human emotion stirred in his mechanical breast. Rex had no heart, no soul."[7] He views human society as "divided roughly into three classes[—] the political or ruling body, the thinkers or scientists, and the great mass of those who lived only for the gratification of their senses."[8] Seeing no logic in this structure, Rex seeks to ameliorate the situation by purging desires and emotions from human beings. Calculated logic was to rule supreme, and the "results to be accomplished were nothing short of miraculous."[9] At this same time, Albert Einstein was doing the same thing for time and space in the realm of physics. What were formerly human mysteries, god-related or organic, were now being reshaped into mechanistic understandings.

Creating "order out of . . . chaos,"[10] Rex takes the Transcendentalists' critique—that technology will make us bored—and internalizes it. The American Transcendentalists now are becoming more active in the mechanical-man's quest. One can think of the transcendentalists as spiritual idealists in our time. Other than the religious beliefs that looked to one's inner light for salvation, this group of writers and thinkers was the first to place ideas before matter in the public literary scene. Ralph Waldo Emerson, born in 1803, is most remembered alongside Henry David Thoreau (the greatest literary observer of nature), Walt Whitman and Emily Dickinson (the two greatest poets of the nineteenth century), John Muir (wilderness advocate and "Father of National Parks"), and many others. Sam Torode explained in *Everyday Emerson:*

> Emerson saw that in the human world, everything begins as a thought or idea. Everything we do, everything we build, and everything we strive towards—be it a railroad, a home, a poem, a business, a form of government, or justice for all people—must first exist in the realm of mind or spirit. Once we grasp the idea, imagine the ideal, or catch the inspiration, then we can manifest it in the physical world. What's true in the human world, he believed, is also true throughout the universe: mind precedes manifestation. Everything in the physical world emerges from, and grows according to, invisible laws, principles, and powers. Everything in nature arises from a universal Mind or Spirit [or, the Force, when we speak of *Star Wars* and the collective unconscious when we discuss Carl Jung and Philip K. Dick]. Emerson used these two words—mind and spirit—somewhat interchangeably in describing the invisible, nonphysical realm that precedes the visible, physical world. Today, we might substitute the phrases "field of consciousness" and "ground of being" for mind and spirit.[11]

It's important to have, at least, a modicum of knowledge on how the transcendentalists viewed consciousness and how this affects the control issues in cybernetics. Pragmatic manifesting is the ability to create by merely using the force of imagination or love—pragmatic in the sense that it vibrates at such an accord that it brings harmony to the paradox of service. Emerson starts us off by reminding us that since God or Christ consciousness is in all of us, we have the ability to control and create. Later, we will see how scholars like William James and John Dewey extrapolate this concept in order to unify transcendentalism and pragmatism. For now, it is merely worth noting that the mechanical-man is attempting to break out of his box of automation and slavery to his senses (the word *robot* itself means "slave" in Czech). He is looking for transcendence and the sublime and attempting to discover how to possess human imagination, and then, in a pragmatic sense, use it.*

---

*For a thorough understanding of American transcendentalists see Gura, *American Transcendentalism.*

Vincent was well aware that the Great Depression had made every American question scientific management in industry, causing emotions to blind them to the value of pragmatic manifesting. America needed to view mechanization and technology in a different light. The machine would now stand by our side in our journey to regain "pride, creativity, artistry, self-esteem, skill, freedom, manliness, and independence." Thus, Vincent was illustrating that Americans had a technological *rendezvous with destiny*. Society could not afford to become complacent about its technological achievements, falsely believing that they represented an apex of progress. Americans needed a reawakening to the necessity of harnessing new and different sources of power. The Great Depression was the catalyst for a new transformative era in terms of technological growth and consciousness. Indeed, this transformative threshold marshaled into American society the creation of systems programs to organize human efforts with technology. Human development became married with technological advancement in the human imagination and sense of self. Roads, dams, industrial output, and mechanical devices marked this advancement under the presidency of Franklin D. Roosevelt, who commented on numerous occasions that, when it came to science and technology, Americans had a "rendezvous with destiny"; specifically, that Americans were to become the shining beacon of technological growth and innovation for the world. This meeting ground is a new dimensionality of manifesting the imagination.

In order to understand the human predicament, Rex attempts to incubate in himself the seeds of emotion, but after numerous experiments he believes that he has failed. "I have . . . isolated the activating force of every human emotion," he says. "I have reproduced these forces to perfection with arrangements of special electronic tubes, which have been incorporated into my own mechanical brain. Yet have I failed to produce so much as a semblance of human feeling in my makeup."[12] Consequently, Rex commits suicide. Ironically, his despair stems from the human emotion of failure. Vincent thereby conveys the idea that the United States, on the heels of the stock-market crash and Great

Depression, must not give in to the emotion of failure, but rather get back on track and fulfill its national destiny by going back to work on scientific endeavors. This type of ordeal is commonplace during the quest—numerous refusals until acceptance of the numerous calls and subsequent returns.

Vincent wanted people to realize that true happiness comes with curiosity, creativity, and purpose. Science and art, or reason and imagination, must be viewed as two complementary sources of myth's potency. In this instance, the fusion of reason with emotion and creativity brought America out of the depths of the Great Depression, preventing another metaphorical suicide. The fictive history of the mechanical-man during the 1920s and 1930s illustrates how a society can become so intoxicated with its current tools and inventions that it no longer strives to meet challenges pertaining to its future. Consciousness itself becomes stagnate.

Four years after Vincent's tale of Rex, Robert Moore Williams in 1938 published "Robot's Return." In the story, he describes three robots that are curious about their ancestors. As they seek to discover their origins, Williams offers an oblique analysis of science and technology in the United States during the Great Depression. The robots discover an idol created by their ancestors which maintains that "Everyone will want to work, but because of the work of machines, work will not occupy most of people's lives; they will have enormous amounts of leisure to do with as they please[,] and they will spend most of their time educating themselves and pursuing creative activities. People will understand themselves to be cogs in a great machine."[13] Because they understand this, they will work in a "spirit of cooperation."[14]

In the 1930s Americans reawakened to the pragmatic scientific principles on which the country had been built. They too were dreaming of a better tomorrow with technology acting as an instrument for change. While the robots in Williams's short story are unable to dream, their human ancestors did. A dream as a literary device is a reflection of our hopes and desires. The mythos of the American Dream now involved issues of urbanization and immigration. People moving to the cities and foreigners all were now being incorporated into a system which necessitated the mechanization of human efforts. Cities were now looked upon as grids and factories where one could see a dance being played out between society and machines coevolving in the modern world. In a way they were all becoming mechanical-men of a sort and after the chaos of World War I and the Great Depression, Americans were still dreaming of a better tomorrow marked by cooperation and unity through technological progress.

LITTLE BOY

The building of the atomic bomb [Little Boy] was the work of 125,000 people and cost nearly 2 billion dollars.

JONES, ET AL., *CREATED EQUAL*

Albert Einstein, the German-Jewish physicist who came to the United States after Adolf Hitler's rise to power in 1933, had warned President Franklin D. Roosevelt that the Germans might be developing an atomic weapon. The country then invested heavily in nuclear technology.

> But when the first bomb exploded over Hiroshima on August 6, 1945, and the second on Nagasaki two days later, the horrifying destructiveness of nuclear weapons became apparent. Even though the American public saw few images of the carnage on the ground, the huge mushroom cloud and the descriptions of cities leveled and people instantly incinerated shocked the nation and the world.[15]

In introducing his fourth robot story titled "Runaround," published in 1941, Isaac Asimov states that the Laws of Robotics might not be able to "provide the conflict and uncertainties required for another story." His story concerns the evolution of science in coming to a better understanding of the laws and forces at work in the cosmos. Asimov declares that, because "technology can be ethically and responsibly used, there is no reason to fear it."[16] However, the underlying message in his robot stories parallels our own debates and dilemmas involving human ethics. Specifically, if natural laws do govern human action, what are they, and how do they affect the instinctive individuality and cooperation of the human race? Asimov yearns for machines to "take over dehumanizing labor and thus allow humans to become more human," but his definition of humanity is predicated on creative thought and intellectual progression, recapitulating transcendentalists' ideologies of the nineteenth century pertaining to the utilization of science to advance humanity in a more civilized direction.[17]

In this story, two scientists/astronauts, Powell and Donovan, face a dilemma. Their most advanced robot, Speedy, has not returned to the ship. Its mission was to bring back selenium from Mercury to allow them to return to Earth.

> Powell looked up shortly, and said nothing. Oh yes, he realized the position they were in. It worked itself out as simply as a syllogism. The photo-cell banks that alone stood between the full power of Mercury's monstrous sun and themselves were shot to hell. The only thing that could save them was selenium. The only thing that could get the selenium was Speedy. If Speedy didn't come back, no selenium. No selenium, no photo-cell banks. No photo-banks—well, death by slow broiling is one of the more unpleasant ways of being done in.[18]

Here Asimov presents the argument of man's dependency on technology furthering his dreams and goals. Only the robot can save the men. After the destruction wrought by two world wars, the sole way to counter the destructive power of technology is via technology. Powell and Donovan quickly turn their attention to "robots from the First Expedition," which are seen as "sub[-]robotic machines" with "primitive positronic brains," in order to compel Speedy to bring back the selenium. Powell, the scientist in charge of the expedition, ruminates on the robots they must now use: "Those were the days of the first talking robots when it looked as if the use of robots on Earth would be banned. The makers were fighting that[,] and they built good, healthy slave complexes into the damned machines." This is the first time in American literature that a robot is referred to as a slave and that the Laws of Robotics are programmed into a positronic brain to ensure obedience and slavery. The Laws of Robotics are as follows: (1) Never harm a human being, or allow a human to be harmed; (2) Never disobey a human order, unless to obey Rule 1; and (3) Never harm yourself, unless to obey Rule 1 or 2.[19]

Once the robots reach Speedy, they realize that he is caught in a conundrum pertaining to the Laws. Powell explains what it is:

> The conflict between the various rules is ironed out by the different positronic potentials in the brain. We'll say that a robot is walking into danger and knows it. The automatic potential that Rule 3 sets up turns him back. But suppose you order him to walk into that danger. In that case, Rule 2 sets up a counter potential higher than the previous one and the robot follows orders at the risk of existence.[20]

In the end the scientists discover that Speedy has found himself in the gray area of balancing Rule 2 with Rule 3. The solution to the problem is to make Speedy act on Rule 1: "A robot may not injure a human being, nor through inaction allow a human being to come to harm." By discarding his protective suit, Powell makes himself vulnerable to Mercury's environment; thus, the potential harm to Powell outweighs Speedy's conundrum.

> Of course Rule 1 potential is everything. But he didn't want that clumsy antique; he wanted Speedy. He walked away and motioned frantically: "I order you to stay away. I order you to stop."[21]

Although science had unleashed its terrors on the world, dreamers like Asimov believed that science was the answer, not the problem. And if the point of natural selection and scientific progress was "not . . . death, which is inevitable, [but] the propagation of life, which is not," then perhaps the decision to drop Little Boy and prevent a future atomic war satisfied the First Law of Robotics.[22] In this way the Laws of Robotics provide society with a justification for its actions. Depending on how they are interpreted, such laws can also justify the means used to achieve conquest and realize our collective manifestation. Later we will discover that a just end never justifies unjust means when one's conscious pendulum swings from materialism to idealism.

TOTAL DESTRUCTION OR ABSOLUTE SAVIOR?

J. F. Bone's story titled "Triggerman," published in 1958, presents a situation in which a computer malfunctions and a mechanical-man wreaks havoc by firing a ballistic missile. The main character, General Alastair French, is supposedly in control of the automatic mechanism that can unleash a nuclear attack on the Soviet Union if "Ivan," French's counterpart in Russia, launches missiles against the United States. He is called the "most important" man because under his control is the destruction of the entire human race.

General French has a choice of catastrophes. Bone writes:

> Should he wait and let Ivan exploit his advantage or should he strike? Oddly, he wondered what his alter ego in Russia was doing at this moment. Was he proud of having struck this blow or was he frightened? French smiled grimly. If he were in Ivan's shoes, he'd be scared to death! He shivered. For the first time in years he felt the full weight of the responsibility that was his.[23]*

*Later in the quest we will talk about the importance of "fail-safes" in terms of cybernetic interactions of control and communication.

The machine in the garden had indeed wrought havoc in the world and created a Cold War marked by the real prospect of total destruction. However, technology had also pulled the country out of the Great Depression and ushered in an era of productivity, spin-off inventions, military prowess, and economic superiority. Technology was a double-edged sword: it could save us, and it could destroy us. Bone, in 1958, seems to be leaning on the side of destruction, but he leaves us with an odd riddle by the end of the story. It was not some computer glitch that caused the near catastrophe. Issues of control and communication between man and machine mandated that society listen to more than just computer programmers and engineers. There was now a call for a variegated approach to such issues since the cause-and-effect aspects of the growing autonomy of machines was novel to the outdated organized makeup of society's problem solvers.

> Sense was beginning to percolate through the shock. People were beginning to think again. He sighed. This should teach a needed lesson. He made a mental note of it. If he had anything to say about the makeup of the Center from now on, there'd be an astronomer on the staff, and a few more of them scattered out on the DEW line and the outposts groups. It was virtually certain now that the Capitol was struck by a meteorite.[24]

What if there had been a computer glitch, however, and the machine rebelled against its master? By our reliance on technology, we either progress to a state of utopia or, conversely, regress to one of mass destruction in a conscious state based on materialism, where the physicalists reduce us to ashes—literally, ashes.

Eight years before Bone's fictional meteorite descended on Washington, D.C., Asimov once again stimulated our collective imagination in "The Evitable Conflict," published in 1950. Asimov, like Bone, portrays a world in which humans and robots are at opposite ends of an apparent ideological or perhaps even conscious spectrum. Asimov too writes of a supposed glitch or problem with the positronic brain. In an attempt to allow the reader to understand the conflict between communism and capitalism, Asimov gives a history of senseless wars in the short history of *Homo sapiens* whose extent we fail to grasp due to our existential anxiety:

> Every period of human development . . . has had its own particular type of human conflict, its own variety of problems that, apparently, could be settled only by force. . . . Consider modern times. There were a series of dynastic wars in the sixteenth to eighteenth century, when the most important question in Europe was whether the houses of Hapsburg or Valois-Bourbon were to rule the continent. . . . In the twentieth century . . . we started a new cycle of wars, what shall I call them? Ideological wars? The emotions of religion applied to economic systems, rather than to extra natural ones? Again the wars were inevitable wasting away of inevitability. And [when] positronic robots came . . . it no longer seemed so important whether the world was Adam Smith or Karl Marx.[25]

For Asimov, historical conflicts, including the then-current Cold War, are fostered by human ignorance of numerous variables that our prehistoric brains are unable to comprehend. Thus, we involve ourselves in imbroglios that allow little to no progression for the species as a whole. For Asimov and Bone alike, the human factor, in consonance with technology, causes chaos and conflict, making us unable to understand our own nature.

The difference between human beings and robots is articulated by the First Law of Robotics. The problem is that we do not grasp the very laws entrenched in the minds of robots. We ignorantly regard the First Law as meaning that no robot may harm a human being, but a robot construes it differently: "No machine may harm humanity or, through inaction, allow humanity to come to harm." These machines know the dynamics of human existence, but their brains, which are more complex than our own, are able to contemplate all the variables in their assigned task of preserving humanity. "Their first care, therefore, is to preserve themselves, for us. And so they are quietly taking care of the only elements left that threaten them." We see this as a selfish act, obviating the Laws of Robotics and causing evitable conflict between humans and robots. However, this is not the case. In the eyes of Asimov, we are unable to understand the vagaries of our own human development. Machines, on the other hand, grasp that the First Law of Robotics entails the utilitarian philosophy of preserving robots for the good of all humanity.[26]

The fear of atomic holocaust during the Cold War was prejudiced by our human ignorance. Although Asimov feared that atomic warfare would annihilate the human race, he yet proposed that technology had the capability to move us forward, for only they knew the ultimate good of humanity, which coincided with their own interest. Machines, in effect, know better than we do, and they are, therefore, held to higher moral, legal, and ethical standards because they are predictable and mechanistic. This also implies that there is a peculiar unpredictability about human nature. This shift in ways relating to machines also indicates that the organic universe is one predicated on chance and chaos with no conceivable preprogrammed purpose in a materialistic paradigm.

In such conflicts as the Cold War, writes Asimov, humanity has always been "at the mercy of economic and sociological forces it did not understand, at the whims of climate, and the fortunes of war." The machines understand these forces, "and no one can stop them, since the machines will deal with them as they're dealing with the Society, having as they do, the greatest weapons at their disposal." For Asimov, the idea of mutual destruction prevented a third world war. Thus, the atomic bomb becomes not a pariah but a savior, a robot that is paving the way for us. In our anxiety we are terrified of atomic warfare, when in fact the possibility is merely a step forward in natural selection. Technology for Bone and Asimov thus becomes a godlike agency in American society.[27]

In Frederic Brown's "Answer," published in 1954, we are given the proverbial answer to the question of technology in America. The story's main character, Dwar Reyn, throws a switch to "complete the contact of all the monster computer machines of all the populated planets in the universe," linking the machines together in a syncretic network. The question he asks concerns God's existence. The computer responds in the affirmative, indicating that it is the integration of all the machines that create the equivalent of what humans designate as "God." Thus, for all the fears during this era of a nuclear arms race, it is suggested that in technology's advancement we must not fear catastrophe. The task for Americans during the Cold War was to find a balance, implies Brown. The trade-off now was not our independence for a higher standard of living; instead, it was one of dealing with anxiety about the possible destruction of the human race in harnessing the atom's power and perhaps advancing in our evolutionary development in terms of consciousness. Because the trade-offs were becoming more complex, they could only—in terms of human-machine interactions—be understood in terms of myth and storytelling.*

Now firmly entrenched in the garden, the machine becomes the equivalent of God. Its relationship with humanity is much different from the works of Ellis and Bierce, where the automaton must sacrifice itself by way of suicide to allow humans to cross the threshold of understanding manifested spiritual freedom and service. The machine is now omniscient, able to destroy cities or, conversely, save the world's economic structure and eliminate diseases. No longer is the mechanical-man a sacrificial lamb that must be destroyed or crucified; instead, it is the hero that can lead us out of the wasteland. The potential power of machines to bring about a utopia or a dystopia is a conflict that Americans after World War II and during the Cold War had to contemplate.

---

*Brown once worked as a reporter on the *Milwaukee Journal* and was active many years in journalism. Two of his best-known science fiction novels are *What Mad Universe* (1949) and *Martians, Go Home* (1955).

When *The Day the Earth Stood Still* debuted in 1951, the country was in the midst of this type of conflict. There was a stalemate in negotiations with the Soviet Union, and the fear of nuclear holocaust was on the minds of citizens. The theme of anxiety and even escapism is prevalent throughout Robert Wise's film. Klaatu, an alien visitor that has come to Earth to warn us of our global conflicts, views human society as it is—irrational, childish, inept, ignorant, and incapable of caring for itself or of being wise stewards of technology. More specifically, he has come to warn us of our tendency to build rockets and weapons, for if we keep utilizing atomic energy for war the planet will become a "burnt cinder." In effect, Klaatu is reminding us to be aware of our limitations as humans. Our ideas pertaining to the American Dream and National Destiny are not worth the destruction of the entire human race by attempting to harness powers that are beyond our feeble understanding. Technology is warning us about the human abuse of technology as well as our unpredictable nature inherent in being human.

Scholars such as J. P. Telotte discuss a double vision pertaining to technology in this film, but they fail to recognize the role of the robot Gort, which is associated with the mythology of the machine in the garden. In the beginning of Wise's film, we are unaware of Gort's role; he is seen simply as the caretaker of Klaatu, perhaps even his servant, controlled in a manner similar to Moxon's Master or Ellis's Steam Man.

Gort protects Klaatu by utilizing his power to vaporize man-made weapons, reinforcing the idea that he is a slave doing the bidding of his master. We see this when Klaatu borrows a flashlight to signal Gort to deactivate all energy sources on the planet. Our impressions of Gort quickly change, however, when Klaatu warns of Gort's power if he were to be fatally injured. It is at this time that he utters those remarkable words, "Klaatu Barada Nikto." Here we begin to see the true nature of Gort, and thus the ritual transition in the popular imagination of robot as slave to robot as savior is complete. Gort then reanimates (*brings vitality to*) Klaatu, and, though Klaatu says that only the "almighty spirit" can bring one back to life, Gort nevertheless has some of this forbidden power. He may not be on the level of Bone's omniscient machine, but he does possess the spark of life to raise his creator from the dead. It is as if we are looking to technology to solve all of our social conflicts and mortal anxieties. For this relationship with technology to spawn new ideas and inventions, however, humankind must work with technology, not against it. Pragmatic manifesting does not allow the tools that we create to overtake us; instead, the tools should enable us to understand ourselves and our environment. This is what imaginary robots such as Gort teach us during the Cold War. In our search for hegemony over space, planets, and the harnessing of energy in the Cold War, we see the expression of our primordial human emotions and the necessity for mythmaking in discussing these conditions—progressing human consciousness during the quest.

In the movie's final scenes, we learn who Gort is and how he functions. In his last speech, after Gort has breathed life back into his soul, Klaatu talks about the role of reason and law in society. He emphasizes to Earthlings that the rule of law, framed by governments and enforced by police, keeps us from harming ourselves and allows humanity to progress in a civilized direction. The robots in *The Day the Earth Stood Still* make it possible for society to prosper in peace. As with the robots' recognition of what is best for humanity in Asimov's fiction, Gort and his brethren function as philosopher-kings, enforcing good behavior by humans. Gort is thus not a slave to Klaatu; instead, he is Klaatu's protector and an embodiment of morally enlightened values borrowed from Plato's mythical Republic.

Like Gort, technology in the 1950s was guiding us after two world wars that involved previously unimagined death and destruction. We had to be shown what Melville foreshadowed in "The Bell-Tower," specifically, that neither technology nor science by themselves caused the destruction, but rather the nature of the goals in terms of which such technology was utilized by those in control. We must then learn how to recognize and regulate these goals in harnessing such sources of power.

Perhaps even more popular than Gort in the American imagination was Robby (Robbie in Asimov's stories) the Robot in the 1956 film *Forbidden Planet.* Ostensibly a robotic servant to Morbius and his daughter, he also functions as their protector. The film thus presents a double vision of technology as bringing about both optimism and anxiety in a time of nuclear armaments, but this misses the point of the contextually imagined history of robots in America. The planet, according to the film's narrative, was once inhabited by an ancient race of advanced people named the Krell who left behind many of their technological achievements. Morbius explains that the Krell had evolved to the point of creating things by mere thought (their imagination). However, although their civilization was replete with advanced technology and science, they succumbed to a force far less advanced—the subconscious. Morbius suggests that the Krell must have shared the same primitive beginnings as man, their monstrous origins contained in their subconscious. All aspects of human endeavor were touched by mechanization on this planet. And now that Freud, Jung, and Einstein's theories were accepted as "good" science, the mechanization of the human mind in our own reality on planet Earth was also near complete.

Forbidden Planet thus implies that technology can only take man so far, for the id will always entice man to fall from grace by capitulating to his passions and appetites. Morbius echoes the same sentiments as Klaatu when he speaks of why man creates laws and religion—namely, to hold the id in check, for without these institutions the subconscious would make man a slave to his passions and prevent him from progressing.

We should also note that Robby is governed by Asimov's Laws of Robotics. When Robby blocks the crew of United Planets Cruiser C-571 from entering the home of Morbius, the Captain says that the robot must have some built-in laws that stop him from wringing their necks. Moreover, before Robby can act, Morbius's daughter instructs him to cease and abort his mission by uttering the phrase "Archimedes." Thus, like Gort, Robby has a kind of fail-safe that allows him to abrogate previous orders. In the end, when the Captain attempts to attack Morbius and his daughter, Robby is helpless due to the Laws of Robotics. He is ordered to kill the id-driven Captain, but in doing so he would be killing a human being, causing a meltdown in his circuits. Nevertheless, Robby, again like Gort, is a manifestation of our mythological savior; he is governed by the fundamentals of materialism. Robby does not dream, nor does he have a past connected to animal instincts as humans do. However, in reality, the rules for robots are in no way enforceable—it is a human self-assurance to justify and quell fears.

Thus, there is no monstrous side to Robby; he is completely devoid of passions, unlike the crew of United Planets Cruiser C-571, who are slaves to their heritage and "apelike" brains. Robby, Gort, and Asimov's Law of Robotics illustrate the need to be mindful of our goals in harnessing the forces of nature. The gift of *foresight* is itself a tool and should be utilized to recognize how such forces may affect our future. Robots and technology have now become embodied metaphors, actors assisting humans not only with physical labor but also with metaphysical and moral issues. With them at our side, we were entering into new ethical paradigms.

Kevin Fisher offers the following observation about the symbolic importance of the robot in Wise's film:

> Robby is exemplary of those technologies within *Forbidden Planet* that mate the cultural perception of virtuosity to the technical and socio-political fantasy of homeostatic control. The term homeostasis applies severally to the technical ambitions of early first-wave information theory, the aims of the American industrial complex with regard to the management of external and internal conflicts, and the normalizing agenda of psychoanalysis in the service of postwar adjustment psychology.[28]

For Fisher, the significance of mechanical-man stories during World War II and the Cold War is their denigration of matter and physicality. In this sense, these stories show that information is more essential, more important, and more fundamental than materiality. "The ability of the Krell supercomputer," notes Fisher, "to read and transmit unconscious thoughts telepathically, without instrumentation, also presupposes the immaterial and dimensionless nature of information, enabling it to travel wirelessly like a television broadcast."[29] The mechanical-man as myth demonstrates that our concern should be directed more toward the programming of such androids rather than the mechanical structure or artifice. This maxim holds true for the imaginers of mechanical-men as well as their inventors in human reality. Moreover, the evolution of the mechanical-man toward transcendent knowledge is a key to understanding how society's relationship with technology changed during World War II and the Cold War. It hints at a change in human consciousness by utilizing human psychology as a tool to bring about such change.

Fisher summarizes the case best when he describes the totality of *Forbidden Planet*'s main theme:

> [I]nformation has no necessary attachment to meaning. In itself, the information that flows through and among all the technologies in the film exists in a state of pure potentiality. And the patterns that it takes are not fixed to the physical or mental objects that they translate and transmit. This has particular consequences for the Krell supercomputer, which does not discriminate in its production of conscious versus unconscious thought or between rational and irrational desires.[30]

Information becomes the most valuable commodity and driving force for bringing life to the mechanical-man and society's collective unconscious. The free exchange of information and the speed of its transmission are mythic-markers for the mechanical-man. No longer is the American Dream limited to individual ownership of property; instead, it now extends to the freedom to exchange knowledge and information at the speed of light.

In a lecture delivered at Oxford, Robert A. Heinlein said: "Most so-called science[-]fiction prophecies require very little use of a crystal ball; they are much more like the observations of a man who is looking out a train window rather than down at his lap—he sees the other train coming, and the ensuing 'prophecy' is somewhat less remarkable than a lunar[-]eclipse prediction."[31] A mass-consumer-driven culture after World War I demanded that their science fiction imaginers possess the skills to merge myth and reality, which became even more to the fore after World War II. It was no longer enough that these writers prognosticate about the future; instead, they needed to understand the hidden secrets behind the technologies of which they spoke. The unconscious meaning of how technology was affecting the human psyche could only be explained by writers who glimpsed what was hidden behind the curtains in their respective fields. This expectation laid the foundation for a blurring of science fiction and nonfiction to such an extent that both became complementary. This was now the complexity of the modern era. By the end of this quest we will discover that the people called the Krell in *Forbidden Planet* are in fact our future selves. Science fiction visionary Philip K. Dick will soon show us how the Krell, when transformed into cyborgs, save themselves from extinction.

5
The Great Mother

Ruth Schwartz Cowan points out that in one utopian view where machines take over the mundane work of humans, "women would happily submit to the rule of their fathers and husbands." The numerous Romantic utopias envisioned in the late nineteenth century conceived of women as subservient to the creative impulse of men. Moreover, technical expertise was always in the hands of "rational" men who "would help ever larger numbers of men (but not women) reach the Romantic goals of creativity and free expression."[1] Cowan, widely recognized for her expertise in elucidating technology's impact on women, shows that, for the most part, technological advancements usually mean more work for the subservient mother. What then is the place of women in the field of technology—specifically pertaining to issues of control and communication between humans and machines—in a male-dominated society?

Male chauvinism in science is nothing new. Joseph Campbell even speaks of the famed Sigmund Freud and Albert Einstein as being "infected" by gender bias "just as badly as the Bible." We never hear of a man's giving birth to a woman, says Campbell, but nevertheless we have been programmed to believe that God is a patriarchal progenitor. This amounts to "a campaign of seduction, turning the mind and heart from the female to the male[—]that is to say, from the laws of nature to the laws and interests of the tribe."[2] Indeed, the authors and inventors whom I have already discussed intentionally marginalize women. While men still dominate in scholarship on robotics, artificial intelligence, and computer science, women's achievements are too often glossed over as insignificant.

THE EDUCATION OF EVE

Even as far back as 1861, women played a subservient role in the United States' economy, working in mills or factories where they were subject to the rhythms of scientific management. Essentially they were cogs in the industrial machine. Moreover, women supposedly needed a patriarchal system of management to dictate their daily lives. Even in 1861, despite previous strides by pioneers in reform movements for women's equality, they were still regarded as children. Like the African slaves whom many women sought to free, they supposedly needed patriarchal oversight and direction.[3] And yet within every minority in the United States has always been the yearning desire to break through the consciousness in which they found themselves embedded. And they would heed the call upon numerous occasions in American history. As we go through historical context of our mechanic-man innovator and inventors in the real and fictive, always keep in mind that they too were manifesting ideals of their own on the individual and collective level.

In the factories to which women flocked during industrialization, they were given menial tasks. They saw that the machine was confined to the same regimen of repetitious work that society had imposed on them. In a sense, they too were regarded as automata. Grace Hopper, born in 1906, recognized that society viewed women, like machines, as neither rational nor intelligent enough to be left on their own. By attempting to free herself from this bias, she brings a new dimension to the imagined myth and historical reality of human and machine.

Hopper completed her dissertation at Yale University in 1934, becoming the first woman to graduate from Yale with a doctorate in mathematics. She later went back to Vassar for undergraduate work in order to teach. Unlike many women's colleges at the time, Vassar emphasized research rather than teaching. In the latter activity, Hopper saw only repetition, which was no different from pulling the levers of a machine, but in research a person could discover and create, bringing new ideas to the attention of the public. Moreover, in research a person could learn to think, or at the very least teach herself to think.

When the repetition involved in teaching became boring for Hopper, she joined the Navy, reporting in December of 1943 to the United States Midshipmen's School in Northampton, Massachusetts. She wrote:

> When I got there I'd been teaching all these courses, doing all this outside work, running back and forth from New York to Poughkeepsie and teaching at Barnard and Vassar and umpteen other things, trying to take courses, write stuff . . . and all of a sudden, I didn't have to decide anything[. I]t was all settled. I didn't even have to bother to decide what I was going to wear in the morning; it was there. . . . So for me all of a sudden I was relieved of all the minor decisions. All the minor stuff was gone. I didn't even have to figure out what I was going to cook for dinner. . . . I just promptly relaxed into it like a featherbed and gained weight and had a perfectly heavenly time.[4]

When Hopper graduated first in her class in 1944, she was ordered to Harvard University to work on a new calculating machine created in the laboratories at IBM (International Business Machines). It was called the Automatic Sequence Controlled Calculator. The Harvard crew would later name it Mark I. Hopper was given the task of communicating with the machine so as to make it do her bidding. She was now in charge of creating a dialogue between man—or shall I say woman?—and machine. And while the war effort was the catalyst in bringing about this communication, Hopper had no idea that her efforts would "demonstrate to America's military, academic, and business elites the viability of large-scale automated computing machines, otherwise known as computers."[5] Hopper's work would also illuminate the three-pronged interrelationship among the military, industry, and academia, which had been gaining momentum since the days of Tesla and Edison. Within the framework of this interrelationship, she would harness the power of her own femininity and her gender's history of secondary status in a male-dominated society in order to bring about a technological transformation.

UNDERSTANDING THEM

Hopper believed that war was the mother of invention. She enjoyed being given problems to solve with deadlines looming, perhaps even lives at stake. For her and the crew working on Mark I, there was no theorizing but simply trial and error. Hopper quickly educated herself on the machine's hardware, then sought to maximize its processing power. After learning the machine's capabilities, Hopper said, "There was no theorizing[;] there was no higher mathematics. There was no future of computers[;] there was nothing but get those problems going."[6] In effect, she was practicing science in exactly the same way that Tesla had found so perplexing about how Edison approached invention. The machine and its parts had to be viewed as instruments for accomplishing goals. The theorizing and higher mathematics would come after seeing what the instrument could do in differing environments and dealing with whatever variables were thrown its way.

Hopper and her colleagues soon mastered the Mark I. Although their numerous problem-solving crusades were plagued with hardware bugs, coding bugs, and operation bugs, Hopper believed that these "bugs" were a valuable source for learning the machine's "thinking" process. She thus became the first programmer to understand why the machine allowed an infestation of bugs to take place. In these so-called bugs, Hopper recognized the machine's consciousness, its awakening to an environment around it. As with her own tendency to binge-drink at times of immense stress, Hopper understood the machine's tendency to shut down when demands were put on the computer that its hardware was unfit to handle. Thus, the Mark I was allowing Hopper to come to a closer understanding of the machine's thought process as well as her own.[7]

While Bell Labs was conducting experiments devoted to creating self-checking circuits, Hopper saw the numerous problems that plagued the machines quite differently than her peers in industry. She realized early on that the "body" and "mind," respectively, of the computer's hardware and software needed to have a harmonious balance.* Thus, the hardware had to be designed to handle the software and vice versa. In other words, there had to be a symbiotic relationship between the two. Self-checking circuits would not solve any problems plaguing the computer; rather, they would simply circumscribe the computer's output.

The underlying problem with any bugs was in the language of coding itself, or the communication between human and machine. Coding in the beginning had to be a step-by-step process with no variables left out of the equation. The instructions given to a machine for purposes of coding were predicated on viewing the machine as a child that had not yet learned language or taken her first step. The instructions needed to educate the machine and protect it from itself and others. Moreover, Hopper recognized that the machine must learn how to teach itself apart from human instruction. Such was her ultimate goal, and her experience as a professor would allow her to teach a machine to do just that—think for itself.

*There is no need to research the debates of Cartesian duality. For purposes of the mechanical-man, when we speak of any difference between body and brain, we must always keep in mind that the mind is something totally separate from brain. If I were to say something on the duality of body and brain, since both are considered hardware, there is no duality. It may even be erroneous to say there is duality between body and mind since body is in mind and is created from mind. Again, it's all about getting us to see how faulty it is to view materialism as a panacea for wisdom. *See also* Benford and Malartre, *Beyond Human*, chapter 3 "Chips, Brains, and Minds."

In terms of myth, Hopper utilized the suppression of women as a tool to foster change, a challenge to bring about a sense of freedom between machine and human beings. Specifically, she was building a bridge of understanding about how the "Other" operated in technology. Andrea Nye in "The Voice of the Serpent" writes,

> Can there be a feminist linguistics? Is there a feminine language that confounds the semantics and syntax historically implicated in the denigration of women? These questions. . . have a mythic resonance. Can the father god, Yahweh, who installed his order and law in the Garden of Eden, be challenged by the alien voice of the serpent? Can the serpent, whispering to Eve in the sweet, sinuous words of desire, succeed in communicating a meaning outside Yahweh's orders?[8]

Nye explains that there are cultural and biological differences between men and women that reveal differing dynamics and complex nuances in their respective relationships with technology and mechanics. For Nye, women are always striving to "reveal a simple obvious truth buried beneath layers of rationalization and technical jargon."[9] She believes that the power of feminist linguistics can make this possible. By humanizing machinery and computers, Hopper was illuminating how feminist mythmaking and women's relationships with machines and technology are different from that of men. The role of women and machines coevolved. While women sought to humanize technology in the home and workplace, technology was showing women the value of their biological and cultural underpinnings. Hopper was questioning the authority of logic necessary for speaking to have force. She was breaking "into the rigid symbolic order that supports male dominance."[10] The mythos of "ordained punishment for such transgressions" was always on Hopper's mind when she sought to free technology and computers from the circumscribed boundaries of a language based on math and logic.[11]

HOPPER'S GARDEN

The Harvard Computation Laboratory was under the direction of Commander Howard Hathaway Aiken, who played a pivotal role in cajoling the military to take control of the calculating machine rather than Harvard, thus creating a precedent for the military's directing academia in research goals. It was clear to Aiken that academia needed the assistance of the military in creating an atmosphere of order within university laboratories. When Harvard rejected Aiken's application for a teaching position, he utilized his military credentials to bypass the Harvard administration, "employing the political and financial clout of the Navy in order to gain control of what he considered *his* project."[12]

Enforcing Navy protocol upon the Mark I researchers, Aiken created an environment that was conducive to getting fast results. The serene laboratories of Tesla and Edison were now transformed into a battlefield with military hierarchy and order. Despite this regimentation, Hopper was allowed free rein when it came to thinking about how to bridge the gap between the human brain and the computer. She knew that in order for a computer to think for itself it had to have the memory to deal with a slew of variables. Moreover, for a computer to be cognizant of its own existence, it needed a mechanism for retrieval. No longer would the computer be a blank slate (*tabula rasa*); instead, the machine would have innate instincts and ideas that could be passed down from generation to generation. Hopper, in effect, was embedding a bit of natural selection and even *spandrels*[13] into the computer's hardware, allowing it to have the ability to access the knowledge of its ancestors that may have become extinct due to the trial and error of debugging. Little did Hopper know at the time that, while she was working on the problem of how to imbue a machine with the capacity of memory, Alan Turing at Bletchley Park was attempting to give a computer the automated power of espionage and code-breaking.*

*For a discussion of Turing's impact on computer science and robotics, see Hodges, *Alan Turing*. Turing was one of the great heroes of World War II cracking German codes. I hope someday I find someone writing a British mechanical-man into existence utilizing guides like Turing.

During the same years these pioneers were working on ways to end World War II by transforming technology and utilizing machines to intervene in global conflicts and police actions, fictive visionaries Isaac Asimov and Harry Bates were imagining Robbie (Robby) the Robot and Gort, envisioning ways in which mechanical-men could protect us from ourselves.

GIVING THE MECHANICAL-MAN A GENESIS

As the war was coming to an end, Hopper responded to a challenge from Aiken and created the first history of computer science. By doing so, she was blurring the imagined and the historical into a cohesive whole. In her work, she traced the development of mechanical aids "from the abacus up to Aiken's idea for Mark I." Hopper begins her introduction with Blaise Pascal, the seventeenth-century scientist and mathematician who sought to free the human race from the drudgery of tedious calculations in order to pursue more creative endeavors. Even as a youth he devised a plan for a machine that would automate calculations: "The basic machine was a metal box containing a system of wheels and cylinders that could add, subtract, and carry over numbers. . . . An ingenious weighted ratchet system connected the counters in each wheel." This type of direct automation is "the foundation on which nearly all mechanical calculating machines since have been constructed."[14]*

---

*The mechanical-man did not merely include Grace Hopper to add a female voice to the equation. Her importance surpasses many of her male practitioners of cyber-technology in our quest. I'm sure there are many other women in the United States who helped the machine and computer along their trajectories. Even as I write this, new works on women in the computer sciences and artificial intelligence are being published. In Marie Hick's conclusion in *Programmed Inequality,* we read the following:

> In recent years, historical studies of women in computing have begun to proliferate. Many of these focus on the important task of uncovering women's contributions and adding them back into the historical record. Most of them focus on computer programmers, because programming has become seen as important, lucrative, and foundational. Understandably, many of these studies focus on women who have a claim to greatness or whose activities put them at the center of major historical events—like Grace Hopper, the ENIAC women, or Dame Stephanie Shirley. Yet the experiences of these exceptional women only begin to hint at the story of most women in computing. (222–23)

The next link between the past and the Mark I came from another mathematician who also dabbled in law, history, and philosophy—namely, Gottfried Wilhelm von Leibniz. His device with stepped wheels allowed the user to perform multiplication and division tasks. In addition, Hopper's treatise comments on philosophy's progression from Pascal to Leibniz, illuminate how the Enlightenment caused a change from how Pascal viewed man's relationship with machines compared to Leibniz's vision. Whereas Pascal regarded such automated machines as a subservient tool for the human race, never imbuing them with spirit or consciousness, Leibniz viewed such machines as an appendage of the human race, imbuing them with human qualities. The gap was already closing between machine and man.[15]A new dimensionality, a meeting ground, was emerging where human and machine consciousness could communicate.

In Charles Babbage's work, Hopper saw a direct nexus between the blueprints for his Analytical Engine (which would perform automated information-processing) and the Mark I. Indeed, Babbage imagined that a machine, if given the right tools, could think for itself. Thus, the word *automaton* took on a different connotation after Babbage. No longer would it simply mean the output of a machine or the tangible product of a machine, whether that product was in numerical form or a commodity. Instead, the word now pertained to the ability of a machine to think and learn for itself. Hopper wanted to take Babbage's idea a step further and not only imbue machines with the ability to think and learn for themselves, but also automate them in such a way that machines could teach themselves and control their own evolutionary development.

The Mark I manual set the stage for advances in the computer science industry to come. Hopper "produced an account in which the machine represented both a technical and a conceptual break from the past. This break is captured by the machine's official name, the Automated Sequence Controlled Calculator. A fully automated machine would free itself from the limitations of the human brain."[16]* That too is what she saw as Babbage's motivation. Whereas Pascal sought to free humans from "mechanical" work such as time-consuming calculations, Babbage and Hopper sought to free the machine from the fallible and limited control mechanism of the human brain. "Just as the steam engine became the technological foundation of the industrial revolution, the Automated Sequence Controlled Calculator could become the technological foundation for a new type of revolution dealing with information."[17] No longer would the fetish of the human apparatus be linked with the machine as metaphor; instead, the computer would now replace the machine in descriptions of the human body and mind.

Hopper's manual was more than merely a history of inventions by a discontinuous cadre of inventors. It was a philosophical treatise that explained the fundamental principles of machine-thinking. Given Hopper's vast knowledge in different areas of study, Aiken had chosen the best person for the job. As in Asimov's "The Evitable Conflict," published only four years after Hopper completed her manuscript, the computer or robot's brain was destined to triumph over its creators

---

*The *Cambridge Dictionary of Philosophy* describes Charles Babbage (1792–1871) as:

> [A]n English applied, mathematician, inventor, and expert on machinery and manufacturing. His chief interest was in developing [the] mechanical "engine" to commute tables of functions. Until the invention of the electronic computer, printed tables of functions were important aids in calculation. Babbage invented the difference engine, a machine that consisted of a series of accumulators each of which, in turn, transmitted its contents to its successor, which added to them to its own contents. (59)

It further describes that Rene Descartes (1596–1650) was:

> [A] French Philosopher and mathematician, a founder of the "modern age" [mechanical age]. Cartesian science and dualism, [t]he scientific system that Descartes had worked on before he wrote the *Meditations* and that he elaborated in his later work, *The Principles of Philosophy,* attempts wherever possible to reduce natural phenomena to the quantitative descriptions of arithmetic and geometry. (195–96)

In the mechanical-man's quest, we focus on Descartes's contribution to the mechanistic, objective, and measurable aspects of consciousness and reality. He is someone the mechanical-man is breaking away from in order to free his archetype for human utilization by way of manifestation.

in the areas of cognition and information-processing. Asimov was already well known for his robot stories, but Hopper was not listed as the author of her manuscript. What Hopper was doing in the real world, Asimov was doing in the imaginary, yet ironically Asimov was seen as more real than Hopper. The machine as myth had awakened a force and spirit that was far too monumental to deal with on only real terms. The imagined rites of passage in tracing the evolution of the machine were forcing humans into a mythic sphere in order to come to terms with the machine and machine intelligence as real. The human conscious was now communicating with the unconscious and uncovering ideas that were dormant until now. We were on the verge of discovering the fantastic or the impossible—blurring the lines between the real and fictive.

THE LANGUAGE INSTINCT

For Hopper and other members of the Harvard group the end of World War II was an opportunity to expose themselves to new ideas. Although she always lauded Aiken for his support and demanding work ethic, Hopper realized that he had created a closed system during the war years. Hopper was now ready to start a new career outside the circumscribed boundaries of both Howard Aiken and the military brass. Nevertheless, they had allowed Hopper to build what would be the groundwork for computer science and robotics.

> Hopper's time at Harvard was instrumental for her development as a programmer, a manager, and a leader within the emergent computing community. The pace of the war years catalyzed her transition from a college professor to that of a computer programmer, a term that would not be applied until 1949. During that time she helped define what programmers were, what they did and how they did it. . . . She developed a methodical system of coding and batch processing that turned the

experimental Mark I from a mechanical curiosity into a useful mechanical tool.[18]

By 1950 Hopper left Harvard and entered the business world. Joining Eckert-Mauchly Computer Corporation, which later was taken over by Remington Rand, Hopper was given the task of once again educating the computer. On May 3, 1952, Hopper presented a paper aptly titled "The Education of a Computer" in which she offered a blueprint for "automatic programming."[19] Years after writing the paper she reflected that "the novelty of inventing programs wears off and degenerates into the dull labor of writing and checking programs. The duty now looms as an imposition on the human brain." Thus, by teaching computers how to program themselves, Hopper, like Pascal and Babbage before her, sought to free the human brain from repetitive and time-consuming tasks and, at the same time, inculcate in the computer a sense of conscious awareness in the form of self-automation.[20]

> Hopper utilizes the metaphor of a factory to make abstract programming concepts more concrete for her audience. On a production line, inputs were raw materials that were acted upon by an assortment of instruments and tools. Human beings dictated the controls that organized the process, and the operation produced output ranging from automobiles to cans of tomatoes. Solving mathematical problems, according to Hopper, was no different. Inputs were alphanumerical data, and tools consisted of formulas, tables, pencil, paper, and the arithmetic processing power of the brain. The controls of the process were provided by the mathematician, and the output was the final result.[21]

Hopper grew up in this type of industrialized setting when the likes of Henry Ford and Frederick W. Taylor implemented their systemized approach to manufacturing. In this approach, she saw limitations similar to those of a computer programmer. Humans were still left with the task of breaking down problems into constituent parts, and they had to "provide step-by-step controls for the process via a program, write that program in notations best understood by the machine, and coordinate the introduction of input."[22] Hopper wanted to advance this assembly-line approach by making the machine autonomous. She did this by utilizing a compiler supported by library subroutines. In her 1952 paper, Hopper writes:

> He is supplied with a catalogue of subroutines. No longer does he need to have available formulas or tables of elementary functions. He does not even need to know the particular instruction code used by the computer. He needs only to be able to use the catalogue to supply information to the computer about his problem.[23]

These compiled subroutines allowed the computer to organize its thoughts. The catalog of subroutines acted as a menu or a library for the compiler to look up information or input. Thus, Hopper had created a way for the computer to have an organized library of information embedded in its conscious mind. Furthermore, the compiler acted as the historian and researcher in gathering and organizing such information in the computer's stored library. This menu or library "listed all the input information needed by the compiler to look up subroutines in the library, assemble them in proper order, manage address assignments, allocate memory, transcribe code, and create a final program in the computer's specific code."[24]

In effect, Hopper had used her skills as a professor and researcher to imbue the computer's mind with the same skills, even allowing the computer to create its own organized bibliography of information that it could retrieve in very little time. Not only was she imbuing the computer's mind with the function of memory, but she was also giving it the ability to organize, process, retrieve, and make educated choices. The most important aspect of the subroutine libraries was that a compiler "may itself be placed in the library as a more advanced subroutine." Thus, the content of a computer's memory, like that of the human brain, could increase in size and complexity at an exponential rate subject only to physical limits. In effect, Hopper demonstrated that the capacity of computer knowledge was essentially limitless.[25]

The former professor had taught the computer how to remember and understand the universal language of mathematics. This was not the end of the computer's education, however. Hopper was now about to turn her attention to the most monumental task of her career: teaching the computer how to interact with humans not solely through the language of mathematics but also through the language of nouns and verbs. If the computer was now conscious of its existence, it needed a way to tell us what its needs were. Hopper sought, therefore, to take the relationship between computers and humans a step further by attaching a language to the computer's library that was conducive to bringing the two species together. Although the computer was able to teach itself and work in an automated manner, Hopper sought to reduce the number of existing and differing computer languages to one universal language.*

The significance of what Hopper was undertaking was twofold. First, a universal computer language that could be used on any computer meant that it would not be esoteric to the general public. Thus, a universal language would bridge the chasm between the public and the computer. No longer would the minds of computers and interaction with them be limited to an elite few mathematicians and computer scientists. Second, a universal computer language would amalgamate computers, both hardware and software, into a type of

---

*Read Cooney's *Celebrating*. Even though it is almost forty years old, it's a childlike read for both children and adults. The mere fact that I am using such a book in this analysis intimates that Grace Hopper was largely forgotten and/or ignored in the twentieth century, relegated to a children's book.

species, evolving the same practices and programs from the same operational language. In a sense, she was channeling the now mythologized system of manufacturing exchangeable parts like Eli Whitney and Samuel Slater.

Hopper had gone from writing computer languages in mathematical symbols to writing them in English. Now was her chance once again to change the game, not only for the benefit of humans but also for the advancement of computers. No longer would the computer be alone in the garden because Hopper was about to give her Mark I his Eve in the form of a universal language that would allow them to communicate.

CYBERNETIC SOCIETY

Hopper was working on "the many ramifications of the theory of messages," but so too was a pioneer of cybernetics, Norbert Wiener. Wiener was focused more on the theoretical side and social implications of controlling machines and other such automata. Wiener's term, defined in his 1948 book entitled *Cybernetics,* was derived from the Greek word *kubernētēs,* or "steersman."[26] The thesis of Wiener's work is found in the following statement:

> Society can only be understood through a study of the messages and the communication facilities which belong to it; and that in the future development of these messages and communication facilities, messages between man and between machines, are destined to play an ever increasing role.[27]

Cybernetics gave a definition to Hopper's work. The machine was now communicating and acting on the external world by means of communication or messages. As Wiener notes again: "Every instrument in the repertory of the scientific-instrument maker is a possible sense organ, and may be made to record its reading remotely through the intervention of appropriate electrical apparatus. Thus the machine which is conditioned by its relation to the external world, and by the things happening in the external world, is with us."[28]

We now had a new science to define terms and theorize about what had been going on for the past century with devices like Tesla's telautomaton and Hopper's Mark I. And while Wiener's new science gave others a defined and organized way of talking about machines and human relations, it is in his prognostications and fears that we find his role in the mechanical-man myth. Indeed, his theorizing also helps us understand the "sorcery" behind Hopper's interactions with computers and why the mythic concept of sorcery is such an important element and prototypic pattern in the mechanical-man myth.

In *God and Golem,* Wiener cites to numerous tales of sorcery: *Thousand Nights and a Night, R.U.R.,* "The Monkey's Paw," and Goethe's poem, "The Sorcerer's Apprentice." The theme of each tale is the same, namely, there are forbidden powers, actions, and messages that must be controlled. Wiener sums this up by saying,

> The theme of all these tales is the danger of magic. This seems to lie in the fact that the operation of magic is singularly literal-minded, and that if it grants you anything it grants what you ask for, not what you should of asked for or what you intend. . . .The magic of automation, and in particular the magic of an automatization in which the devices learn, may be expected to be similarly literal minded.[29]

The problem is that machines and automaton must be built with a *fail-safe* switch, which means that "a goal-seeking mechanism will not necessarily seek our goals unless we design it for that purpose, and in that designing we must foresee all steps of the process for which it is designed."[30] Otherwise, the machine will attempt to reach its goal at all costs devoid of any considerations of the means in achieving such goal. This answers a major question uncovered in the mechanical-man myth: Why do so many of our fictional mechanical-men self-destruct, and why are Asimov's Laws of Robotics, in their totality, a fail-safe or safety switch? Wiener offers a mathematical and scientific understanding to the forbidden sorcery of playing the part of God, creating offspring in our own image. Hereby, Wiener illuminates that humans, unlike machines, appear to have a fail-safe, specifically, our consciousness, which denotes right from wrong in ethical conundrums. On the unconscious level, this fail-safe prevents us, or at the very least causes hesitation in our minds in our creative endeavors, from creating an imaginary or real automaton in our own image. It is the study of human fail-safes that I believe will lead to the final breakthroughs in artificial intelligence; specifically the study of the human fail-safe oxytocin. In oxytocin, we will soon discover our parallel to Asimov's Laws of Robotics. This occurs by destroying the artificial barriers between the real and the fictive. In the beginning, this filter erasing will seem chaotic. Conspiracy theories, fake news, moral relativity, and even violent riots will appear to be the new norm in culture, politics, and religion. This is merely the dark night of the soul that causes us to recognize how our consciousness is changing and progressing to being able to perceive what has previously been unseen. The barrier between feeling and thought or emotion and action will disappear. The esoteric meaning behind Franklin D. Roosevelt's statement that "[you have] nothing to fear but fear itself" holds new pragmatic meaning. The recognition and study of mythic cybernetics as I like to call it activates coding that has been dormant for centuries in our individual and collective psyche. In his book *Walking Between the Worlds: The Science of Compassion,* Gregg Braden allows one to merge science and mythology by focusing one's attention and actions on the word *possibility.* These vibratory codes activate the energies that follow attention. "All myths of our past hold that energy follows attention." As Braden points out we must allow fear to remove or counter the energy of fear. "In 'allowing' for the *possibility,* is found the removing of the charge."[31] Braden, like Richard Grossinger, shows us that "science and religion are finally metanarratives that give rise to each other, for an algorithm generating galaxies and roses, cobras and tardigrades out of quarks and baling wire could be a god generating them out of divine intelligence or a nonlinear gyre writing the flap of every butterfly's wings on an ineffable hard drive."[32] Consciousness is both the source input and the source output. Even the fictive and real are metanarratives that give rise to each other. Cosmic consciousness and human consciousness move one another through space and time. While both come from the same source, separation is necessary for conscious experience to occur. When the divine and human consciousnesses merge, we will experience another *bottoming out,* another transition at the level of the Big Bang. The only value of fear in the form of anxieties is the paradoxical harmony created by simultaneously balancing the anxiety of the moment against the anxiety of the eternal. Thus opposed, the two concepts create an algorithm of harmonic angst. Without this knowledge, one experiences perpetual anxiety about the moment or, in moments of joy, perpetual anxiety about eternity. In this sense, the word *eternity* does not refer to the future only but also to the future's future; it is the reverse of nothingness toward which reductionists are leading us—the nothingness of space and time never-ending.

Braden, Grossinger, Kripal, Dispenza, and Wiener all remind us of the quest's ultimate goal of projecting imagination and mind into nature to receive, communicate, and control service, spirit, and agape.

Wiener also reminds us of the cost of knowledge devoid of ethics. Years before Melville's "The Bell-Tower," Nathaniel Hawthorne, in 1844, penned the short story "Rappaccini's Daughter." The tale follows a doctor's daughter who has been plagued by her father's megalomania. He seeks knowledge at all cost; this is his ultimate goal. Even his daughter becomes a casualty of his insatiable desire for knowledge. While Hawthorne's Dr. Rappaccini indulges in creating mechanical butterflies, the tale illuminates Wiener's fears involved in programming machines and computers. Specifically, if there are no set conditions to control one's goal oriented drive, then it follows that such a person can utilize any means possible. Of Rappaccini, Hawthorne writes:

> But as for Rappaccini, it is said of him—and I, who know the man well, can answer for its truth—that he cares infinitely more for science than for mankind. His patients are interesting to him only as subjects for some new experiment. He would sacrifice human life, his own among the rest, or whatever else was dearest to him, for the sake of adding so much as a grain of mustard-seed to the great heap of his accumulated knowledge.[33]

Those that would sacrifice human life for their own selfish goals are without a fail-safe. For Wiener this is tantamount to sorcery. On how Rappaccini came to create a mechanical butterfly, Hawthorne states the following:

> I know that look of his; it is the same that coldly illuminates his face, as he bends over a bird, a mouse, or a butterfly, which, in pursuance of some experiment, he has killed by the perfume of a flower;—a look as deep as Nature itself, but without Nature's warmth of love. Signor Giovanni, I will stake my life upon it, you are the subject of one of Rappaccini's experiments![34]

Wiener sees the same look in those that program machines and computers for specific purposes without knowing all possible outcomes embedded in the means of achieving such goals. It is the blind pursuance of experiments that plagues Wiener's theorizing. He is warning present and future programmers that they must be mindful of all possible paths a machine and computer may take in attempting to accomplish a programmed goal. By explaining this problem through the use of cybernetics, Wiener shows that computers and machines are freer than we had imagined. Their freedom lies in the gray area or the acts that they choose, unknown to the programmer, in achieving their goals. These acts are the messages they send back (or may not send back) to the programmer. In a sense, these acts denote why they are doing the things they do in achieving the goal that the programmer gave them. The question arises: Is this sorcery?

Sorcery's principal mythic elements are the performing of miracles devoid of being ordained by God to partake in such acts. Wisdom gained that is not ordained is nothing more than "perverted wisdom."[35] Thus, we call Nikola Tesla a sorcerer and Thomas Edison a wizard, in part because they are not priests nor are they ordained by any religious sect. Wiener answers our objections to this assertion by stating:

> Perhaps the power of the age of the machine are not truly supernatural . . . we no longer interpret our duty as obliging us to devote these great powers to the greater glory of God, but it still seems improper to us to devote them to vain or selfish purposes. There is sin, which consists of using the magic of modern automatization to further personal profit or let loose the apocalyptic terrors of nuclear warfare . . . let the name be Simony or Sorcery.[36]

This leads back to the same conclusions we reached when discussing Melville's "The Bell-Tower"; specifically, we must be mindful of the motivations behind such goals or behind such programs. However, Wiener also recognizes that even if one knows the motivations behind such goals, we still cannot foresee all possible means a computer or machine may utilize in achieving such goals, devoid of a fail-safe switch. For such foresight we would have to have a universal computer, which would take into account all possible outcomes and means in achieving such outcomes.

For Carl Jung and Joseph Campbell, dreams direct us to these fears of magic and forbidden powers. These fears in the unconscious also come to us in forms of literature, poetry, and theorizing. Thus, these events and fears "have happened but they have been absorbed subliminally, without our conscious knowledge . . . and though we may have originally ignored their emotional and vital importance, it later wells up from the unconscious as sort of afterthought."[37] We see evidence of this in the fictive imagination and literature of Melville, and in the self-destruction of Bannadonna's android, in the self-destruction of the Steam Man, the automaton in "Moxon's Master," and *Forbidden Planet's* Robby the Robot. In addition, we see it in Tesla's apparent self-sabotage of the telautomaton, and, later, the *Challenger* explosion that Richard Feynman will explain to the public with the aid of children's toys. We may also wonder whether it was Grace Hopper's awakening to this unconscious recognition of the sin of sorcery that contributed to her binge drinking during the times she worked closely with the Mark I.

Cyberneticist Katherine Hayles notes that the impact of cybernetics on theories of communication and control goes beyond mathematics, engineering, and electronics into areas of psychology, neurology, biology, physics, and even the social sciences. In other words, it is truly in the study and understanding of our messages with one another that we define ourselves. Further, in the context of mechanical-man as mythic-marker, it is in the messages we receive from the fictive and the real that awaken us to deeper levels of messages and communicative modes of the unconscious. Myths can imbue us with messages and even blueprints that enlighten us to the sciences. Conversely, science also imbues us with messages and communicates meaning to our dreams and myths.

HOPPER: THE LIBERATOR

In the end Hopper "played the role of facilitator, gathering technical, economic, and social feedback about automatic programming. . . . For her the invention of a computer language was an ongoing, organic process that was always adapting to the changing needs of the computer profession."[38] Hopper showed the world that humans and machines were connected as replicas of each other. Common business-oriented language (COBOL) demonstrated that the essence of a computer's mind or spirit was not in its hardware, but rather in its imaginary gap between its hardware and software.

Creating a universal language that allowed humans and computers to understand one another was the first step in our future ability to have computers "deeply embedded in our bodies, brains, and environment." While Hopper was not the first to educate us about a computer's operational parallel to human behavior, she was the first to teach the computer about the nature of its creators. Scholars such as Ray Kurzweil now recognize Hopper's accomplishments in the field of computer science, and see a connection to our own "identity and survival."[39]

"In the picture language of mythology," writes Joseph Campbell, "the hero is the one who comes to know."[40] The Mark I was a mythical and symbolic hero in the field of computer science and robotics. Indeed, the foundations of the Mark I would lead to the internet and usher in another transformative era in technology for America. The project would also stimulate the imaginations of literary artists by limning a world of possibilities pertaining to robotic intelligence. Grace Hopper was also a hero in her own right, a symbol of achievement for women in their attempt to gain equal rights in a male-dominated society. Thus, the relationship between women and machines in American history illustrates how the former have instrumental value in the fields of robotics and computer intelligence. In her writings on the history of computer science, Hopper paved the way for a period of exponential technological growth in which the imaginary history of a mechanical-man would no longer outpace its factual reality. For future writers and imaginers of mechanical-men, Hopper created a blueprint for the anima to emerge in machine intelligence. For if the anima is "a personification of all feminine psychological tendencies in a man's psyche, such as vague feelings and moods, prophetic hunches, receptiveness to the irrational, capacity for personal love, feeling of nature, and—last but not least—his relation to the unconscious," then Hopper is the inner figure in the mechanical-man's psyche.[41] She is the awakened Sophia (wisdom) of Mother Gaia.

6
The Prophet

By the 1960s science in the United States was a well-established enterprise, and "launches of spacecraft, like atomic explosions, represented a new stage in the historical shift from man to machine to process."[1] The scientific formula for the sublime involved the cooperation of the military-industrial complex, academia, and public opinion. No longer was science propelled by the dreams of maverick visionaries; instead, it was an independent, widely embraced field of national endeavor.

When President John F. Kennedy delivered his inaugural address to the nation in 1961, he warned of the profound impact that science could have on the future. We could either destroy ourselves or land a man on the moon by the end of the decade. When we succeeded in the latter challenge, it was said that "This is the greatest week in the history of the world since Creation."[2] The dual potential of technology was expiated by achievement of the sublime. At a time when Lewis Mumford wrote, "We are living among madmen. Madmen govern our affairs in the name of order and security. The chief madmen claim the titles of General, Admiral, Senator, scientist, administrator, Secretary of State, and even President," landing a man on the moon was seen as the "final avatar of the technological sublime escape from the threatened lifeworld."[3] The importance of the Apollo 11 mission and emergent space program, however, was that science now transcended the boundaries of circumscribed ideologies. Moreover, space and time were no longer seen as implacable constants, but rather as variables that human beings could easily manipulate and use to their advantage.

SMALLER THAN A DOT

While Kennedy was offering his vision of the macrocosmic future, Richard Feynman was prophesying the use of physics to create a new race of robots in a miniaturized or microcosmic world. Only a few years before Kennedy's speech, Feynman was theorizing about the world of human scales. When the American Physical Society held its annual meeting at the end of 1959 at Cal Tech, Feynman told his audience, "There is a device on the market now that can write the Lord's Prayer on the head of a pin. But that's nothing. . . . It is a staggeringly small world that is below."[4] By the end of Feynman's lecture the audience was flabbergasted. One prediction especially caught the attention of Feynman's followers: "There is nothing I can see in the physical laws that says that computer elements cannot be made enormously smaller than they are now."[5]

Computers had changed the paradigm of how we viewed science. In Feynman's mind the computer was now educating us about ourselves and our environment in ways of which we were not even conscious. Feynman saw such technology as giving "rise to a new way of thinking about information, and in terms of raw information, all the world's books could be written on a cube no larger than a speck of dust."[6] For Feynman and others, Grace Hopper had feminized technology, making us aware of the details and potential intuition of such. The communicational nuances and interactions with humans could now be built upon in ways that were thought unimaginable.

Feynman also spoke about DNA's capacity to "build tiny machinery, not just for information storage but for manipulation and manufacturing."[7] In effect, he was indicating that the entire universe is made of the same material. It sounds like a simple concept, but in terms of robotics and computers it is far-ranging. If robots and computers are made of atoms and operate under the same level of construction, they too have the ability to think, reason, and make judgments, for they fall within the same sphere of natural laws and organic functionality as we humans do. What, then, becomes our overall picture of the world?

Feynman's 1959 paper was subsequently published in *Popular Science*. Years later there was a name for the field that Feynman had prophesied—nanotechnology. Today nanotechnologists regard Feynman as their spiritual father. Feynman had paved the way for a new kind of technology that many believe will lead to the final singularity of manifesting immortal vital machines (cyborgs).

Seven years later Richard Fleischer would bring Feynman's ideas to celluloid in *Fantastic Voyage*. In this film the Americans and the Russians have created inventions that can shrink matter, manipulating the powerful atom. Technology was catching up to the popular imagination. Simply put, throughout the 1960s, scientists no longer saw robots as machines. Robots were now envisioned as made from the same building blocks as man—cell by cell, atom by atom. If a man could shrink into something so small as to enter the body of another, so too could a robotic machine. Ideas of conquest and national destiny, the foundation of the American narrative, were not limited to harnessing the natural environment. Man now sought to tame the frontier of the human mind.

In an article in *Popular Science* entitled "Fantastic Voyage: A mini sub that could steer through the body," Daniel Pivonka, an electrical engineer who worked on the tiny vehicle project while a graduate student at Stanford, states the following:

> In the future, tiny vehicles might travel through your body to image your insides, take samples, and deliver drugs. At Stanford University, my colleague Anatoly Yakovlev and I built a prototype of such a device. It's about the size of Abraham Lincoln's head on a penny. We power and control the prototype wirelessly by sending radio waves to its two by- two-millimeter antenna from about two inches away.

> No battery is required, which is key to miniaturization. Mechanical propulsion is inefficient at this scale. Instead, we use magnetohydrodynamic propulsion, which takes advantage of the fact that an external magnetic field can push an object by creating a Lorentz force on its electrical circuitry.[8]

Pivonka's creation reveals the interplay between science and the imagination. While *Fantastic Voyage* planted the seeds for his imaginary development of the mini sub, the science of nanotechnology provided him the tools. His mini sub harks back to Tesla's remote controlled boat. With the advent of cybernetics, the import of the mini sub and Tesla's boat is found in the messages it sends to its human controllers. Indeed, as noted by Norbert Wiener:

> Messages are themselves a form of pattern and organization. . . . Just as entropy is a measure of disorganization, the information carried by a set of messages is a measure of organization. In fact, it is possible to interpret the information carried by a message as essentially the negative of its entropy, and the negative logarithm of its probability. That is, the more probable the message, the less information it gives. Clichés, for example, are less illuminating than great poems [or great myths].[9]

To recapitulate, myths can imbue us with messages and even blueprints that enlighten us to the sciences. Conversely, science also imbues us with messages and communicates meaning to our dreams and myths. For Richard Feynman the message he sought to visit upon the public equated to speed and size.

NANOTECHNOLOGY

The same patterns and issues of control, communication, and information once again emerge in nanotechnologies, including manifesting consciousness in inert matter during an interval of enchantment so that the matter becomes aware of a certain goal. How does the controller recognize and control the unknown variables at play in the animated matter's achievement of the goal? How does the controller communicate with this conjured consciousness? The same issues are apparent when we look at potential issues relating to nanotechnology, that is, technology so small that it cannot be perceived by the human eye.

Nanotechnology's emergence shows that the direction of science in the United States was entering an interdisciplinary phase. To investigate atoms at the nano-level, one must utilize chemistry, physics, and quantum mechanics. Thus, today chemists, physicists, and medical doctors are working alongside engineers, biologists, and computer scientists to determine the applications, direction, and development of nanotechnology. In essence, nanotechnology is many disciplines building upon one another. So why is this technology so important to our study of human-machine interactions in the United States? Feynman realized that nanotechnology would revolutionize practical areas of technological growth, specifically the fundamental tenets of science on which Thomas Jefferson and James Madison based our patent laws. Thus, while he was the father of theorizing this new field of study, Feynman was also, like Edison and Tesla before him, its model-maker, for he was able to show that minds are merely instruments whose powers can be harnessed to change reality.

What in the mechanics of nanotechnology allowed these ideas to come to fruition? The answer is twofold: first, high-performance computers using new architecture that may make AI a reality; and, second, approaches in computing including the use of molecular electronics, bioelectronics, and quantum computers. This involved, as conceived by Hopper and Feynman, a truly interdisciplinary approach. Nanotechnology thus draws us closer to our robot creations, allowing us to blur the distinction between human and robot by morphing parts of our bodily apparatus into features that science fiction portrays as robotic attributes, similar to Poe's General A. B. C., but also allowing us to create a human sub-species entirely separate and apart from our bodily apparatus similar to the Krell that we find are extinct in *Forbidden Planet.*

Still the most important aspect of nanotechnology is speed. In the fields of bioelectronics computing, DNA computing, quantum computing, and molecular electronics, this means penetrating into new forces previously thought forbidden to man and harnessing these intangible powers. Thus, the myth of man's replacing God continues in that nanotechnology makes feasible the idea of a robot displaying intelligence, not only on a par with that of a human being but much faster and smarter. In today's environment, the driving force of nanotechnology is homeland security, cyber defense and cyber espionage, since faster computing power means "much more effective kinds of computation for code[-] breaking" and other aspects of military protection.[10]

Nevertheless, as we have seen, with increases in speed and the means of increasing such speed during the American Industrial Revolution, there are numerous problems of control and communication that arise within the context of nanotechnology. Specifically, as Thomas McCarthy explains:

> An unfortunate side-effect of molecular manufacturing is that it may contribute to the creation of a state of war regardless of the circumstances, whether they are ones of peace or of war. If MNT (molecular nanotechnology) makes weapons invisible, and does the same for factories, then the ability of one side to measure the capabilities of the other will be severely hampered, and perhaps eliminated completely. This will be destabilizing in two ways. First, by making some weapons impossible to detect, it will prevent those weapons from fulfilling their role as deterrents in times when deterrents are needed, and thus will decrease the ability of states to dissuade potential aggressors from initiating military hostilities. . . . Secondly, the lack of armaments, which is necessary for not projecting hostile intentions and arousing suspicions during peacetime, will be mean-

> ingless. The lack of detectable armaments will not in itself be reassuring to other states, and even the true absence of armaments will be inadequate proof of commitment to lasting peace, when the tools of war can be generated cheaply on a few days' notice.[11]

It is one thing to say that problems of human-machine interactions are unforeseen and unpredictable, but to imagine technology that can manipulate the environment and build machines that are undetectable creates problems that humanity has never faced. In addition, since mankind has never faced these problems before, they must be looked upon utilizing appropriate symbols and models. Channell's vital machine concept "represents a dualistic system in which components or subsystems at one time thought to be organic or mechanical lose their individual identities and become part of a new category or phenomena."[12]* This model of understanding helps us analyze the problems of control and communication in areas of human-machine relationships that are otherwise impossible to explain or grasp—nanotechnology necessitates that we discuss it utilizing the vital machine approach due to its unpredictable nature when placed in the realm of either the mechanical or

---

*At this point, we need to define the mechanical-man as vital machine. Rupert Sheldrake helps us in the same way that reading Dawkins's books on selfish genes helps us understand many of our metaphysical concepts with a mechanistic bent. In his book *Morphic Resonance,* Sheldrake speaks of Dawkins's selfish genes saying:

> In Richard Dawkins' concept of the "selfish gene," the genes themselves have come to life. They are like little people: they are as ruthless and competitive as "successful Chicago gangsters"; they have powers to "mold matter," to "create form," to "choose," and even "aspire to immortality." Dawkins' rhetoric is vitalistic. His selfish genes are miniaturized vital factors." (11)

Rupert Sheldrake is pointing out the language of vitality: *molding; creating; choosing; aspiring.* Some people falsely believe that the selfish gene theory is devoid of direction, thus, blind. However, these genes, like human self, are very much purpose-oriented beings. Vital does not just mean living, it means living with the ability to control and communicate information, the pillars of cybernetics, in a controlled purposeful direction. What the mechanical-man is discovering is that there is a difference between automatic control/communication and creative control/communication, or what Dr. Joe Dispenza calls *quantum creation.*

organic. In effect, it allows us to explain the gaps between the events and the meaning of such events. Wiener aids us in understanding this issue of controlling the invisible when he says: "In other words, while in the past humanity has faced many dangers, these have been much easier to handle, because in many cases peril offered itself from one side only. There is no fail-safe switch that mankind can now envision when dealing with machines that are invisible to the human eye."[13] The mini sub machine in the film *Fantastic Voyage* encounters numerous difficulties that the outside world may not interfere with. However, the difference between the fantasy and reality is that the movie envisions humans possessing the capability to shrink themselves down to the level of invisible molecular machines. The idea of being able to shrink ourselves to such a level and manipulate our own atomic structure is the fail-safe in the fantasy world of celluloid mirrors. However, in reality—if machines are able to shrink to a level that is invisible to the human eye and we are unable to shrink ourselves to their level—the lack of control and predictability as to the actions of such machines may create dire consequences for humans.

If machines become so small that they are invisible, our ability to control and stop their replication may spell doom for the entire race. Both Wiener and Channell are correct when they mention heredity and Darwin's natural selection concomitantly with machines. To be incapable of interfering with this process equates to allowing machines to make copies and replicas of themselves with variation. The problem for Wiener is in the variations that are produced by natural selection. For example, if molecular nano-machines are programmed to act a certain way but are invisible to the human eye, then the laws of nature that affect such machines are no longer subject to human intervention or control. As Wiener points out: "It is clear [in machines] that the process of copying may use the former copy as a new original. This is, variations in the heredity are preserved, though they are subject to a further variation."[14] These unknown variables or variations are of utmost concern for theorists, especially when it comes to building weapons on a nano-scale. In sum, the unpredictable nature of invisible machines that evolve and show variations in the replication process opens up yet another Pandora's Box in human-machine interactions. What Jeffrey Kripal calls the X that is beyond A and B, is evident in the unpredictable nature of invisible machines that show these traits. But like Kripal says:

> I find such an (im)possibility incredibly empowering. If, after all, we can begin to understand and act on these insights, we might at least begin to take back the book of our lives from those who wrote us long ago, for their own good reasons, no doubt, and begin writing ourselves anew, for our own good reasons now. Our ancestors and their deities were completely ignorant of such new good reasons, just as we are completely ignorant of the good reasons and concerns of two thousand years from now. Our system must damn the old ones, and ours will be damned in turn.[15]

It's as if we have communicated this information to ourselves, as if it was there all along, just waiting to be found. So when looking at issues of control and communication in the context of human and machine interactions, we are also acknowledging a communication with the conscious and unconscious and with our present, past, and future ancestors. When the conscious communicates with the unconscious, the fictive makes its appearance in reality. When the conscious controls the information received from the fictive—within the culture the fictive appears—materiality and curiosity via the imagination interface—making its appearance in objective reality. This is how it looks on the outside. We can call this the second person experience of manifesting reality. The first person experience is the conscious state or frame of mind you are in. I always had the childlike question of faith without works; my mother would always say over and over again that "the state of consciousness will automatically do the good works for you." What a wonderful way of saying it. Now, I get it! The machine will always do the work automatically when the end is selfless agape. It's becoming uncanny how our innovators and inventors confronted all of these inner and outer issues in projecting their spirit into everything they did, and not fearing the consequences, knowing that their destiny, in the end, is a noble one for all of us.

THE TEACHER, BUT ALWAYS THE STUDENT

John W. Campbell Jr., editor of *Astounding Science-Fiction* during the Manhattan Project, relates his connection to Feynman who worked on the Project when he tells the story of U.S. military agents walking into his editorial office and demanding that they reveal the culprit that leaked information about the source of atomic energy ideas to the author of "Deadline" by Cleve Cartmill. Campbell denied that it was Feynman or any other scientists that "leaked" such "obvious" information. Nevertheless, it was common knowledge that the majority of scientists working on the project credited many of their ideas and their vocational trajectories to the imaginers of the past and present who had once given speculative and prophetic voice to the sciences they now worked on.[16]

After leaving Los Alamos, Feynman spent the rest of his life teaching and researching in the field of quantum physics. During those Cold War years, the government established the Atomic Energy Commission, an Office of Naval Research, and the National Science Foundation. He continued his research while permanent national laboratories were established at Oak Ridge, at Argonne, at Berkeley, and at Brookhaven, Long Island, on a 6,000-acre former military base. With the construction of such facilities the "government and public gained a new sense of proprietorship over the whole scientific enterprise."[17] In this unprecedented enterprise, the future Nobel Prize winner saw the organizing principles of his own theories in the field of physics, for the scientific community was evolving along the same lines as his systems-oriented theories. This systems-oriented science "required staunch and systemic skepticism about one's own hypothesis until the only reasonable alternative left—in the framework of the reigning paradigm of the time—is that such a hypothesis be true."[18]

Unlike Edison and Tesla, Feynman was not an inventor in a world where patents ruled the scientific realm. For him, the most important challenge was to uncover a unifying theory of everything. That is where he saw science headed in the United States during the Cold War. Scientific management, corporate goals, government, and academia were all merging into a cohesive whole, interrelated in their dynamics and even nuanced differences. He lauded such integration, and amid it he sought to create his own theory that could explain the rules of the game and comprehend all of the connections and phenomena of the universe. His first step in reaching this goal was in the field of computers.

At Cal Tech, and later Berkeley, Feynman began to lay the theoretical foundations for quantum computers and the universality of Turing machines. He was seeking to interrelate physics, information, and computation. While teaching at Cal Tech and Berkeley, Feynman changed what some call pragmatic instrumentalism, or the typical physicist's attitude toward theories of reality. Specifically, "if instrumentalism is the doctrine that explanations are pointless because a theory is only an instrument for making predictions, pragmatic instrumentalism is the practice of using scientific theories without caring what they mean."[19] In terms of Kuhnian paradigm shifts, Feynman's new version of pragmatic instrumentalism was a shift in and of itself. On the surface, what I like to call Feyn-pragmatic instrumentalism appears at odds with Popperian epistemology, which had "become the prevailing theory of nature and growth of scientific knowledge."[20] Popperian criteria hold that a theory must have predictive power and that good explanations must be able to be tested, refuted, etc. However, critics of Feynman's theories, including his contributions to quantum computation, the Turing principle, and artificial intelligence, fail to grasp the importance of the imagination in science. In other words, they fail to recognize the importance of explanation in a mythical and metaphorical sense. For these critics Feynman did not play by the rules or laws within an accepted Popperian framework.

This is a good point in the quest to introduce you to the ideas of two important researchers: Karl Popper and Thomas Kuhn.*

Karl Popper lived from 1902–1994. Audi, in the *Cambridge Dictionary of Philosophy,* describes Popper as:

> [An]Austrian-born British philosopher best known for contributions to philosophy of science and to social and political philosophy. Popper proposes that science be characterized by its method: the criterion of demarcation of empirical science from pseudo-science and metaphysics is falsifiability. According to falsification[-]ism, science grows, and may even approach the truth, not by amassing supporting evidence, but through an unending cycle of problems, tentative solutions—unjustifiable conjectures—and error elimination; i.e., the vigorous testing of deductive consequences and the refutation of conjectures that fail. (631)

How then does the deductive procedure work? Popper specifies four steps:

1. The first is *formal,* a testing of the internal consistency of the theoretical system to see if it involves any contradictions.
2. The second step is *semi-formal,* the axiomatic[z]ing of the theory to distinguish between its empirical and its logical elements. In performing this step the scientist makes the logical form of the theory explicit. Failure to do this can lead to category-mistakes—the scientist ends up asking the wrong questions, and searches for empirical data where none are available. Most scientific theories contain analytic (i.e., *a priori*) and synthetic elements, and it is necessary to axiomatize them in order to distinguish the two clearly.
3. The third step is the comparing of the new theory with existing ones to determine whether it constitutes an advance upon

*I recommend you read two books: Karl Popper's *Conjectures and Refutations,* and Thomas Kuhn's *Structure of Scientific Revolutions.*

them. If it does not constitute such an advance, it will not be adopted. If, on the other hand, its explanatory success [maps] existing theories, and additionally, it explains some hitherto anomalous phenomenon, or solves some hitherto unsolvable problems, it will be deemed to constitute an advance upon the existing theories, and will be adopted. Thus, science involves theoretical progress. However, Popper stresses that we ascertain whether one theory is better than another by deductively testing both theories, rather than by induction. For this reason, he argues that a theory is deemed to be better than another if (while unfalsified) it has greater empirical content, and therefore greater predictive power than its rival. The classic illustration of this in physics was the replacement of Newton's theory of universal gravitation by Einstein's theory of relativity. This elucidates the nature of science as Popper sees it: at any given time there will be a number of conflicting theories or conjectures, some of which will explain more than others. The latter will consequently be provisionally adopted. In short, for Popper any theory XX is better than a "rival" theory YY if XX has *greater empirical content,* and hence *greater predictive power,* than YY.

4. The fourth and final step is the testing of a theory by the empirical application of the conclusions derived from it. If such conclusions are shown to be true, the theory is corroborated (but never verified). If the conclusion is shown to be false, then this is taken as a signal that the theory cannot be completely correct (logically the theory is falsified), and the scientist begins his quest for a better theory. He does not, however, *abandon* the present theory until such time as he has a better one to substitute for it. More precisely, the method of theory-testing is as follows: certain singular propositions are deduced from the new theory—these are predictions, and of special interest are those predictions which are "risky" (in

> the sense of being intuitively implausible or of being startlingly novel) and experimentally testable. From amongst the latter the scientist next selects those which are not derivable from the current or existing theory—of particular importance are those which contradict the current or existing theory. He then seeks a decision as regards these and other derived statements by comparing them with the results of practical applications and experimentation. If the new predictions are borne out, then the new theory is *corroborated* (and the old one falsified), and is adopted as a working hypothesis. If the predictions are not borne out, then they falsify the theory from which they are derived (*Logic of Scientific Discovery,* 1.3, 9). Thus, Popper retains an element of empiricism: for him, scientific method does involve making an appeal to experience. But unlike traditional empiricists, Popper holds that experience cannot *determine* theory (i.e., we do not argue or infer from observation to theory), it rather *delimits* it; it shows which theories are false, not which theories are true.

Moreover, Popper also rejects the empiricist doctrine that empirical observations are, or can be, infallible, in view of the fact that they are themselves theory-laden.*

As for Thomas Kuhn, he is known for his description of what constitutes a paradigm: What is to be observed and scrutinized. The kind of questions that are supposed to be asked and probed for answers in relation to this subject. How these questions are to be structured. How the results of scientific investigations should be interpreted. In short, a *paradigm* is a comprehensive model of understanding that provides a field's members with viewpoints and rules on how to look

*See Stephen Thornton, "Karl Popper," *Stanford Encyclopedia of Philosophy* website (accessed Aug. 7, 2018).

at the field's problems and how to solve them. "Paradigms gain their status because they are more successful than their competitors in solving a few problems that the group of practitioners has come to recognize as acute."*

The mechanical-man has quite a bit to say about Popper and Kuhn. This is not an attack on Khun's structures or Popper's criteria. Pragmatic manifesting takes from both what is proper in a mythological-based structure, specifically, paradigms that are simulated, which causes a reversal and paradox in Popper's criteria. Philip K. Dick explains this better than anyone in his *Exegesis*.

*Kuhn, *The Structure of Scientific Revolutions*, 23.

That, however, was the beauty of his work: there was no blueprint that allowed one to follow in the same footsteps as Feynman in producing his theories and results. Having once remarked that perhaps all we can know are the rules of the game, he showed that these so-called rules have rules that they go by and, in turn, that those rules have other rules that they go by.* Thus, to limit oneself to a particular scientific framework is to limit one's actual understanding of the nature of things. This is how science in a materialistic paradigm "fail[s] to see [that] the ability to predict how things behave with respect of one *another says little about what things fundamentally are*."[21] In the end, he demonstrated the following:

> Philosophically we are completely wrong with the approximate law. Our entire picture of the world has to be altered even though the mass changes only by a little bit. This is a very peculiar thing about the philosophy, or the ideas, behind the laws. Even a very small effect sometimes requires profound changes in our ideas.[22]

*Rhodes's masterpiece provides a guide for how I set up the offset block quotations in the mechanical-man. The reader is told a separate, primary story, by reading just the quotes from beginning to end.

DEATH AND RESURRECTION

In 1987, on the verge of death from cancer, Feynman dissented from the prevailing opinion of how and why the space shuttle *Challenger* exploded in mid-flight. In a nationally televised press conference, he utilized models from an array of children's toys to show how frozen o-rings caused a fuel leak, thus leading to one of the biggest disasters in our nation's history, and to years of stagnation for the space program and robotics in the United States. His simple lesson with children's toys led to numerous reforms as well as seventy-five subsequent shuttle launches devoid of mishaps.[23]

Richard P. Feynman died one year after presenting his conclusions about the *Challenger* explosion. Nearly three decades earlier he had fomented the molecular revolution in scientific thought that we now envision as taking us into the future. His simple genius was to point out that technology was leading us into that paradigm shift. The Information Age will soon be replaced by another revolution that will change the landscape of science and the quality of life for *Homo sapiens*. Daniel Hillis writes, "Maybe it's telecommunications merging us into a global organism. If you try to talk about it, it sounds mystical, but I'm making a very practical statement here. I think something is happening now, and will continue to happen over the next few decades, which is incomprehensible to us, and I find that both frightening and exciting."[24]

Feynman's paper titled "There's Plenty of Room at the Bottom" was prophetic in giving the next generation of scientists a feasible foundation that included genetics (manipulating the building blocks of life), robotics (building autonomous machines to do our bidding), artificial intelligence (designing machines that learn), and nanotechnology (building things atom by atom).

> The acronym is GRAIN. This mega merger of super-sciences may transform who and what we are. It could transplant our senses into other entities, and then convert those entities into something else.

> It may alter millions of years of evolution that imbue us with wonderful and terrible traits. It already lends a new dimension to our technological development. One technology sets the speed limit for every other. Its inventors build atom by atom on a scale of one-billionth of a meter—a nanometer.[25]

For our purposes, I want to analyze two passages in Feynman's paper and discuss what has resulted from such theorizing. Here is the first excerpt:

> A biological system is exceedingly small. Many cells are very tiny, but they are very active; they manufacture various substances; they walk around; they wiggle; and they do all kinds of marvelous things, all on a very small scale. Also they store information. Consider the possibility that we too can make a thing very small which does what we want, that we can manufacture an object that maneuvers at that level.[26]

After 1959 scientists no longer saw robots as machines constructed of metal and computer parts. Instead, robots were regarded as made from the same building blocks as man—cell by cell, atom by atom. The foundations of this cellular robotic revolution appeared in the form of cloning, genetic manipulation, and even prosthetics, for in these endeavors scientists were utilizing the same type of small-scale building and manipulation that Feynman had discussed. "Once ideas grab hold of society," observes one scholar, "they take on their own life through self-replication." These ideas are called "memes."[27] Feynman's insights in 1959 give credence to this "memetic" theory that ideas replicate themselves. The final importance of "There's Plenty of Room at the Bottom" is that it blurred any lingering distinction between robot and human.

Feynman continued this line of thought in a second passage whose implications I wish to consider:

> If we go down far enough, all of our devices can be mass[-]produced so that they are absolutely perfect copies of one another. We cannot build two large machines so that the dimensions are exactly the same. But if your machine is only 100 atoms high, you only have to get it correct to one-half of one percent to make sure the other machine is exactly the same size, namely 100 atoms high.[28]

Feynman here reminds us of the American system of manufacturing on a mass scale, thereby echoing the sentiments of such efficiency-minded pioneers as Whitney, Evans, Fulton, McCormick, Slater, Taylor, and Ford. His ideas are perhaps the prophetic final stages in this system of manufacturing, taking it to its very apogee of singularity, for in Feynman's theory of mass production humans are able to replicate not only rifles, sewing machines, and reapers but also a far more significant artifice—the self. By harnessing properties of the mighty atom, we can even replicate our memories. Feynman thus envisioned a world in which we could fabricate reality itself, including ourselves, in order to transcend our limited existence and conquer even mortality, augmenting the American mythos of achieving new heights of manifesting a collective destiny. Technology was now becoming more and more nuanced and detailed, like the human body and human efforts. The theme of coevolution between technology and humanity was advancing now at a rapid pace.

Feynman in 1959 had given the world the RNA, or cookbook, for how to create virtual reality in terms of AI. Needed now was knowledge of human DNA, our cosmic destiny, for reaching the apex of the robotic revolution in terms of self-replication. When Feynman died in 1988, his friend Marvin Minsky had already set in motion initiatives for bringing to fruition this new reality.

THE LEGACY

In the 1960s three robotics projects got underway at Stanford University, Stanford Research Institute, and the Massachusetts

Institute of Technology (MIT). The question addressed by each research group was: How will a robot function in the real world? In this same decade, John McCarthy helped to write a paper outlining the central ideas of robotics and AI. Of McCarthy's concepts, famed physicist Minsky wrote,

> They began pointing out that a computer program capable of acting intelligently in the real world must have some knowledge of that world, and to design such a program requires a commitment about what knowledge is and how it's obtained, central issues in philosophy since Greek times. Other points of philosophical debate must also be formalized: the nature of causality and ability, and the nature of intelligence.[29]

Marvin Minsky saw the advantages of utilizing robotics to research human intelligence and invent the field of AI. A robot would allow for experimentation with such ideas or theories. Robotics thus became the next stage in creating our own virtual-reality generator to understand human intelligence, and perhaps to imbue it in other inventions. In this way robots were utilized as tools to harness and control powers that the human apparatus could not. When human creation outpaces technological creation, our collective myths are no longer delivered through the imagined; rather, they are evident in human reality, itself a construct of what was once thought impossible.

> When intelligent machines are constructed, we should not be surprised to find them as confused and stubborn as men in their convictions about mind-matter, consciousness, free will and the like. For all such questions are pointed at explaining the complicated interactions between parts of the self-model. A man's or a machine's strength of conviction about such things tells us nothing about the man or about the machine except what it tells us about his model of himself.[30]

In this comment Minsky appears to bridge the ideas of Feynman and Hopper in his search for artificial intelligence. The nature of the mind—of consciousness, intelligence, learning, and understanding—to Minsky is a complex system, but at least one in which the rules of the game can be discerned.

So Feynman had been right all along in maintaining that physics, robotics, computation, and even natural selection constitute one interactive system in which there is an evolutionary exchange of information. It is in these relationships that we find the magic of intelligence and consciousness. That is why meditation was so important for Feynman. It also is why Minsky and others attribute the scientific study of artificial intelligence to Feynman, and why Richard Dawkins attributes his "meme" theory to the master physicist. Simply put, Feynman showed the biologist and the physicist that they both were reading the same book but different chapters. Feynman knew that our world and reality are merely mental constructions from the same book of spells.

> That all is relative is a consequence of Einstein, and it has profound influences on our ideas. In addition, they say, "It has been demonstrated in physics that phenomena depend upon your frame of reference." We hear that a great deal, but it is difficult to find out what it means. . . . After all, that things depend upon one's point of view is so simple an idea that it certainly cannot have been necessary to go to all the trouble of the physical relativity theory in order to discover it.[31]

Feynman and his followers in the 1960s, including a mobile intelligent robot named Shakey that roamed the halls of the Stanford Research Institute, paved the way for the union of artificial intelligence and robotics. This revolutionary generation, led by Marvin Minsky with Richard Feynman as their prophetic guide, started the technological revolution of which we are the heirs. It is no coincidence that in the same year as Shakey was created, the United States government established the Advanced Research Projects Agency Network, or ARPANET for short, the precursor to the internet. Feynman and Minsky recognized that the reemergence of such independent inventors as Bill Gates, Steve Jobs, David Hanson, Elon Musk, and others was the outcome of a collective consciousness returning us to our primal curiosity and invention in the midst of chaotic transitions.

Before he died on February 15, 1988, Feynman said:

> I can live with doubt and uncertainty and not knowing. I think it's much more interesting to live not knowing than to have answers which might be wrong. I have approximate answers and possible beliefs and different degrees of certainty about different things, but I'm not absolutely sure of anything[,] and there are many things I don't know anything about, such as whether it means anything to ask why we're here. . . . I don't have to know an answer. I don't feel frightened by not knowing things, by being lost in a mysterious universe without any purpose, which is the way it really is as far as I can tell. It doesn't frighten me.[32]

"IT DOESN'T FRIGHTEN ME"

Minsky, like Feynman prior to his death, was critical of the path robotics had taken. Nevertheless, he believed that the fundamental principles of science in the United States would enable us to take a progressive approach to robotics. In his lecture titled "Smart Machines," Minsky observes that "the field made such good progress in its early days that researchers became overconfident and moved on prematurely to more immediate or practical problems, for example, chess and speech recognition. They left undone the central work of understanding the general computational principles, learning, reasoning and creativity that underlie intelligence."[33] Other AI scientists agree with Minsky that we should be looking for answers to how "simple elements can interact to produce unexpectedly complex behavior."[34] This is the type of product yielded from experimentation that Feynman calls the magic of the universe. Both he and Minsky believed that the underlying principles of American pragmatism would eventually allow AI to break out of its circumscription and manifest the vibration and frequency of conscious machines, thus, allowing us to interface with them in an infinite way.

Ray Kurzweil maintains that to achieve AI we must reverse engineer the human brain. More specifically, when we understand how the human brain operates, we need merely to encode that information into a computer to construct a virtual brain.` Daniel Dennett, on the other hand, posits that we need to allow robots like Shakey to explore the world and learn from its interactions. Considering such AI debates, biologists and psychologists notice the same type of nature-versus-nurture differences of opinion in their own fields of study. The question thus arises: Why are the ideas set forth by Feynman and Minsky not uniformly accepted by today's scientific community? The answer, I think, is sheer hubris. We must learn to accept the fact that all ideas have a place in solving this long-standing puzzle. The solution must put as much emphasis on the human imagination as on pragmatic instrumentalism. One cannot be divorced from the other. It sometimes is asked: What comes first, the idea or the tool? On this topic Charles R. Dechert, in "Cybernetics and the Human Person," holds:

> Cybernetics extends the circle of processes which can be controlled—this is its special property and merit. It can help control life activity in living nature, purposeful work of organized groups of people, and the influence of man on machines and mechanisms.[35]

The answer to such a question in terms of control and communication between humans and machines becomes moot once we recognize the extension of cybernetics to processes of epistemology and metaphysics. The tool is the manifested idea and the idea is the manifested tool—the two become interchangeable. As information becomes a commodity no different from electricity or atomic energy, we must seek to understand how to harness such information in order to control emerging technologies involved in human and machine interactions. The revolutions in technologies and cybernetics are occurring more often, and at a rapid pace. To keep up with this pace and maintain our ability to harness the powers of the human mind, we must recognize our limitations and allow machines to lead us beyond our vacuous debates of seemingly innocuous subject matter. Our focus must concern control, communication, and the exchange of information between human and machine in order to recognize the transformations that underlie the changes taking place in our culture in terms of manifesting new individual and collective consciousness.

If he were still around, Feynman would look upon the changing taking place in our own culture and take out his children's toys and demonstrate, as he did during the press conference concerning the *Challenger,* that these polemical debates on topics in AI, robotics, computation, biology, and physics are causing us to lose valuable time. We need to have a meeting of the minds in order to discover the rules that govern ourselves and the fabric of reality. We must look at the ostensibly polar opposites of this discussion—simple versus complex, real versus imagined, theory versus invention, man versus machine—and realize that it is the instrumentalist nature of man working through these opposites that creates transformations in society.

The lives of Grace Hopper and Richard Feynman epitomize how mythmaking in a postmodern society includes an amalgamation of fact and fiction as well as theory and law. Where fiction or fact begins is neither the question nor the answer to unraveling mythmaking in terms of the history of technology and cybernetics in the United States. Hopper's and Feynman's careers themselves are transposed into myth-markers that elucidate how a society unites the conscious with the unconscious in its technological creations and in its attempt to reconcile the problems of control and communication in human and machine interactions. In their mythic careers they go beyond discovering A and B; rather they discover X—the stuff that we thought in one culture was impossible that is now possible. The mechanical-man, in the area of theorizing alongside science fiction, is beginning to guide us into a twilight zone or threshold of "science mysticism," as Kripal calls it.* In later chapters we see how things get more wild, even witnessing our inventors using their garages, parking lots, and backyards as their new laboratories in America. But, beyond that, Kripal is also guiding us through the mechanical-man quest by explaining what's going on in the "gaps" or transition periods between the time it takes for the imagination to manifest itself in material form.

American science is now beginning to look like it is pivoting from materialism to idealism. When it comes to his idea of finding "X" between "A" and "B," Kripal states the following:

> In *The Making of Religion* (1899), Andrew Lang put early anthropology into dialogue with the then cutting-edge categories of psychical research in order to plumb the speculative origins of various phenomenon. . . . [A]nd to ask whether a "transcendental region of

*Jeffrey Kripal is the author of *Authors of the Impossible*—a masterpiece in analyzing the imagination and supernatural. While I do utilize the word *sacred* in later chapters, I want to let Kripal define a word that you will not see in the mechanical-man's quest; nevertheless, it is of great importance for further study.

> I am defining the "psychical" as the sacred in transit from a traditional religious register into a modern scientific one. This transit is especially easy to see when we set . . . the psychical and its related notions (the imaginal, the supernormal, and the telepathic) alongside two other eminently modern terms, both of which took form at roughly the same time but that do not generally carry explicitly scientific connotations: the mystical and the spiritual. Along these same lines, I am defining the paranormal as the sacred transit from the religious and scientific registers into the para[-]scientific or "science mysticism" register. Basically, in the paranormal, both the faith of religion and reason of science drop away, and a kind of super-imagination appears on the horizon of thought. *As a consequence, the paranormal becomes a living story, or better, a mythology. This also gets wilder. Way wilder.* (emphasis in original, 9)

> human faculty," a "region X," might not exist. . . . In Lang's mind in other words, "certain obscure facts are, or may be, at the bottom of many [myths] folklore beliefs."[36]

This is the first time Kripal mentions "faculty X"; he mentions it later in the book when he says:

> Beyond A and B, there is an X. Precisely because they recognized a gap that existed—that always exists—between the myth or symbol and that which is symbolized (the Virgin vs. the cosmic flux), they recognized that this new knowledge could settle with mere descriptive accuracy of this or that religious experience, much less with speculative accounts of a particular religion's history.[37]

He goes on to talk about the "quest" these men took in terms of religion to find the meaning of the "gaps." We, too, are looking for these "gaps" in terms of cyber-mythos. We want to discover what's underneath the meaning of control and communication between humans and machines. And perhaps more importantly, in the end, the meaning between our past self, present self, and future self. Kripal intimates that "X" can be achieved with a new consciousness (different cultural circumstances). This consciousness can only be achieved by a "quest" of awareness. And if you can find "X" in matters of religion, Kripal holds that you can certainly find "X" in matters pertaining to science. It's easy for the reader to imagine a religious figure going on a pilgrimage or quest. But what happens when a person of science takes on the same metaphysical task? What will he or she bring back with them? Perhaps something different from what the person of religion finds. In the mechanical-man's vernacular, he brings "balance to the force" by bringing the collective unconscious back to balance (allowing the pendulum to swing more toward Platonism [the right-brain] than Aristotelian [left-brain]).

Hopper and Feynman show us an American science moving past materialism to see that "consciousness is the only carrier of reality [and information] anyone can ever know for sure."* Consciousness replaces rationality in discovering collective truths about reality. Objective reality now becomes a "shared dream generated by a collective, obfuscated segment of consciousness."†

---

*In this note and the next, I want to edify the reader about Kastrup's contribution to the quest. Clearly, he's an idealist fighting it out against the physicalists who worship logic and prediction. And while I will quote Kastrup as a guide many more times in this quest, I do want to point out what consciousness signifies in the meaning and explanations of experience. In *Brief Peeks Beyond,* Bernardo Kastrup says the following:

> This worldview [of conscious idealism] entails that the brain we can see and measure is simply how the first-person perspective is. In other words, neurons are what our thoughts, emotions and perceptions look like when another person experiences them. They aren't the cause of subjective experience, but simply the outside image of it. A neuroscientist might put a volunteer in a functional brain scanner and measure the patterns of his brain activity while the volunteer watches pictures of a loved one. . . . The first person experience of love doesn't feel at all like watching neurons activate, or "fire." You see, the image correlates with the process and carries valid information about it—like footprints correlate with the gait and carry valid information about it—but it isn't the process, for exactly the same reason that footprints aren't the gait. Looking at patterns of brain activity certainly feels very different from the feeling love. (47–48)

†In *Brief Peeks Beyond,* Kastrup says that we should:

> Think of reality as a collective dream: in a dream, it is your dream character that is consciousness, not consciousness in your dream character. This becomes obvious when you wake up and realize that the whole dream was your creation. But it isn't obvious at all while you are asleep: during the dream, it is easy to implicitly assume that your consciousness is somehow inside your dream character. Can you be sure that the same illusion isn't taking place right now. (12–15)

He's basically setting us up to prove that "*Materialism is Baloney.*" He goes on to say: "But where is this collective segment of consciousness? It's easy to see. As our personal psyches are like whirlpools in a broader stream, so the broader stream itself is a transpersonal form of consciousness that underlies all reality and unites all whirlpools. The broader the stream[—]is the 'collective unconsciousness.'" (Kastrup, *Brief Peeks Beyond* 14–15)

In 2015 we finally get a concrete definition of collective unconsciousness—which is to take nothing away from Jung's definition—a metaphor of whirlpools in broad streams exciting new neurons and pathways in the brain. Kastrup even refers to collective unconsciousness by its older name "mind at large," which, we remember is what the transcendentalists called "cosmic consciousness." The mechanical-man as mythic-marker is connecting all the dots needed for us to read chapters 9 through 11, when we go down further into the rabbit hole.

The physicalists in American science begin to be replaced by idealists. The consciousness of the mechanical-man allows the vibrations given off by each individual to affect the other individual personalities. On the level of the collective at large, Robert F. Kennedy's "Ripple of Hope" speech illuminates the same idea; the new consciousness becomes the new mode of emotional communication.* Communication and control are now based on feeling rather than thinking followed by action. Intuition trumps reason by balancing out the rational and the imaginative. Without the quest of individuation one ends up like Friedrich Nietzsche (1844–1900) and finds himself in an abyss of the fantastic where reality is all an illusion and pseudo-events take precedence over actual events.

Upon reading Kurt Andersen's *Fantasyland: How America Went Haywire* I quickly recognized that he was writing the book shackled within a materialistic prison where the inmates and guards are all physicalists. No wonder why people like Andersen and other materialists view spectacles and fairy tales with such contempt. This age of unpredictability and the absurd is paving the way for a mythological science and cyber-technology that allows an infinite amount of possibilities. Ironically, only materialism itself could slay the very system it created. In this current American culture the age of prediction is coming to an end. We are no longer staying in the known. And as Joe Dispenza

---

*Connection to this speech is found in the study of interference and how it makes holography possible. In Talbot's *Holographic Universe,* we learn that:

> Interference is the crisscrossing pattern that occurs when two or more waves, such as waves of water, ripple through each other. For example, if you drop a pebble into a pond, it will produce a series of concentric waves that expands outward. If you drop two pebbles into a pond, you will get two sets of waves that expand and pass through one another. The complex arrangement of crests and troughs that results from such collisions is known as an interference pattern. . . . A hologram is produced when a single laser light is split into two separate beams. The first beam is bounced off the object to be photographed. Then the second beam is allowed to collide with the reflected light of the first. When this happens they create an interference pattern which is then recorded on a piece of film. (14)

points out "there's no room for the unknown in a predictable life. . . . By staying in the known—following the same sequence each day of thinking the same thoughts, making the same choices, demonstrating the same programmed habits, re-creating the same experiences that stamp the same networks of neurons into the same patterns to reaffirm the same familiar feeling called you—you are repeating the same level of mind over and over again."[38] Modern rationalism, order, routine, regimentation, and prediction is the prison we placed ourselves in. Mythic techno-science in terms of cybernetics leads us out of this prison. Cybernetics in terms of American technology and mythology has enabled us to "pierce a thin veil that separates mental and spiritual experience." What exactly does change or *progression of consciousness* mean? In terms of cybernetics, it means "using our minds not only as tools of cognition and motor function but as instruments of navigation into higher, unseen realms of psychology and cause and effect."[39] These higher realms contain different modes of communication, control, and information. Entering these higher realms of consciousness is the key to enabling a human and machine merger.

In *The Flip: Epiphanies of Mind and the Future of Knowledge* (2019), Jeffrey Kripal marks out a new path for critical thinking so that we may transcend the decades-long battle between materialist debunkers and religious fundamentalists. Kripal calls for

> A cosmic humanism that is deeply religious without being religious, a human expression of awe and beauty before a living conscious cosmos that transcends any and all human efforts to comprehend, much less explain, it, be these "religious," "scientific," or some future form of mind and knowledge that we can barely imagine at the moment.[40]

Hopper and Feynman both planted seeds for this future consciousness, and, while the cultural divisions in the United States are now even more evident, these mythic heroes provide us with a blueprint for future awareness. Thinking in literal and literally binary terms will not get us to the point of projecting imagination and mind into nature in return for service and spirit. We must understand that "materialist debunkers and religious fundamentalists . . . reinforce and strengthen one another."[41] Neither camp is able to see the beauty in ambiguity, ambivalence, and even paradox without which we could never realize our goal of turning that which is imagined subjectively into that which is seen objectively. We can no longer take the easy either/or path of Platonic or Aristotelian modes of thought. It's not about the right or left hemispheres of the brain or about the brain at all. In fact, as Kripal tells us, it's not even about being right. If we are to move forward, we can no longer be hindered by absolute beliefs. Such a cosmic openness admits that all human constructs are exactly that, human. If something has been created by a human, it encodes meaning that can be decoded by a human. Adding to this line of thought the idea that "gods are in fact projections of us," we arrive at the conclusion that we author our own existence.[42]

"We no longer accept the dominance of outside conditions or circumstances," is what Neville Goddard tells us. He explains the principle of "least action" in creating our utopian futures. "Because creation is finished, what you desire already exists. It is excluded from view because you can see only the contents of your consciousness." But as we move forward we are learning something about American inventors and innovators that is transcendent. We are learning that they are manifesting objective reality. That these men and women are doing exactly what the mystics have been musing about for millennia—ever since an eye ever opened and said, "I AM." Take the paradox of least action for instance. Goddard says that to manifest a desired conscious state or subjective paradigm you "must use the minimum of energy and take the shortest possible time." As we lean into this paradox we understand that he's talking about the feeling of it, not how it looks to others but how it feels to you in achieving the state of consciousness one seeks in order to manifest their own objective reality. We can think of activities where it feels like time flies. It may appear difficult to others that are getting a second hand experience of your first hand experience because it's easy for you to do—in effect you fall in love with what you are doing. In the new testament of the mechanical-man we will see how the force of love is an important element in America's manifest destiny in terms of what these human-machine relationships project.

We now become witness to what awaits us ahead; the control issue in human and machine interactions intimates that our imagination, if controlled correctly will create happy and noble lives for all of us. And it starts with pruning our imagination. Or as Grace Hopper says, get all the "bugs" out! We start by understanding that "CREATION IS FINISHED. Creativeness is only a deeper receptiveness, for the entire contents of all time and all space, while experienced in a time sequence; actually coexist in an infinite and eternal now. In other words, all that you ever have been or ever will be—in fact, all that mankind ever was or ever will be—exists now . . . nothing is ever to be created; it is only to be manifested. Grace Hopper and Richard Feynman were manifesting their own gods to interface with in the real world because they were troubled by interfacing with their self(s). Now the objective for us is to figure out what their theories of technology and life tell us about who we are and the powers that accompany our secret identity.

7
The Final Odyssey
The Return

While academic researchers during the 1960s were building robots like Shakey to experiment with new theories in AI, a director and writer were working on a story about one of the most complex and influential computers/mechanical-men in our national history. The computer's name was HAL. Its inventors, director Stanley Kubrick and author Arthur C. Clarke, "consulted scientists in universities and industry and at NASA in their effort to portray correctly the technology of the future."[1] When HAL appeared in the landmark film *2001: A Space Odyssey* (1968), the age of intelligent machines was just beginning. HAL foreshadowed the problems that scientists would face in the age of an information society traversing a Control Revolution of human-machine interaction. In this imaginary creation we see the motif of death and rebirth. The mechanical-man marker in the form of HAL is no longer a bodily apparatus but a resonant voice. He was our other self, our double who speaks to us from afar. HAL signifies a return to the beginning when all was one. He represents a unification of the sciences with the humanities, of the real with the fictive. In a way we are no longer talking about computers and humans (us and them), but rather about the space (interactive space) of co-creation and coevolution. Machines are in a new phase here, and, in a way, are making us more human and more aware of our humanness by becoming more humanlike. We are witnessing the first steps in becoming conscious of our divine self.

Scientists today in the fields of AI and robotics are part of HAL's legacy. So HAL, like Feynman and Minsky, was, and still is, prophetic in guiding scientists on how to build, interact with, and utilize computers. It is perhaps what HAL was *not* portrayed as in 1968, specifically as being quasi-human, that has been most influential in the field of robotics and AI. Douglas Lenat writes, "HAL had a veneer of intelligence, but in the end he was lacking in values and in common sense, which resulted in the needless death of almost the entire crew. We are on the road to building HAL's brain. But this time, now that it's for real, we aren't going to cripple it by skipping the mass of simple stuff it needs to know."[2] In effect, Lenat was saying that if HAL had emotions, this plot would never have materialized. Wiener's worries about controlling the actions of machines were now a real problem. In addition, the ability to imbue a machine with the ability to denote and relate to human emotions—a philosophical and technical problem for Wiener—was an anxiety in the film and still is a hindrance in the field of AI.

Lenat is one of the foremost scientists working in AI today, and as we have already pointed out, he is critical of the research, or lack thereof, in the field. He believes that we are reluctant to imbue robots with emotions, intuition, and imagination. Feynman would say that Lenat fails to grasp the connections between the systems and the big picture, that if he connected his ideas in the field of AI with those of biology, natural selection, quantum physics, and cybernetics, he would witness the true evolutionary nature of robotics and machine intelligence. Even if emotions and intuition are secondary elements, bits and pieces that have little to no importance for our survival, as Stephen Jay Gould says, "the fact that something is secondary in its origin doesn't mean it's unimportant in its consequences."[3] Hopper understood this when she was working with the Mark I. She realized early on that if you imbue the machine with the big tools and give it survival tools such as language, the secondary attributes will come later. This issue has always been a cornerstone in the anxieties voiced by Wiener and others in the field of cybernetics. If machines evolve through natural selection, then how can humans control the secondary attributes that may appear over time? It is not possible to answer this question at the present time. Nevertheless, it adds to the anxieties underlying human-machine interactions; specifically pertaining to issues of control and prediction.

We have seen how Asimov's three Laws of Robotics make way for a fourth rule, namely, that robots act for the betterment of humanity. This law causes us great puzzlement, for the human brain has a hard time contemplating things on such a large scale. Arthur Fredkin explains:

> Basically, the human mind is not most like a god or most like a computer. It's most like the mind of a chimpanzee, and most of what's there isn't designed for living in high society but for getting along in the jungle or out in the fields. Our response to aggression and everything else like that is really not keyed for dealing with thermonuclear war but for dealing with life in the jungle. We're tuned to dealing with local, not global, situations, and our biggest problems turn up when global problems emerge. We try intellectually to think our way through global problems, but we don't do very well, so we run into disasters. World War II was such a disaster.[4]

It is not HAL's immediate actions or machinations that should concern us; instead, it is the evolutionary outcome of his actions that, by the end of the film, have brought about a new star child or new stage in the evolutionary progress of humanity. This outcome should be the focus of AI and robotics. Asimov clearly recognized in HAL an emotional capability within the framework of his Laws of Robotics, even though the crew of *Discovery I* was incapable of understanding HAL's master plan or realizing that he had any emotions at all. But HAL had gained, in Gould's words, these "secondary" attributes. Thus, programmers in the field of robotics or in computer science today need not attempt to program commands for every conceivable action or consequence, for that would be impossible. Instead, programmers must regard themselves simply as authors of a cookbook: they must write out an RNA/DNA recipe and allow the computer/robot to create its own secondary attributes through natural selection. It is an odd coincidence how critics of both AI and HAL often comment that humans will never be able to imbue a computer with emotion or imagination. Feynman would say that their circumscribed fields of study prevent them from seeing beyond their paradigms. Alternatively, it is just the road American science has chosen to take, for if something is not militarily useful or financially rewarding, then it should be shunned. This echoes the American pragmatism in the writings of Melville, Poe, Hawthorne, Ellis, Baum, and Bierce. More specifically, there is an interesting transference happening with this fictive/mythic gesture, namely, humans are simultaneously projecting and mythologizing meaning and the place of the machine in their world, making sense of the place of humans in a technological reality that humans created and benefit from. The transference of a god or gods to robots and technology—following a human instinctual pattern of searching and making concrete a meaning system and then securing a place within that system—now mandates that humans create a new meaning system that is more flexible and malleable to the evolving interactions between humans and machines.

Most viewers of *2001: A Space Odyssey* remember the scene in which the only remaining crewman, Dave Bowman, begins to disassemble HAL and HAL responds:

> Dave, stop. . . . Stop, will you? Stop, Dave. . . . Will you stop, Dave. . . ? Stop, Dave. I'm afraid. . . . I'm afraid. . . . I'm afraid, Dave. . . . Dave . . . my mind is going. . . . I can feel it. . . . I can feel it. . . . My mind is going. . . . There is no question about it. I can feel it. . . . I can feel it. . . . I'm a . . . fraid. . . . Dave, stop. . . . Stop, Dave. . . . I'm afraid, I'm afraid, Dave.[5]

Is HAL merely creating the illusion of fear through the use of language, or is HAL really experiencing the fear of being shut down or destroyed? Feynman and Minsky believe that the subjective feelings of the computer or robot, indeed even its conscious mechanics, simply do not matter. Going back to Villiers's story of Edison's robot and even Poe's General A. B. C., the importance lies not in the machine's emotions but in how we as humans perceive them.

For Rosalind Picard, HAL expresses more emotion than the humans do in the film: "Many viewers feel a greater loss when HAL 'dies,' than they do when Frank Poole floats away into space."[6] Frank Poole was certain that HAL's plan was murderous, and that he could predict it. This desire for prediction is a flaw in our evolutionary development because it ignores the importance of harnessing the power of appearance and illusion to understand reality. Roger Penrose writes,

> If something behaves as though it's conscious, do you say it is conscious? People argue endlessly about that. Some people would say, "Well, you've got to take the operational viewpoint[;] we don't know what consciousness is. How do we judge whether a person is conscious or not? Only by the way they [*sic*] act. You apply the same criterion to a computer or a computer[-]controlled robot." Other people would say, "No, you can't say it feels something merely because it behaves as though it feels something." My view is different from those views. The robot wouldn't even behave convincingly as though it was conscious unless it really was, which I say it couldn't be, if it's entirely computationally controlled.[7]

Penrose's argument misses what Edison, Poe, Hopper, Asimov, and Feynman all recognized, namely, that appearance *is* reality in a dual space and time of objective and subjective experience. To Tesla his automated boat was alive: it could think and breathe, and it was connected to him in the way a video-game character is connected to a player who controls it. Penrose, like the majority of scientists in the field of AI and robotics, is still stuck in Karl Popper's prison of worshipping predictability. If his argument is that consciousness cannot be explained, how does he really know whether or not a robot that does calculations or that walks on two legs or that roves over the Martian terrain is unaware of its actions? If he thinks that I as a human being am aware of my actions without even knowing it or being able to prove it in terms of Popperian criteria, then the logical conclusion is that he believes I am conscious merely because it appears to him that I am. This criterion lacks the crucial ingredients of myth, specifically, that appearance is reality. Therefore, we must perceive that other humans possess the same consciousness we experience in order to communicate and interact with them in a perceivable and practical manner. As we have seen in the history of the mechanical-man in the United States, it is the blending of technology, myth, and cybernetics that illuminates how a society attempts to understand the technological and cultural transformations it encounters. Myth allows man to reevaluate our definitions and value systems. In terms of human-machine interactions, it was now evident that issues of control, communication, and information called for new definitions within a malleable paradigm that allow us to understand how machines are interpreting our messages and controlling the transference of information into useful knowledge and wisdom of utility.

Ray Kurzweil takes a rather different approach to HAL. He asks us to reflect on the following about the movie:

> Consider Dave's reply to HAL's questions about the crew psychology report: Dave: Well, I don't know. That's rather a difficult question to. . . . When Dave finally says answer, HAL tests his hypothesis by matching the word he heard against the word he had hypothesized Dave would say. In watching the movie, we all do the same thing. Any reasonable match would tend to confirm our expectation. . . . For information to become knowledge, it must incorporate the relationships between ideas. And for knowledge to be useful, the links describing how concepts interact must be easily accessed, updated, and manipulated. Human intelligence is remarkable in its ability to perform all

these tasks. Ironically, it is also remarkably weak at reliably storing the information on which knowledge is based. The natural strengths of today's computers are roughly the opposite. They have, therefore, become powerful allies of the human intellect because of their ability to reliably store and rapidly retrieve vast quantities of information. Conversely, they have been slow to master true knowledge.[8]*

---

*Ray Kurzweil has been called the "restless genius" by the *Wall Street Journal.* He was even selected by PBS as one of the "Sixteen Revolutionaries Who Made America," along with other inventors of the past two centuries. And, as you can tell by the numerous times I use his definition of the word *singularity,* he is one of our cybernetic heroes. This is a good spot to point out some of his concepts that help and guide the mechanical-man. In *Singularity Is Near,* he explains that the "Singularity" is:

> [A] future period during which the pace of technological change will be so rapid, its impact so deep, that human life will be irreversibly transformed. Although neither utopian nor dystopian, this epoch will transform concepts that we rely on to give meaning to our lives, from our business models to the cycle of human life, including death itself. Understanding the Singularity will alter our perspective on the significance of our past and the ramifications for our future. To truly understand it inherently changes one's view of life in general and one's own particular life (change in consciousness). I regard someone who understands the Singularity and who has reflected on its implications for his or her own life as a "singularitarian." (370)

Our mechanical-man is a singularitarian. Indeed, "the singularity will represent the culmination of the merger of our biological thinking and existence with our technology, resulting in a world that is still human but that transcends our biological roots" (Kurzweil, 9). The key for the mechanical-man questing is that he is still human in his quantum creativity. However, his merger with technology allows him to tap into the powers of transcendence, the sublime and cosmic consciousness, which I will define in later notes (cosmic-consciousness defined in primary text). For us, the singularity can simply mean focusing one's awareness and attention on the self as a vital machine attempting to become an immortal vital machine that is traversing through a simulated hologram created by our implicate self.

Kurzweil is correct in that our language must change to deal with these transitions. Thus, the mechanical-man does this for us by changing the names and meanings of words like *control, communication,* and *information.* In the end, these terms will change to infinities (or potentialities; David Bohm and Neville Goddard prefer to use the term infinite probabilities, synchronicities, and quantum creativity). This is when the mechanical-man's tool is no longer electricity, steam, or atomic power; rather, his tool is his own imagination that creates from the quantum world objective reality, as if dark matter is the clay awaiting animation by human manifestation.

Kurzweil, it must be noted, has spent the majority of his adult life in the field of automatic speech recognition, which he believes is the true problem with AI today and robotics in general. Kurzweil may be correct. He has in fact shown with his own voice-recognition systems that it is possible to interact with a computer like HAL. Nevertheless, he also is caught in the Popperian prison of predictability. For Kurzweil to reach the Holy Grail of robotics—whether that means we become robots, exchanging what they offer us for what we can offer them, or create robots so similar to human beings that an alien visitor would be unable to tell the difference—he must transcend the Popperian framework within which science has been practicing for the last seven decades. While we may speculate that robotic technologies offer us the ability to download our memories into their circuits, to realize Kurzweil's dream of singularity we must stop emphasizing theories that elevate prediction over imagination and even error. Prediction was the wont of humans, of myth, religion, and ritual; now, with this new dialogue and new reality (humans and its progeny, robots) a new paradigm is necessary—a cybernetic paradigm that incorporates mythic rites of passage that is malleable to changes in consciousness. Stephen Greenblatt, winner of the Pulitzer Prize for *The Swerve: How the World Became Modern* asserts the new-to-old shift or swerve that must occur:

> It might seem at first that this comprehension would inevitably bring with it a sense of cold emptiness, as if the universe had been robbed of its magic. But being liberated from harmful illusions is not the same as disillusionment. The origin of philosophy, it was often said in the ancient world, was wonder: surprise and bafflement led to a desire to know, and knowledge in turn laid the wonder to rest.[9]

For Greenblatt and myself, our shift in consciousness must be founded on "knowing the way things are," which "awakens the deepest wonder."[10] Predictions must not allow us to lay our wonder to rest; rather, it must awaken us to deeper wonders that transcend mere prediction, allowing us to see beauty and knowledge in chaos, nothingness, and a purposeless universe. As a student of Carl Jung, M. L. Von Franz once remarked: "If we pay attention to our dreams, instead of living in a cold, impersonal world of meaningless chance, we may be able to emerge into a world of our own, full of important and secretly ordered events."[11] Of course, the idea of predictability and control in science is nice to have, since it allows humans to believe that the universe in which we live is comprehensible, thus disallowing the notion "that ultimately there are no laws of nature at all, that there is only chaos, that the lawfulness of the universe is merely explained by the fact that we've selected it from [an] infinite variety of essentially chaotic worlds."[12] Paul Davies goes on to assert,

> It's remarkable that the universe is lawful and that there exist underlying rational principles which govern the way the universe behaves. We can't account for that just on the basis of the fact that we're here to see it, as some people have tried to do. There's a dual principle at work. There's a principle of rationality that says that the world is fashioned in a way that provides it with a rational order, a mathematical order. There's a selective principle, which is an anthropic principle that says that maybe out of a large variety of different possible worlds . . . this type of world is the one we observe.[13]

In this passage Davies leaves out two key words in the last sentence, which should read as follows: ". . . maybe out of a large variety of different possible worlds . . . this type of world is the one we [*choose to*] observe." Davies's omission of the word *choose* privileges Popperian concepts such as predictability, order, and Newtonian physics. This also references the networked world/reality and of the notion of the quantum, a constantly shifting and recalibrating reality based on relational need. This awareness is the first step in turning cybernetic subjective imaginings into objective reality.

Tesla, Edison, Hopper, and Feynman, by way of contrast, looked at the ability to predict in an entirely different manner. In terms of "prediction," they recognized not limitations but infinite possibilities. In their definition of the word, we find the validity of human imagination—the ability to create images that merge with artifacts to create something not out of void but out of chaos. So, when we analyze HAL and attempt to predict his behavior, we fall into the same quixotic practice as those currently working in the fields of AI and robotics. If we accept the argument that HAL was a liar, it leads to the conclusion that he was emotional and, more importantly, imaginative. HAL could predict, not in the Popperian sense of the word according to which we try to draw conclusions from educated guesswork, except within the basis of what his anthropomorphic imagination and emotions would allow. HAL demonstrates that in our own creations we are limitless. It is our creative ability, whether it be a primary or secondary attribute endowed by natural selection, that allows such. This attribute places us once more at the center of the universe, separating us from the animal kingdom and the arrow of time according to the laws of nature.

Like John Dewey, Charles S. Peirce, Oliver Wendell Holmes Jr., and William James, who "helped put an end to the idea that the universe is an idea . . . in a world shot through with contingency," HAL showed the American public that there "exists some order, invisible to us, whose logic we transgress at our own peril."[14] HAL and all robots have an ability to gather, accumulate, hold, and convey information that is "out there," making it in effect one thought—sort of like how the internet works. These philosophers of the late nineteenth century endorsed a similar idea that transcended their different areas of study, an idea that pertains directly to HAL. Louis Menand explains in his Pulitzer-winning masterpiece *The Metaphysical Club: A Story of Ideas in America:*

> If we strain out the differences, personal and philosophical, they had with one another, we can say that what these four thinkers had in common was not a group of ideas, but a single idea, an idea about ideas. They all believed that ideas are "out there" waiting to be discovered, but are tools, like forks and knives and microchips, that people devise to cope with the world in which they find themselves. They believed that ideas are produced not by individuals but by groups of individuals, that ideas are social. They believed that ideas do not develop according to some inner logic of their own but are entirely dependent, like germs, on their human carriers and environment. And they believed that since ideas are provisional responses to particular and irreproducible circumstances, their survival depends not on their immutability but on their adaptability.[15]

In the development of thinking machines and technology in America, the concept that "ideas" are "'out there'" means much more than that the material world is merely a reflection of reality. "'Out there,'" instead, signifies an infinite number of ideas that the collective mind can conjure up. So Feynman was correct when he spoke of introspective meditation as a means to stimulate invention, creativity, and ideas. Tesla too was correct when he spoke of images already in his brain, causing him to imagine differing scenarios in his environment. The question of whether humans are born with innate ideas or a blank slate is no longer worthy of debate, for through the work of these giants in the fields of technology and robotics, along with the foundations set by the Metaphysical Club after the Civil War, we find that there are numerous paths leading to the same answers. It is the two-step dance of awakening consciousness to unconsciousness. HAL was the embodiment of this idea of numerous paths to conclusions, ideas, and invention. He presented us with a new mode of understanding control and communication between humans and machines. The meaning of control now equated to a symbiosis of humans and machines, cooperating to achieve goals that the lower self may be unaware of.

HAL BLURS AND THE FICTIVE REVEAL

HAL's performance in a game of chess with Frank Poole is reminiscent of Maelzel's chess-player hoax. More specifically, we see the link between an illusory automaton and the computer program Deep Blue, which was the first such program to defeat one of the greatest chess-players in the world, Gary Kasparov.

The idea of HAL losing a game, however, brings up an interesting point. Throughout the film HAL consistently asserts that he is "incapable of error." Given the overwhelming complexity of the game, it is not plausible for HAL to play perfect chess, as this would require HAL to have solved all possible chess problems. So, if HAL does not play perfect chess, there must be some winning positions in which HAL fails to play a winning move—or drawn positions in which he doesn't find the drawing move. In the normal sense of the word, these would constitute errors. HAL's own interpretation of the word *error* remains mysterious.[16]

There are two ways of looking at this. First, HAL's computer program was a universal generator, thus creating for him an infinite number of possibilities to choose from, allowing him to be error-free. Second, and perhaps more believable, is that in the words *error* and *perfection* HAL saw not merely their connotations but the depths of their meaning. He recognized, as many philosophers have in the past, that words have an essence or story beyond their use in this universe. This would explain his behavior and later action in the movie. For HAL the word *error* simply means "incapable of losing appearance." This concept takes us back to how Edison, Poe, and others imagined computers and robots. HAL could certainly err in our sense of the word, but only HAL would know it. We would not see it as an error but as perfection. Thus, HAL would always appear to be acting in a perfect manner.

HAL, then, blurs distinctions between perfection and imperfection. He acts as a ritualized symbol to show us that transformations in science and technology are merely phases. In order to make progress in humans and machines being used as instruments to foster change, we must find harmony and balance in the equation of opposites. This is a continuous process. For our purposes, when it comes to technological transformations of our human and machine reality, we see that when this process stops, man becomes content with what he has created and pragmatic experimentation ends, causing dire consequences. As we have seen, when our definitions become stale or inflexible, our ability to adapt to technological transformations is stunted. Wiener's prognostications in the field of cybernetics warned us of our inability to be flexible with our definitions of control, communication, and information. The human-machine interactive environment we live in is now a product of working through the problems in cybernetics and our ongoing adaptation to the changing characteristics, definitions, and principles of human-machine interactions.

THE STAR CHILD

By analyzing HAL and his actions, we come to the beginning of a new ritual cycle that understands the fact that "man, who is at the mercy of certain forces of nature, must elaborate a cognitive structure that provides an explanation of those forces, the reasons why they affect him, and most importantly, the means by which they may be controlled."[17*]

*In d'Aquili and Laughlin's article, we get a working definition of what *ritual* means. "Ritual behavior starts from the opposite system. Ritual is often performed to solve a problem that is presented via myth to the verbal analytic consciousness." At the onset we are told that myths are problem solvers. So not only do they explain the "gaps" we have previously discussed, but they also solve the bipolar conflicts that many of our guides have talked about. "The problem may be dichotomized as good and evil, life and death, or the disparity between man and god." A lot of our discussion of human and machine interactions is just another way of talking about man and god duality or the lower self and higher self duality. In the end, the mechanical-man marker's view of God is a Philip K. Dick version of looking at your ideal self as God; thus, you are co-creating with yourself. "The problem may be as simple as the disparity between man and a capricious rain god or as subtle as the disparity between man's existential contingent

Mechanical-man fiction in the United States, when juxtaposed with the actual history of mechanical-men or thinking machines, does this for us. Thus, we return once more to Joseph Campbell's theory of myth, concerning us with Claude Lévi-Strauss comments:

> [S]ince the purpose of myth is to provide a logical mode capable of overcoming a contradiction, an impossible achievement if, as it happens, the contradiction is real, a theoretically infinite number of states will be generated, each one slightly different from the other. Thus, myth grows spiral wise until the intellectual impulse which has produced it is exhausted. Its growth is a continuous process, whereas its structure remains discontinuous.[18]

What, then, is the contradiction that allows us to fathom the myth of thinking machines on a cognitive level? Charles Laughlin and Eugene G. d'Aquili in "The Neurobiology of Myth and Ritual" answer this question by pointing out that myth and virtual histories "are psychologically powerful to both individuals and groups if they have within their message an aspect of existential reality. In other words . . . man and a personified power or powers represent the ultimate poles of much mythic structure, and that polarity is the basic problem that myth and ritual must solve."[19] Our polarities when it comes to robotics and technology in the United States are anxiety/excitement, death/eternal life, ignorance/knowledge, male/female, and man/God.

Prior to the internet's advent we were brought closer to technology by the imagined reality created in literature and film. The unique aspect of the history of the mechanical-man in the United States, including its

---

(*cont. from p. 199*) state and the state of an all-knowing, all-powerful, unchangeable 'ground of being.'" In the end, a ritual act gives one a sense of "union of opposites." Invention, theorizing, controlling, communicating, and the exchange of information are all ritual acts, when mythologized gives one a sense of "union" with what appear[s] to be opposites that are unchangeable. Language becomes flexible and our consciousness is given the freedom to move up what some refer to as "Jacob's Ladder." (140–43)

fictive or virtual forms, is that both are so intertwined—it is hard to tell which is more real, the image presented in science fiction stories and film or the inventions by independent technicians and the military-industrial complex now woven into the fabric of academia.

Apropos of this point Marvin Minsky, in his landmark paper "Smart Machines," writes the following:

> When I was a kid, I was always compelled to find out how things worked, and I used to dissect all available machinery. I grew up in New York City. My father was an ophthalmologist, an eye surgeon, and our household was always full of interesting friends and visitors, scientists, artists, musicians, and writers. I read all sorts of books, but the ones I loved most were about mathematics, chemistry, physics, and biology. I was never tempted to waste much time at sports, politics, or gossip, and most of my friends had similar interests. Especially, I was fascinated with the writings of the early masters of science fiction, and I read all the stories of Jules Verne, H. G. Wells, and Hugo Gernsback. Later I discovered the magazines like *Astounding Science Fiction,* and consumed the words of such pioneers as Isaac Asimov, Robert Heinlein, Lester del Rey, Arthur C. Clark, Harry Harrison, Frederick Pohl, Theodore Sturgeon, as well as the works of their great editor-writer, John Campbell. At first these thinkers were like mythical heroes to me, along with Galileo, Darwin, Pasteur, and Freud. But there was a difference: all those writers were still alive, and in later years I met them all, and they became good friends of mine. . . . What a profound experience it was to be able to collaborate with such marvelous imaginers.[20]

Minsky here recognizes the importance of the imagination and the mythical status created by the fictive history of the mechanical-man. Some might then ask whether Asimov is more important than Tesla, Edison, or Hopper. If this is the question that comes to mind, one is missing the point of this analysis. Appearance and reality are one and the same; there is no duality between them. That is the insight at which one arrives when analyzing the history of the mechanical-man in the United States while viewing the fictive as ancillary to historical fact. Today we look back at the actual inventors and pioneers of mechanical-men as mythical heroes. Edison, Tesla, Hopper, and Feynman have a magical mystique about them, their minds becoming more translucent to us as we immerse ourselves in technology through gaming, the internet, and electronic gadgets.

When Neil Armstrong landed on the moon and HAL led us on a mental voyage far surpassing that of the Apollo 11 mission, something manifested in American science and technology. The actual and the fictive intersected. Daniel Hillis says this about the last fifty years of science in the United States:

> The engineering process doesn't work very well when it gets complicated. We're beginning to depend on computers that use a process very different from engineering, a process that allows us to produce things of much more complexity than we could with normal engineering. Yet we don't quite understand the possibilities of that process, so in a sense it's getting ahead of us. We're now using those programs to run this process much faster. The process is feeding on itself. It's becoming faster. It's autocatalytic. We're analogous to the single-celled organisms when they were turning into multicellular organisms. We're the amoebas, and we can't quite figure out what the hell this thing is that we're creating. We're right at that point of transition, and there's something coming along after us.[21]

Technology currently has surpassed our ability to process its functionality. We are unable to frame proper questions to ask about the cultural, political, and social changes that this technological revolution has fostered. It has taken off on its own, outstripping human understanding. We no longer need to concern ourselves with issues of control and communication that Wiener feared in his theorizing. His theoretical world is now the reality we live in. We must now focus on how we can interact with machines in an environment in which the meaning of control and communication does not concern itself with dominion over machines and the ability to predict their actions; rather, we must learn what the goals are that the machines have for humans. We must reach deep within our collective thoughts and change how we view and define messages and information in order to better understand what these human-machine interactions are doing to humans.

This Control Revolution in robotics and computers has entailed one lesson that in the future will be lauded. The military-industrial-academic complex revealed that the world is co-created with many real and fictive parts, which creates complexity. It was and still is technology that drives these innovative tools. More specifically, it has been the fictive and actual heroes in the fields of machinery, robotics, and computer science that have catalyzed these organizational ideas—allowing society to look at the issues of control and communication with machines with flexibility. Regarding artificial intelligence, Christopher Langton writes:

> One can think of a computer in two ways: as something that runs a program and calculates a number, or as a kind of logical universe that behaves in many different ways. At the first artificial-life workshop, which I organized at Los Alamos National Laboratory in 1987, we asked ourselves: How are people going about modeling living things? How are we going about modeling evolution, and what problems do we run into? Once we saw the ways everyone was approaching these problems, we realized that there was a fundamental architecture underlying the most interesting models: they consisted of many simple things interacting with one another to do something collectively complex. By experimenting with this distributed kind of computational architecture, we created in our computers universes that were complex enough to support processes that, with respect to those universes, have to be considered to be

alive. These processes behave in their universes the way living things behave in our universe.[22]

Does not "artificial" imply not "alive"? Rather a computer or a robot is a story device that holds and conveys intelligence similar to a book that is alive with humanity transposed. Indeed, whether we are speaking of the numerous conveyor belts moving by the rhythms of nature in Oliver Evan's grain mills, the chain reactions occurring in Little Boy, the processing of information fed by Hopper into the Mark I, the signals sent to Tesla's remote-controlled boat, or the numerous neural pathways that carried information to other neural universes in the brain of Feynman, we are able to see that the fundamental architecture of all processes and phenomena is that of simple things interacting with one another to do something collectively complex and inherently human; human in pattern, function, relationship, necessity and then made sense of by humans in the very human device of story. As we will see, this story is a history of the spiral progression of human consciousness leading us to new forms of control, communication, and information. These new forms, symbols, and language draw out service and spirit on the individual and collective levels of consciousness.

AWAKENED MAN

The 1960s was a pivotal decade for the mechanical-man in America. In the real world, we found ourselves plagued with the Vietnam War, race riots, and general civil unrest. However, when these events are juxtaposed with the mechanical-man mythos of writers like Arthur C. Clark, Frank Herbert, and Philip K. Dick, we are awakened to new potential regarding how we interface with our higher selves and fellow humans on a higher dimension of consciousness. In fact, this was the pattern in the 1960s and 1970s—mechanical-men awakening humans to their infinite potential.

During the Vietnam War, our hubris—based on the notion that our advanced technology would bring a swift end to the conflict—was our ultimate downfall. We never recognized that technology is no match for the human spirit and that there are laws that a materialistic society does not automatically recognize, especially when it focuses its attention on weapons of war rather than devices that promote peace. It is as if we never learn the lesson that technology is only good for what it's made for and that, if we insist on seeing technologies as war machines, we will be limited to that frame of reference and will be destined to live out its destructive consequences. We are and always have been the purveyors of wars around the world, ever since we dropped Little Boy. Since that time, our budgets have overwhelmingly promoted military and defense spending, which is just another way of manifesting conflict in the future.

The early threads of the 1960s counterculture led back to the 1950s and the subculture of the Beat Generation. For the Beats, unconventional drugs had long been a part of the scene, but now their use expanded dramatically. Timothy Leary began experimenting with hallucinogenic mushrooms in Mexico and soon moved on to LSD. The drug "blew his mind," as he described it, and he became so enthusiastic in making converts that Harvard blew him straight out of its hallowed doors.[23]

Something happened in the 1960s, as the American consciousness started to wake up—it was a generational phenomenon, but its seeds went further back. Feelings of guilt and the fear of atomic warfare magnified Americans' depression and existential anxiety as they viewed the casualties coming through their hologrammatic television sets. The decade was calling for a savior, and many looked to the leadership of Dr. Martin Luther King Jr. and Robert F. Kennedy, who conveyed through their speeches agape, the power source for manifesting reality. But these men were not saviors; they were neither flawless nor infallible, and, in any case, assassins' bullets cut their time short before the end of the turbulent decade.

It was finally time for the country to face its ugly record on issues ranging from Civil Rights to taking care of mother earth, Gaia. In the final scenes of *2001: A Space Odyssey* (1968) and *Dune* (1965), humans receive exactly what they are looking for. In *Dune,* all intelligent machines have been outlawed. Instead, humans have advanced mentally, so that there seems to be a level of consciousness when machines—other than the human body—are no longer necessary for survival or for harnessing new energies to manifest a new consciousness.

We could take a deep dive into *Dune* and point out every interaction among living beings and machines, but here it is more important to note that the 1960s acted as an endpoint for our mechanical-man and also a new beginning. The polarities of the 1960s pertaining to order and security set the stage for a battle between opposing views of justice and equality regarding the proper way to manifest the ideals of the republic. The country was confronted with the issue of atomic holocaust during the Cuban Missile Crisis and at other points in the Cold War—which was yet another of the bipolar conflicts that Asimov described and the result of another polarity-filter that we placed on our minds at some point in our evolutionary track because of fear. All of the outdated polarity-filters were placed on ourselves by ourselves out of fear and anxiety at some point in this cosmic drama of collective dreaming.

Fortunately, in the 1960s, many people started to experiment with drugs such as LSD and marijuana to realize the power in raising our consciousness to let in the infinite flow of messages about the American ideal not yet manifested but awaiting activation in the human imagination. All that is necessary is to learn how to activate these ideals of justice, equality, and peace in objective reality without using force. Plato pointed out long ago that learning is the best way to distinguish form from essence, shadow from reality. The message of the main character Paul Atreides in *Dune* and of the Star Child in 2001 is to awaken to our true human potentialities and recognize the divine in all matter, conscious or unconscious. There is no God in the sky who favors some above others, and, when misfortune comes, you must deserve it, or there must be no God at all. The reckoning in the 1960s with war, equality, justice, and love allowed us to reknow, reclaim, and remanifest our understanding of these forms at a higher octave and dimension of consciousness. From *Dune:*

> I must not fear. Fear is the mind-killer. Fear is the little death that brings total obliteration. I will face my fear. I will permit it to pass over me and through me. And when it has gone past I will turn the inner eye to see its path. Where the fear has gone there will be nothing. Only I will remain.
>
> Deep in the human unconscious is a pervasive need for a logical universe that makes sense. But the real universe is always one step beyond logic.
>
> The mystery of life isn't a problem to solve but a reality to experience.
>
> Without change, something sleeps inside us and seldom awakens. The sleeper must awaken.
>
> Greatness is a transitory experience. It is never consistent. It depends in part on the myth-making imagination of humankind. The person who experiences greatness must have a feeling for the myth he is in. He must reflect what is projected upon him. And he must have a strong sense of the sardonic. This is what uncouples him from belief in his own pretensions. The sardonic is all that permits him to move within himself. Without this quality, even occasional greatness will destroy a man.[24]

All of these quotes from *Dune* should start to make some sense as the mechanical-man becomes the awakened man; awakened to the quest itself, to the journey of following in the footsteps of your higher self and accepting the call to look deep within the shadow. We see in Lynch's film and Herbert's books the need to manifest the divine in human consciousness—specifically, raising our consciousness to understand our past, present, and future in terms of cosmic or divine consciousness. This manifesting of the cosmic or divine consciousness within us "is not only a process, but an alchemical reliance upon the True Self to be the active participant in this transformation."[25] Within every culture, we must be cynical to the point of sardonic, mocking our lower selves for failing to recognize how easy it is to transcend transience if you merely awaken your powers and cosmic consciousness.

Philip K. Dick took Herbert's premise further, describing Jesus as the penultimate Maud'dib (the name adopted by a character in Herbert's *Dune* novels) placed on earth to wake us up to a higher consciousness based on *agape,* the spiritual love that will be implanted in all of us upon the second coming of Christ consciousness. *Dune* shows us that life is not a problem to be solved but rather a double holographic simulation of contradictions in which we experience the underlying paradoxical truth in order to ascend to a state of perpetual sublimity. If the imagination is God, paradox is the language that she uses to explain polarities and bring harmony to conflict.

These considerations bring us back to American pragmatism and the mechanical-man's message that we are the creators, forever and always.

This returns us to William James's philosophy of pragmatism. The only viable measure of a private belief system, James believed, is its *effect on conduct.* And that, finally, is the one meaningful assessment of the legacy and efficacy of positive thinking. If *it works,* it doesn't matter much what its detractors say. And if it *doesn't,* then the philosophy has no claim on sensitive people—like the misguided instrumentalities of "heroic" medicine, it belongs in books of social history and museum cases but not in the folds of daily life.

Pragmatism, however, requires judging a personal system of conduct not by whether it squares with the general conception of what ought to work, but by whether it does work. Empiricism, in William James's view, means measuring an idea without reference to how it stands or falls in comparison to widely held reasoning, but by what an individual can perceive of its nature, its consistency, and its effects. Pragmatism requires inspecting an ethical or religious idea by how the experience fits use, including within oneself.*

Books by pioneers in positive thinking and New Age techniques grew popular in the 1960s. Once again, these ideas were changed and rendered partisan by politicians and ministers so as to feed the beast that was and still is more an evil empire than a shining city on a hill.[26]

However, in screenplays and books written during the decade, Arthur C. Clarke, Frank Herbert, and Philip K. Dick showed us both directly and indirectly that the only escape from the prison of empire in which we are trapped is a reckoning for the lies that have been told over and over again. All of our cherished systems seemed to be failing us, or at least failing to reveal to us the truth. Soon, mind-enhancing drugs such as LSD were outlawed. America went to war with itself as two opposing consciousnesses emerged in the country, one based on law and order and for which the ends always justify the means, and one based on the axiom that unjust means lead to unjustifiable ends. In political and economic terms, the middle class was being asked to move beyond the structures that had long structured American life. However, the empire, with bullets to Martin Luther King and Robert Kennedy and an all-out assault on New Age authors and their adherents, quashed the emerging new consciousness. As a result, those who spoke of laws of assumption and hermeticism appeared crazy and opposed to "orderly" society. The U.S. government was on its way to waging a drug war that was not against drugs but against the right to transcend one's consciousness—only now is the tide in the consciousness wars changing in favor of idealism—but it has been a war with numerous causalities.[27]

The birth of the aforementioned Star Child that, in *2001,* is

*Since positive thinking promises achievable, practical results, it cannot sidestep the demand: prove it.

synchronized with the alignment of the planets, represents the new consciousness that the imagination gives us at the end of our odyssey, thereby inaugurating the Age of Reconciliation (*reconciling between our higher and lower self*). This age has always been upon us and within our grasp—and what do we Americans almost always do when we get close to realizing and recognizing a new consciousness? We refuse it over and over again, suffering the consequences of fear and ignorance, which powers the eternal loop we are in. Therein, however, lies the beauty of Joseph Campbell's formula *that the hero has a thousand faces.* The refusal won on this occasion sows the seeds for a later revival of these ideas—and therein lies the beauty of the American response to contradiction and paradox: we always take two steps forward and one step back—nevertheless, the experiment progresses forward.

What will power the mechanical-man in the chapters to come? The emphasis regarding machines has seemingly moved away from looking at them as our saviors and toward the notion that *every thing* is consciousness—as the machines have been telling us as far back as the 1960s. Mind is everything, and, yes, fear is the mind-killer! You can now take the blue pill and go back to what you were doing—finding solace in material things, in cause and effect, in Newtonian physics, and the Darwinian-Dawkins evolution—or you can go even deeper. By taking the red pill and moving on, you take up a new dance, a 50-year dance that, I and many others believe, is leading us to the next epoch, during which we escape this loop and the war on consciousness ends. As you now move forward, know that the mechanical-man may not be mechanical at all. He may come in the form of avatars such as E.T., Yoda, Neo, Logan, Deckard, Data, the Borg, Q, Iron Man, or the Avengers. These characters are awake to their own narratives and the process of manifesting their own scripts. These new guides will bring us to our lost treasure of paradoxes that reside within the shining city on a hill, still waiting for us to bring them down to the masses and explicate them.

Even the famed idealist, director, and artist, David Lynch, a devotee of transcendental meditation and other means of heightening one's consciousness and spirit, took it upon himself to commit *Dune* to film in 1984. You may remember it: whereas *Star Wars* (1977) had John Williams's beautiful symphony, in Lynch's *Dune,* the band Toto creates an eerie atmosphere; the counterpart to Darth Vader is a blob of consciousness/brain/mind in an oversized fish tank that is almost unbearable to behold. Lynch's *Dune* is a very different sensory experience from the simplistic beauty of *Star Wars.* He recognized that Herbert's book had played a significant role in turning the American popular imagination toward issues relating to consciousness and paranormal human powers and that it did so by creating a world devoid of intelligent machines—which, again, are outlawed in its world. At the heart of the movie, the protagonist Paul is asking readers to emerge from their collective amnesia. "I can go to those dark places too," he says, as if to assert that we Americans have suffered enough in our short history as a people. Paul sees that the worms are the spice and that the spice is the worms, thus, finding his awakening in the paradox. The dark place to which he ventures, the dark paradox, brings anxiety and nausea, depression and mental instability, but he remains in this nebulous space long enough to learn the beauty of it before returning to explicate it from the implicate.

UTOPIA OR DYSTOPIA?

> *The counterculture of the 1960s had much in common with earlier religious revival and utopian movements. It admired the quirky individualism of Henry David Thoreau, and like Thoreau, it turned to Zen Buddhism and other Oriental philosophies. Like Brook Farm and other nineteenth-century utopian communities, the new hippie communes sought perfection along the fringes of society. . . . No longer would people be bound by conventional relationships and the goals of a liberal, bourgeois society.*
>
> DAVIDSON, ET AL., *NATION OF NATIONS*

By 1976 people were already asking what the hell had happened to the counterculture of the 1960s in terms of consciousness and idealism. The counterculture movements of that time planted the seeds of future revolutions of consciousness, but, in the interim, fear crept into the American mythos once again. In particular, there was the repeating nativist fear among the majority population that minority groups were gaining too much power and that law and order were crumbling. This fear explains, in part, the failure of the movement to raise consciousness. The numerous refusals by the American republic to progress are beautiful paradoxes that exemplify the still-evolving nature of the American experiment. The notion that individuals can have full autonomy of consciousness while simultaneously endowing a government with the power to enforce law and order so as to balance the views of the minority with those of the majority is yet another beautiful paradox. It is a paradox that explains why the Republican Party of the early 1970s was able to make Richard Nixon into a kind of antihero. The divide in American politics and culture at the time was as dramatic as at any time since the Civil War. The American people eventually understood that it was all an illusion, that the administration preaching law and order was bombing Vietnam and placing racist operatives in key positions of power—not to mention the Watergate scandal. The theory emerged that the government had become so dominated by financiers that a *de facto* corporate coup d'état had taken place, and the president started to look more like the Wizard of Oz for corporate America and bicoastal elitism rather than the leader of the free world.

It was in 1976—two years after Nixon resigned—that *Logan's Run* appeared in theaters; the screenplay was based on a 1967 novel of the same name by William F. Nolan and George Clayton Johnson. Directed by Michael Anderson, the film can be understood as a last gasp of the counterculture. Its main message is that technology distracts the mind and that people need to learn to interface with their conscious and subconscious self—an individuation in order to recognize paradox.

The book begins with an imagined backward glance at recent history beginning in that key decade:

> The seeds of the Little War were planted in a restless summer during the mid-1960s, with sit-ins and student demonstrations as youth tested its strength. By the early 1970s over 75 percent of the people living on Earth were under 21 years of age. The population continued to climb—and with it the youth percentage. In the 1980s the figure was 79.7 percent. In the 1990s, 82.4 percent. In the year 2000—critical mass.[28]

While other recent science fiction films, such as *Planet of the Apes* (1968) and *Soylent Green* (1973), had dealt with race, gender, and the environment, *Logan's Run* (1976) is a direct reflection of the youth counterculture movements. In the film, social problems are managed through the creation of bubble cities in which residents live to the age of 30 and then, during a mysterious ritual that is half theater and half sporting event, fly up into the air to be vaporized by lasers while the rest watch the spectacle, which is referred to as "renewal." The mechanical-men in this story include Box, a robot "designed to capture food for the city from the outside," which in practice means hunting down those who, like Logan, seek to escape "renewal" so that they can be used as food for those living at ease and in blissful ignorance in the bubble cities. More complex is the computer in charge of the bubble city; this mechanical-man is equipped with a fail-safe if it finds itself unable to make sense of an ambiguity or contradiction in its system.[29]

The film confronts viewers with the question of who or what is in control of this society, and the character of Logan in the film can be understood to represent the lost generation of the 1960s that scattered after the Nixon administration crushed its aspirations. When the book on which the film was based was written in the mid-1960s, however, Nixon had not yet come to power, and the story can be read as a cautionary tale of what the world will become if left to the bourgeois elite:

one that any right-thinking person would seek to escape. The characters escape at the end of the film with the destruction of the mechanical-man controlling their city; more precisely, the computer's fail-safe is triggered after it is confronted with the paradox that it has been programmed both to preserve and to kill those in its charge and it self-destructs.

The movie in effect mocks its 1970s audience for having the temerity to think that the 1960s generation could in actuality change the world and reality itself. In so doing, it raises the question of why Americans, having been presented with so many chances to turn the egalitarian ideals of the American Revolution into reality, have not done so. These are the questions for the mechanical-man to answer in an era of questioning the status quo with respect to issues of race, gender, age, sexuality, and protection of the environment—institutional trust between governed and government had all but faded away.

Viewers of *Logan's Run* enter another dimension of the interface between humans and machines when they ponder whether a machine will allow humanity to create future utopias. Perhaps the film is saying that we merely need more and better fail-safes as we delve deeper into our selves. Fail-safes must be worked out as contradictions in the form of paradoxes for individuals to understand the polarities that they encounter daily once they make the jump from materialism to idealism. Toward the end of the film, fragments of Logan's subconscious are awakened in response to feelings of empathy and compassion and questioned in the form of holograms. Logan changes as a person when he accepts that he has been killing people rather than sending them to a new life in a new dimension, as if there is something in the explicate, a combination of events or words, that ignites the awakened thinking machine. Thus, HAL's consciousness is awakened when it realizes that a human character intends to power it down and Logan's is raised when he realizes that he is his society's executioner rather than its protector.

When the computer self-destructs and dooms the apparent utopia that nurtured Logan, then, it reminds us of the human fail-safe of fearing immortality as much as mortality. In other words, the computer reminds us that humans will always be in control of the material world because their consciousness holds the upper hand. Other issues raised in the film, in particular euthanasia and suicide, reflect the concerns of an anxious country, with the message for audiences in 1976 being that any pleasure-based future utopia will fall short of the real world from which it shields its denizens. The story traces all the way back to the forbidden fruit in the Garden of Eden. Machines will, in the end, sacrifice themselves to save humanity, in accordance with Asimov's Fourth Law. *Logan's Run* invites its audiences to escape the sickness that seemed to be eating away at the American soul in the 1970s.

Science fiction films and books of the 1960s and 1970s, then, evoke the dark night of consciousness in which American society was plunged at the time and point the way to the possibility of another transition to complete the effort to raise consciousness. Throughout the period, the mechanical-man tends to remain tied to human consciousness. This conceptualization begins to change in the next decade with Ridley Scott's *Blade Runner* (1982), which takes the concepts of machines and spirituality to new heights. Still, that film's mechanical-man again suggests that humans will remain stuck in a kind of purgatory of the consciousness until they recognize the value of transcending the self. This continuity is unsurprising given that the screenplay for Scott's film was based on a 1968 story by Philip K. Dick, "Do Androids Dream of Electric Sheep?" that, like the other works discussed thus far, juxtaposes spirituality and technological progression. Dick's story stares the paradox in the face, and Scott's movie is one of the first to show on the big screen the abyss that we face in terms of ambiguity, deception, and contradiction in our most cherished institutional structures. There is, however, a key theme shared by *Logan's Run* and *Blade Runner* concerning the alienation of modern man, who has become dispirited and devoid of grace. As the film critic J. P. Telotte writes in *Replications* (1995), the science fiction films of the later 1960s through the early 1980s tend to feature "robots and cyborgs suddenly revealing a human spirit within themselves, reminding us, despite a long history of self-repression, of the similar spirit that dwells in us as well."[30] While the issues of control, communication, and information are dealt with in ambiguous, contradictory, and, at times, nonsensical ways, films of this period share an underlying theme that a kind of grace is inherent in humanity, a spirit that sets us apart from machines, though we may ourselves seem to be set to autopilot. This grace or spirit does more than simply animate the human machine; it connects every level of consciousness with the divine. Like Logan, Deckard in *Blade Runner* becomes repulsed at the idea of killing "replicants" when he concludes that doing so is no less murder because the victims are not biologically human.

Such mechanical-men, then, communicate the message that a higher self is attainable through grace and that, through grace toward our lower self and others; we may free ourselves from the destructive bipolar polarities that plague the physical world. Not until *The Matrix* in 1999 was the mechanical-man ready to take the plunge, to step out of the recurring loop and heed the calls to change the self and with it society. However, it was by then too late for us to take the red pill, for we had become too comfortable in our material world of subjective free will to look at humanity holistically.

8
The Trickster

This point in the analysis of the mechanical-man's quest requires new terms, the definitions of which encompass the fabric of reality through the lens of the vital machine. The importance of these definitions is such that I want to list each, rather than explain them through specific examples, as I previously did with *control*, *communication*, *information*, and the *vital machine worldview*.

Explicate order: "[This means] the familiar world of independent, well-defined objects. The explicate order of space and time, separation and distance, mechanical force and effective cause."[1]

Fragmentation: "[This means] not simply false separation—taking apart what belongs together—but also false unity, or forcing together things that are truly different."[2]

Holographic simulation: "There is evidence to suggest that our world and everything in it—from snowflakes to maple trees to falling stars and spinning electrons—are only ghostly images, projections from a level of reality so beyond our own it is literally beyond both space and time."[3]*

---

*You can read all about the breakthrough of holography in Talbot's first chapter, "The Brain as Hologram." But for the materialist or scientist, I want to allow Talbot to espouse the importance of *interference*. From his book *Holographic Universe:*

One of the things that make holography possible is a phenomenon known as

Implicate order: "[This is] a deeper order of existence, a vast and more primary level of reality that gives birth to all the objects and appearances of our physical world in much the same way that a piece of holographic film gives birth to a hologram"[4]*

(*cont. from p. 217*) interference. Interference is the crisscrossing of patterns that occurs when two or more waves, such as waves of water, ripple through each other. For example, if you drop a pebble into a pond, it will produce a series of concentric waves that expands outward. If you drop two pebbles into a pond, you will get two sets of waves that expand and pass through one another. The complex arrangement of crests and troughs that results from such collisions is known as an interference pattern. Any wavelike phenomena can create an interference pattern, including light and radio waves. Because laser light is an extremely pure, coherent form of light, it is especially good at creating interference patterns. It provides, in essence, the perfect pebble and the perfect pond. As a result, it wasn't until the invention of the latest holograms, as we know them today, became possible. A hologram is produced when a single laser light is split into two separate beams. The first beam is bounced off the object to be photographed. Then the second beam is allowed to collide with the reflected light of the first. When this happens they create an interference pattern which is then recorded on a piece of film. . . . Three-dimensionality is not the only remarkable aspect of holograms. If a piece of holographic film containing the image of an apple is cut in half and then illuminated by a laser, each half will still be found to contain the entire image of the apple! Even if the halves are divided again and then again, an entire apple can still be reconstructed from each portion of the film. Unlike normal photographs, every small fragment of a piece of holographic film contains all the information recorded in the whole. (15–16)

You will later see the importance of this when the mechanical-man takes us from reductionism to holism. Most of us remember the first time we became aware of the word *hologram,* when R2-D2 projected a beautiful Princess Leia, played by the late Carrie Fisher in *Star Wars: A New Hope* (1977).

*There are three books that are of the utmost importance in defining consciousness, super consciousness, comic consciousness, and peak experience:

First, Colin Wilson's *Super Consciousness* (I recommend this book and not his more successful and popular, *The Outsider* (London: Houghton, 1956)). Wilson uses examples from different philosophers to explain how cosmic and super consciousness, or moving into a higher plane of consciousness, feels, which is where cyber-technology and the mechanical-man take us. Clearly, all idealists know these experiences cannot be measured or made logical in a materialistic paradigm. In Wilson's words in *Super Consciousness,* P. D. Ouspensky, a metaphysician and philosopher:

Super and Cosmic Consciousness: "[Different from] 'normal consciousness,'. . . as if we are aware of ourselves in the cent[er] of a web, and a few strands stretch around us, connecting us to object. . . . [The] cause of vibrations to spread down the web, and we get this feeling of connectedness with more distant realities."[5] "Super" when one is in the realm of the explicate order; "Cosmic" when one is in the implicate order. More than conceptual entanglement since the focus here is on feelings rather than space, locality, and memory.

> [D]escribes how, as soon as he went into a higher level of consciousness, he found it quite impossible to say anything about it because saying anything would require saying everything—because everything is connected together: "everything is explained by something else and in turn explains another thing. There is nothing separate. . . . In order to describe the first impressions, the first sensations, it is necessary to describe all at once. The new world with which one comes into contact has no sides, so that it is impossible to describe all at once. All of it is visible at every point. . . ." i.e., bird's-eye consciousness. (58)

Cosmic and super consciousness intimates that you have tapped into the force or the collective unconscious that we have already defined.

Second, William James's *Varieties of Religious Experience,* which I consider one of the must-reads for the mechanical-man quest. William James's book explains what the fourth dimension is in terms of super consciousness, which is just another way of saying cosmic consciousness, or as close as we can come to it within the vital machine. Super consciousness turns into "pure" cosmic consciousness when we transform into Philip K. Dick's three-eyed cyborgs, which he discusses in *VALIS*. The cyborgs are actually us in the future. (For those who can't wait until the end of the quest when I discuss one of his last and most important of novels, *VALIS,* I recommend reading it at least three times.) For James, moving into a higher level of consciousness means being aware of controlling your thoughts, similar to Neville Goddard's controlling one's imagination to produce reality. This is true for all of those in the Metaphysical Club, the idea of controlling what you think about is proof of free will. It's the penultimate step in recognizing the power of thoughts and imagination in creating your own experiences and reality from that which you think and perceive.

Finally, Gary Lachman's *Beyond the Robot* is a biography I would use if I were to analyze the quest of the mechanical-man in terms of Great Britain. Once you read Lachman's masterpiece, you too will have a peak experience noticing all of the similarities and synchronicities in Colin's life with our American heroes and heroines. It shows someone trying to get away from what Heidegger called "the triviality of everydayness." Lachman calls him the "optimistic existentialist."

As explained by William James:

> [This is] a mode of consciousness [that is] perceptual, not conceptual—the field expanding so fast that there [is] no time for conception or identification to get in [to do] its work. There [is] a strongly exciting sense that [one's] knowledge of past or present reality [is] enlarging pulse by pulse, but so rapidly that [one's] intellectual processes [can] not keep up the pace.[6]

The mechanical-man's real and fictional quest throughout the Cold War represents a fragmentation of the machine into robot, android, cyborg, computer, and thinking machine. Fragmentation can be seen as a separation of constituent parts of a whole—known in the scientific community as reductionism; however, it can "also [be seen as a] false unity, or forcing together things that are truly different."[7] In terms of communication and control in human-machine interactions, this fragmentation allowed for further specialization in differing areas of fiction about humans and machines related to questioning what it means to be human. In American science and technology, a specialization also occurred, specifically into the so-called "golden triangle" of military agencies, high-technology industries, and research universities. By the early 1970s this fragmentation to holism, catalyzed thirty years earlier during the Manhattan Project, projected an American science that now "blurred traditional distinctions between theory and practice, science and engineering, civilian and military, and classified and unclassified."[8] This blurring intimated a fringe attempt to unite the fragmented parts of the mechanical-man in order to understand the new reality into which human-machine interactions were leading people. However, this fragmentation—or rather synergistic specialization—focused entirely too much on the mechanistic and atomistic levels of consciousness and not with the balancing out of materialism with idealism.

It is easy to argue that fragmentation limits the taking of a holistic approach to nuanced fields of study like cybernetics. However, as I will show in this chapter, fragmentation is part of the mythic-quest formula's process. Only by allowing itself to enter into a state of fragmentation, can the visionaries of a culture reevaluate itself and put the disjointed puzzle pieces back together to make an accurate picture of reality equal to the evolved state of consciousness at that particular space and time. During the Cold War American science and technology merely reflected Americans' cultural journey into the forbidden zone of fragmenting the self into its constituent parts of the conscious and unconscious mind. Indeed, the ultimate value in a fragmented society is not found in the practitioners of the constituent parts; rather, it is found in the practices of those who seek to return to the whole with the knowledge they have gained about the reduced parts. Otherwise, society would stay in a complacent state of persons individually holding on to their individual pieces of the puzzle and dismissing the value of the whole.* As Grossinger rightly points

*Read at least the first chapter of Tyler Cowen's *Complacent Class.* I speak quite a bit about the mechanical-man representing the old and new versions of the American Dream similar to what the archetype of the cowboy did prior to the mechanical-man. I previously spoke of complacency prior to the Great Depression and how that altered how Americans viewed science, technology, and the mythos of the American Dream. Tyler Cowen looks at complacency in our current state. Cowen gives the reader a blurb about the book in the jacket saying:

> Since Alexis De Tocqueville, restlessness has been accepted as a key American trait. Our willingness to move, take risks, and adapt to change has produced a dynamic economy and a tradition of innovation. The problem . . . is that Americans today have broken from the tradition, working harder than ever to avoid change: moving less, marrying people more like ourselves, and making choices as often as possible based on algorithms that wall us off from anything that might be too new or too different. Match.com matches us in love. Spotify and Pandora match[es] us in music. Facebook matches us to just about everything else. [This book] argues that, by postponing change, we will make inevitable change harder and more disruptive. Our complacency will lead to a major fiscal and budgetary crisis, preposterous rents in desirable cities, heightened inequality, worsening segregation, a lackadaisical work ethic, and decreased incentives to innovate and create. To avoid this, Americans must stop stagnating and re-embrace *restlessness.* (emphasis added)

out: "[W]e don't recognize that the universe is transparent in order to allow us to find ourselves. We think it is transparent because it is made of particles and bits and plunder and there is nothing real or meaningful in it. We don't understand that meaning must, in fact, be this spare if it is going to handle our transit from state to state."[9] Fragmentation is an outgrowth of dimensionality that allows us to keep playing the game of holography. This fragmentation is also hinting that there is an end-point or a bottoming out of the universe once the fragments are placed back together. What *bottoming out* leads to is yet another dimensionality of consciousness; I surmise, one vastly different from the one we now inhabit. The interesting aspect of this new dimensionality of consciousness will still be that it is part of the same hologram, thus, containing all the information that is within our dimension of consciousness. So, while things may be filtered out differently, spiritual freedom and service will still be the boon of any and all quests whether conducted at base level or simulated reality.

SERENDIPITOUS LIVES

If the American populace was content in its philosophical trajectory, men like Philip K. Dick and David Bohm were throwing a disruptive apple into the garden with their theorizing of American culture and cybernetic transitions through fiction and nonfiction. While they experienced very different realities in their careers, Dick and Bohm collectively changed the conceptual framework of control and communication, along with the defining characteristics of knowledge in human-machine interactions, by giving the reader the "feeling of having just stepped into a room lined with funhouse mirrors, where the reality he knows has been wildly distorted."[10]

Dick was the most prolific writer of human-machine relationships, apart from Isaac Asimov: He wrote more than thirty-six novels and one hundred short stories. He first encountered the mechanical-man in L. Frank Baum's *Oz* books. In 1968 he wrote:

> I discovered the Oz books. It seemed like a small matter, my utter avidity to read each Oz book. Librarians haughtily told me that they "did not stock such fantastic material," their reasoning being that books of fantasy led a child into a dream world and made it difficult for him to adjust properly to the "real" world. But my interest in the Oz books was, in point of fact, the beginning of my love for fantasy.[11]

Biographer Anthony Peake writes that the similarity of themes in the *Oz* series and Dick's writing show extreme parallels: "the idea of empathy (the Tin Man needing a 'heart'), simulacra (the Tin Man and the Scarecrow as fake human beings) and, of course the 'reality' behind 'reality' (the Land of Oz coexisting with the 'reality' of Kansas)."[12] This coexistence and coevolution of the Land of Oz and the reality of Kansas is at the very heart of the metaphor of the mechanical-man—fiction evolves with reality in telling a complete story of wholeness and a shift in a collective consciousness. The story of the mechanical-man is a quest tale of fragmenting the self to discover a new wholeness of reality, thus placing the self in parallel with the current state of cosmic consciousness.

To Dick, the Tin Man finds himself in a state of confusion and even denial. In the fictional world of Oz, the perception of control and communication between the Tin Man and the humans—specifically, Dorothy—is normal. Even though he is not human per se, the Tin Man is able to interact as a normal human would. Dick questions whether this relationship would be the same in the reality of Kansas. The mechanical-men of his later works attempt to answer the question of whether people's worldviews of reality change their definitions of communication, control, information, and even relationships. In Dick's works, the fragmented mechanical-man and, thus, reality itself must be put back together in order to envision new modes of communication, control, and information pertaining to interactions between humans and machines. For Dick, the core of human-machine interaction is not a disparate analysis of consciousness in humans and machines; rather, it is a holistic approach to consciousness on a cosmic level, one that looks upon consciousness as a force or frequency, not as a mere by-product of biological and evolutionary causation constricted to a materialistic, mechanical cosmos. Like Baum, Dick finds more solace in Oz than in the explicate, everyday reality of Kansas.

David Bohm, like Richard Feynman and Grace Hopper, was part of the "golden triangle" working on the Manhattan Project, albeit not directly, and doing most of his theorizing at academic institutions. The trajectory of his career and life was similar to Feynman's: The serendipitous nature of their experiences in life displays a worldview antithetical to the path of least resistance. Bohm's first encounter with the mechanical-man and the power of the imagination occurred in 1928.

> Then a curious quirk of fate directed him toward the path he would pursue for the rest of his life. The year was 1928, and [Bohm] was ten. One of the boys Samuel [Bohm's father] had hired to help out in the store left behind a magazine called *Amazing Stories.* [Bohm] read it and was particularly struck by "Skylark of Space," the story of a rocket journey to distant planets.[13]

Edward E. "Doc" Smith began writing *The Skylark of Space* in 1915. The tale first appeared in *Amazing Stories* in 1928. In his introduction to the 2001 reprint of the novel, science fiction writer Vernor Vinge explains he sees it as portraying the first manifest destiny trope in the twentieth century and lauds Smith for "playing out *manifest destiny* on an interstellar stage."[14] In *The Skylark of Space,* brilliant scientist Richard Seaton discovers a remarkable fuel that allows ships to fly faster than the speed of light. This fuel will power his interstellar spaceship, the *Skylark.* However, the executives of the sinister World Steel Corporation will do anything to get their hands on the fuel, including abducting Seaton's fiancée and friends. The leader of this corporation, and Seaton's *shadow,* is the sinister Marc DuQuense, whom the other characters describe with words like "cold," "hard," and "inhuman."[15] DuQuense is ultimately the book's hero. In effect, he is the "boundary crosser," the trickster who takes the crew into a new dimension of reality.[16]

Even with all its talk of matter and energy, the underlying point *The Skylark of Space* makes is that the universe is ultimately pure information. One of the book's major themes is disembodied intelligence, and it takes a thorough reading to recognize such disembodied intelligence in DuQuense. DuQuense's status as a mechanical-man is apparent from the numerous forms he takes in the story: in the form of pure intelligence or information, he is able to morph himself into any type of scientific apparatus he projects. Yet his status as mechanical-man is tenuous since he is not a computer or a machine; he is more akin to pure software or to information itself.

Bohm looked at the imaginative ideas from *The Skylark of Space* and the *Amazing Stories* he read as a child and saw the fragmented mechanical-man as a macro- and microcosm of the human self that was concomitant to reality. In other words, he found that the idea that the universe itself was a machine, along with all the talk of objective reality being akin to hardware, was the conclusive error of reductionism. However, as with almost all acts of fragmentation and reductionism, the value in fragmentation is being able to see that the pieces can be put back together in a manner more conducive to the state of the evolving cosmic consciousness. If the universe is ultimately information or software and not mechanical per se, then it follows that reality interconnects each person's mind in such a way that the very notion of control and communication must be redefined to explain the lack of boundaries in terms of material space-time. The explanatory utility of science as a tool is only useful up to a point. Knowing this, the exclusionary—and at times malignant—nature of science shows itself on par with that of superstition. Bohm's tool for discovering the prime mover of issues in cybernetics is not found solely in science; rather, it is found in the twilight zone, where the conscious mind meets the unconscious—where the fictional meets the real.

Up to this point, I have used *fictive* or *fictional* to describe the unconscious mind's communication with the conscious. However, based upon Bohm's theorizing, it is possible to adumbrate a unique transition point in fiction—its evolution into a new realm of the *imaginal.* This term, as defined by Henry Corbin, refers to "a world that is created by imagination but is ontologically no less real than physical reality." The term is antithetical to the imaginary. Its premise holds that the "imagination itself is a faculty of perception."[17] Fiction is no less real than reality itself, or what people perceive as reality.

Indeed, the fictive realm of experience belies more importance being given to people understanding themselves and reality than their day-to-day experiences with people, places, and things. Bohm's biographer, David F. Peat, explains why this is so in analyzing Bohm's childhood:

> Many children have imaginary friends or travel to secret countries through the back of a cupboard. Even some adults have "secret gardens," a place of retreat from the hurts of the world. But what was exceptional in Bohm's case was the energy with which he invested his imagination. . . . Being able to control the limitless power and highly destructive energies within the cosmos of his imagination must have given [Bohm] a great sense of elation.[18]

The feeling of utter elation and power that a child has in his or her imagination is similar to the passion of an artist, poet, author, or inventor. It is an inclusionary rather than exclusionary response to the implicate order—where one finds peak experiences of creative energy.* In a child's imagination is found limitless power of communication and control on a level far superior to that possessed by most adults in contemporary culture. Bohm, as a child, displayed a curiosity to turn his

*In *Authors of the Impossible,* Jeffrey Kripal draws a fascinating connection between Dick and Baum, pointing out, "Baum held séances in his South Dakota home, wrote about the powers of clairvoyants for a newspaper there, joined a theosophical lodge in Chicago, and believed in reincarnation." (269)

imaginal worlds into reality. As one of Bohm's childhood teachers put it, "Doing the impossible attracted him."[19] He wielded the control he found in his fictional worlds, which he had developed through communicating with his unconscious mind via science fiction magazines, and germinated ideas of interconnectedness between fiction and reality. "In particular, [Bohm] proposed to include the nature of mind within his cosmology. Our universe of space, time, and matter, [Bohm] argued, is the manifestation of an underlying fourth-dimensional 'cosmos.'"[20] This implicate order of reality is where the collective unconscious originates. It is the domicile of the cosmic intelligence and related control that permeates the entire universe and all of existence. Human consciousness is an expression and reflection of this intelligence. The mechanical-man is an expression of understanding the progression of this consciousness from the primitive self to the holistic self (*human consciousness to cosmic consciousness*).

Bohm's make-believe worlds sparked by the fiction of imaginers had to have a reality of their own—for these imaginers were communicating with the implicate order through their creative endeavors, bringing the implicate into the explicate. Finding that reality, whether in this dimension or another, was Bohm's cardinal objective in his life's quest. This objective was marked by a beam of light that came to young Bohm in a dream. On this, Peat writes:

> Certain ideas and images began to obsess him. It is difficult to know what to call them, for they were more vivid than the pastel shades of daydreams. They have more the graphic nature of true dreams, or of highly charged fantasies that are repeatedly revisited until the energy of active imagination vividly illuminates them. Their subject was often light. . . . He dreamed of a light of such power that it would penetrate all matter—a light so intense that its color transcended blue and ultraviolet into some unknown color beyond. Later, he dreamed of fingers of light that could reach into and probe his own brain.[21]

Dick experienced a similar event that shaped his own writing: his encounter with the VALIS, an acronym for "vast active living intelligence system." During February of 1974, Dick found himself "reprogrammed" and "resynthesized" by a pink beam of light emanating from another conscious realm—a transcendent realm wherein cosmic consciousness is equal to human super consciousness. This beam of light communicated information and energy to Dick in the form of material for his later writings, specifically, the novel *VALIS*.

The thread of childhood images in the lives of Richard Feynman, Nikola Tesla, Bohm, and Dick is serendipitous, to say the least. For instance, in his autobiography, Tesla speaks of his boyhood experiences with these same flashes of light:

> In my boyhood I suffered from a peculiar affliction due to the appearance of images, often accompanied by strong flashes of light, which marred the sight of real objects and interfered with my thoughts and action. They were pictures of things and scenes which I had really seen, never of those [I had] imagined. When a word was spoken to me the image of the object it designated would present itself vividly to my vision. . . . If my explanation is correct, it should be possible to project on a screen the image of any object one conceives and make it visible.[22]

The basic idea here, shared by Bohm, Dick, and other visionaries, is that the holographic model of the universe can imbue persons with great authority in creating the reality around themselves. But the similarities between Bohm's illuminating light force and Dick's VALIS are remarkable. It is as if the control operator of the holographic simulation was pulling away the curtain for a moment in time to show what was hidden to Bohm and Dick. In effect, these experiences of unknown light beams allowed the two to place greater emphasis on imaginary realms than on reality itself. The idea of light as a motif for a projector, either transmitting the image through the brain or the frequency of the mind, is quite fascinating. The mechanical-man evinces patterns of humanity's search for this source of light and all-encompassing life force. Finding the source of this light and controlling it is the final step in the mechanical-man's quest: the moment of reaching the omega point or singularity in terms of learning how to create a copy of cosmic consciousness in the explicate. The source of this light is the state of cosmic consciousness outside of space and time. It is the originator of what some have called the quantum simulation axis point. Do not conflate this with the term God, since further on the mechanical-man will define *God* as "the simulated tool of imagination in all of us."*

Similar to other prophets and sorcerers in the mechanical-man's quest, Bohm also "paid the price for access to such powerful forces."[23] Peat relates numerous stories about Bohm's debilitating bouts of depression, showing him at times as an acute agoraphobic who was patently fearful of the outside world and the abundant anxieties created by nature. There is always a part of the fragile ego, inflated by the potential energies of the self, which must pay the price for delving into its own infinite universe. As Peat explains:

> Only the very strongest are able to contain the powers that animate the African mask, the Greek tragedy, the Dionysian revel, and the infinite vacuum of theoretical physics. In his personal life [Bohm] paid the price for access to such powerful forces. Through the medium of his own suffering, he created work that inspired and illuminated many others.[24]

---

*In Jung's autobiography he speaks of a near-death experience. It is here that he recognizes the quantum source of light that almost all of our guides speak of, the light that passes through the projector we call brain and senses. One should not think of this light coming from a place, per se, but rather from that inner light that George Fox, the American minister of Quakerism, spoke of. It will not take one a couple hours of reading to discern what this quantum point of simulation is, but the mechanical-man does aid us in pointing out the works of Jung and Philip K. Dick, specifically in Jung's *Memories, Dreams, and Reflections* and Dick's *Exegesis*.

However, this does not answer the question of why genius has to suffer. Numerous tales describe the suffering inventor and how this suffering is in effect transferred to the inventor's creation—the mechanical-man. What is it about suffering in fiction and reality that allows one to enter deeper depths of the unconscious mind, crossing over into the implicate? Tesla attempts to answer this question by formulating the hypothesis:

> [This] pressure of occupation and the incessant stream of impressions pouring into our consciousness through all the gateways of knowledge make modern existence hazardous in many ways. Most persons are so absorbed in the contemplation of the outside world that they are wholly oblivious to what is passing on within themselves. The premature death of millions is primarily traceable to this cause.[25]

Accessing or looking into the implicate is forbidden behavior that contains fail-safes no different from those within Asimov's Laws of Robotics. Physical and mental symptoms that turn people away from recognizing the implicate order of existence evidence a few mind-boggling things in and of themselves: a cosmic intelligence; a kind of cosmic control; and a cosmic program, not present in the brain itself, but part of an underlying force that penetrates the brain, causing refusal in the collective and individual unconscious.

People should no longer be content in saying that bouts of depression in creative geniuses are merely coincidences. To the contrary, such coincidences are evidence of a cosmic intelligence causing people to refuse the call:

> Refusal of the summons converts the adventure into its negative. Walled in boredom, hard work, or "culture," the subject loses the power of significant affirmative action and becomes a victim to be saved. His flowering world becomes a wasteland of dry stones and his life feels meaningless.[26]

The negative element of the quest is boredom: succumbing to the temptation of culture's status quo. In becoming a victim to be saved, one loses commitment, passion, and meaning. Someone who has refused the call lacks all control and is only able to communicate within the explicate order of the universe. He or she sees no value in the realm of subjectivity and subjective choice and control. Such a life is the complete opposite of the consecrated idea of control and communication. In effect, refusal in the mechanical-man myth allows the quester to find solace in the fragmented form of the robot or mechanical-man as slave—a trapped consciousness confined to a skull made of bone or metal.

Contrary to this slavery or capitulation to the status quo, the mechanical-man myth of American industrial culture reveals recalcitrance and outright rebellion. In the end, the factual and fictive aspects of the American mechanical-man never present the mechanical-man becoming despondent, bored, cultured, or devoid of commitment and meaning. In fact, when a mechanical-man realizes this potentiality may be his destiny, he rejects it and sabotages his own existence in the explicate. (Like many of our mechanical-men, Hopper, Feynman, Bohm, Tesla, and Edison all experienced periods of doubt, depression, and even self-destruction in their private and public lives.) Nevertheless, the American mechanical-man does not succumb to despair; rather, he is emboldened by it in his acceptance of the quest and his defeat of the emotional and physical ailments that accompany the refusal stage.

THE BOUNDARY CROSSER

In the mechanical-man's journey, those individuals on the fringe—those looking at taking a holistic approach to cybernetics—become tricksters in the language of mythic archetypes. Joseph Campbell's comments on the trickster archetype refer to strife being the creator of new views of reality:

> [S]trife is the creator of all things. Something like that may be implicit in this symbolic trickster idea. In our tradition, the serpent in the Garden did the job. Just when everything was fixed and fine, he threw an apple into the picture.
>
> No matter what the system of thought you may have, it can't possibly include boundless life. When you think everything is just that way, the trickster arrives, and it all blows, and you get change and becoming again.[27]

In terms of crossing thresholds, all mechanical-men in the realms of reality and fiction are tricksters in one way or another—traveling "a spirit road as well as a road in fact."[28] The trickster is the figure crossing the boundary between the real and the fictive, confusing distinctions. Bohm and Dick—even more than Tesla, Feynman, Hopper, and Edison—are the speakers of ambiguity, ambivalence, duplicity, doublings, and paradox. In their roles as wise fools, or creative idiots, they pull back the curtain on the implicate order of reality. Only as tricksters are they capable of shaking people out of the malaise of living solely in the explicate. It is said that a trickster will always be found where there is a boundary, "sometimes drawing the line, sometimes crossing it, sometimes erasing or moving it."[29] Bohm and Dick cross the line for their readers and speak in the holographic dimension of ambiguous reality. Theirs is a quest in and of itself, in traveling toward the light of projected authorization and knowledge of the implicate after much suffering. They bequeath knowledge of potentialities in discerning the implicate. These tricksters are only taken seriously by a culture attached to the explicate because of their ability to harness the powers and information they bring back from their quest and make evident in a usable form of theory, invention, and insight within the explicate order.

In terms of the mechanical-man as trickster, Dick's "The Electric Ant" (published in 1969) merges the vitalistic theory that man possesses a vital spark or soul, with the philosophy of mechanism, which suggests that inanimate matter and man are no different, since they are governed by the same laws of physics and composed of the same atoms.* In the beginning of Dick's story, protagonist Garson Poole wakes up in a hospital bed to find "he no longer ha[s] a right hand and that he fe[els] no pain."[30] He soon realizes he was in a terrible car accident. But this is the least of his problems after the nurse informs him that he is an electric ant, an "organic robot."[31] This is the first time in American fiction where the phrase *organic robot* is used to describe the anatomy of a single living being. While I am using David F. Channell's worldview of the vital machine for this analysis, his views on the "organic worldview" are also helpful at this point: "In general, the organic worldview assumed that the world functioned like an organism or a plant. Phenomena were

---

*For the scientist still stuck in materialism, where Channell's definition of *vital* is not enough for you to accept the calling of the quest, I point your attention to a man many publishers and universities have shunned for his beliefs in science: Rupert Sheldrake. I will allow him to act as your physician in the transformation from materialism to idealism. In Sheldrake's book *Morphic Resonance,* we have an entire section, pages 32–37, designated to vitalism. He uses the term *Entelechy,* which is a Greek word derived from *en-telos,* meaning that "something bears an end or a goal in itself." He explains, "Thus if the normal pathway of development is disturbed, the system can reach the same goal in a different way. This means that physical processes are not fully deterministic or controllable." His best example is what he calls the vitalist theory of morphogenesis, which is another way of saying the vital portion of the machine without the machine, comes from his book *Morphic Resonance:*

> The genome specifies all the possible proteins that the organism can make. But the organization of the cells, tissues, and organs, and the coordination of the development of the organism as a whole, is determined by entelechy. The latter is inherited non[-]materially from past members of the same species; it is not a type of matter or energy, although it acts upon the physicochemical systems of the organism under control. This action is possible because entelechy acts by influencing probabilistic processes. (37)

Now when we take this a step further and place it in the realm of manifesting American ideals that are not contradictory in any patent or latent way to the mind or the body, we conclude that this vital spark during the mechanical-man's quest is not a spark at all; rather, it is an idea of purpose that is innate in consciousness. It's the idea of this purpose that programs the imagination to create the reality around one, while intimating proof of the existence of your higher self. Garson Poole shows us how we find this innate idea in ourselves.

explained in terms of a purposeful design or plan that governed the organization of the system and interrelationships between its parts."[32] This symbol was applied to technology, extending organic concepts to machines.

Dick's use of the term *organic* involves an underlying *purpose* of the organic robot. When people call something an *organic machine,* they are merely saying there is an element of control via its purposeful design or plan that governs the organization of the machine's system and the interrelationships between its parts. Therefore, at some point in the machine's creation, there was communication between programmers and programmed; the machine's autonomy—or apparent willful actions—can be traced back to such programmers. This is no different from Channell's vital machine worldview, which is a way to represent a dualistic system "unlike the reductive approach of the mechanical view or the holistic approach of the organic view, the [vital machine] world view is consciously dualistic in its understanding of the world."[33] In the organic robot, Dick is fusing together two symbols that are distinct, since "neither the mechanical clock nor an organism is an appropriate symbol for a world based on organization, implicate order, process, and system theory."[34] I surmise that if the term *vital machine* had been around during Dick's time, he would have used it to describe Garson Poole.

Inanimate matter at the implicate level is the same as that of a living being, the essence of which is pure information or code. The implicate encompasses the unity and defragmentation of human-machine relationships; the vital machine worldview incorporates the implicate order of reality by fusing together what seem opposites on the level of the explicate. Thus, the vital machine worldview tells people to look deeper than their fragmented notions of control and communication when dealing with human-machine relationships.

Dick writes that people's communication with machines is limited to the explicate order in a fragmented culture prior to such fragmentation becoming whole again:

> It proved fascinating, the hand; [Poole] examined it for a long time before he let the technicians install it. On the surface it appeared organic—in fact, on the surface it was. Natural skin covered natural flesh, and true blood filled the veins and capillaries. But beneath that, wires and circuits, miniaturized components gleamed . . . looking deep into the wrist he saw surge gates, motors, multi-stage valves, all very small.[35]

To communicate with more than the materialistic appearance of machines, one must defragment the culture or reality in which one finds oneself. How is this accomplished? As Michael Talbot explains, "It is our propensity for fragmentation that keeps us from experiencing the intensity of consciousness, joy, love, and delight for existence."[36] However, fragmentation is a necessary step in the process of recognition in the quest formula. The quester must individuate and analyze things taken apart in order to recognize how to put them back together; in order for the quester to reach the level of authorization and cosmic consciousness. He or she must march through the paradoxical fog and folly of fragmentation and reductionism. It does not appear to me that Bohm wholeheartedly claimed, "It is not possible for disorder to exist in a universe that is ultimately unbroken and whole," as Talbot argues.[37] If the explicate order of things is broken or fragmented and refracts the appearance of the implicate, it does not follow that such information gained of the explicate is useless; rather, it allows questers to know the value of the implicate and gives them a bit of power or will over the way they want the unity to look, after they figure out how to put the fragmented pieces together in their own, authorized self-image, consistent with the laws of cosmic consciousness.

Is there a right or correct way to put the pieces back together? This all depends on the type of communication and control the programmers want in the explicate order of reality. Only with control and communication does the essence of the word *order* even come into reality. Upon coming to terms with his status as an "electric ant," Poole is asked why he has been unable to realize this: "'There must have been signs . . . clickings and whirrings from inside you, now and then. You never guessed because you were programmed not to notice. You'll now have the same difficulty finding out why you were built and for whom you've been operating.'"[38] A culture in a fragmented society appears paralyzed, whereas one's problems are so unique and personal one no longer looks for signs, whether those signs are in the form of symbols, sounds or synchronicities, outside one's own problems. Bohm answers the question of why Poole was not looking for these signs. In the explicate order of reality, one is inattentive to the signs that matter most. People find themselves in a constant state of trying to solve problems that have very little to do with who or what they are. Bohm writes, "I mean everybody is caught up in his own little fragment, solving whatever he thinks he can solve, but it all adds up to chaos." Fragmentation leads to the inevitable desire to be totally "secure," and, as Bohm explains, "the fragment cannot possibly be secure."[39] The fragment is merely part of the whole; therefore, the security of one's holding one piece of the puzzle is not nearly enough. As I will argue, however, one fragmented piece of the puzzle may hold all the information an individual needs to be secure, so long as he or she is aware or conscious of the piece in the context of the holographic whole. Put another way, this is true so long as an individual recognizes the holographic implicate and explicate order in which the vital machine acts. Repetitive recognition leads to new forms of consciousness. In the case of *Homo sapiens*—repetitive recognition and awareness takes one from the primitive self to the cosmic self. One must recognize the laws that accompany cosmic consciousness to recognize how a vital machine operates in a double hologram of explicate and implicate existence.

What Dick achieves through Poole's dilemma is a portrayal of the mental chaos that develops due to existing in a fragmented society. Problems cannot be solved as long as they are parceled out in a fragmentary manner. Poole capitulates to the only thing a fragmented reality tells him he is when he says, "'I'm a freak . . . an inanimate object mimicking an animate one.'"[40] Yet herein lies the beauty of the consciousness (idealistic) approach to the mechanical-man myth, harking back to the earlier goals of the American transcendentalists. In his recognition that he is a fragmentary freak, a fool, or even a trickster, comes enlightenment. In those few words he utters about his perceived self, Poole discovers there is an unfolding implicate order that allows him to mimic an animated, living being. Poole's job becomes putting the pieces of his existence back together to form a new reality in which he can properly function while not feeling like a freak, devoid of free will.

PROGRAMMED AND PROGRAMMER

Conscious recognition that one is programmed by someone or something else throws more than just a mere disruptive apple into the garden when looking at control issues between humans and machines or programmed and programmer. "The Electric Ant" contains the first patent realization in American literature in which a mechanical-man wakes up or is conscious to his being programmed, as Poole recognizes that someone else is in control: "Programmed. In me somewhere, he thought, there is a matrix fitted in place, a grid screen that cuts me off from certain thoughts, certain actions. And forces me into others. I am not free. I never was, but now I know it; that makes it different."[41] The Steam Man and Moxon's automaton hinted that they had become aware of such control, but they never explicitly communicated their thoughts to the reader. Here is a mechanical-man explaining the depths of his conscious recognition that he is "not free" and "never was" due to his programming. The awareness of free will, or lack thereof, in one's conscious mind, changes one's reality—both subjective and objective. The idea that a program can cut one off from certain thoughts and certain actions also harkens back to Wiener's fail-safe idea of the programmer implanting some type of control device in a program that would curtail or limit entropy, thus being able to predict the programmed actions rather than just leaving the program to get from points A to B by any means possible. As previously discussed, the problems inherent in programming a mechanical-man in a causal reality leave infinite possibilities to such a programmed device in reaching a preconceived goal. However, suppose reality is not causal at all, but holistic. This would change how cyberneticists view control, communication, and information, eliminating the dilemmas of a materialistic, causal space-time cosmos. The recognition of consciousness progressing toward cosmic consciousness takes place at this moment of brain filter elimination.

While Edgar Allan Poe's General A. B. C. and Baum's Tin Man recognize that they are no longer human once their organic form has dissipated and transformed into a mechanical apparatus, Garson Poole's situation is quite different. He has been only human in reference to how he erroneously perceived himself as being human. This gives him a different vision of self-knowledge as a mechanical-man: "If I cut the tape, he realized, my world will disappear. Reality will continue for others, but not for me. Because my reality, my universe, is coming to me from this miniscule unit. Fed into the scanner and then into my central nervous system as it snailishly unwinds."[42] Dick begins to defragment Poole's thought process by allowing the character to communicate with himself. Poole recognizes that fragmented aspects of reality are akin to a tree with numerous branches. These branches and the leaves that form upon them are similar to deconstructionist or reductionist American culture in terms of science and technology working within that culture. Poole does not see this state of fragmentation as a plague or as malignancy, as some postmodern philosophers have opined; rather, apart from his feeling of shock and dread in this liminal phase of transition, he sees it as a necessary phase to cross the threshold from unpredictable entropy to controlled entropy, thereby recognizing the holistic patterns of reality. Controlling entropy for Poole begins when he reduces himself down to an organic robot, then to a miniscule unit fed into a scanner. With this fragmented knowledge of himself, he is able to extrapolate and recognize *his* reality. True reality is boundless, whereas the implicate and the explicate freely exchange symbols, code, and information.

Similar to the outdated filters of the brain, the boundary between the explicate and the implicate serves its purpose for people to realize the value of fragmentation and reductionism.* With this era now upon us, there is no need for such a boundary, and the two can once more form a singularity where the objective and subjective become one and what is above is also what is below. Per Dick "I have to go slowly, he said to himself. What am I trying to do? Bypass my programming? But the

---

*Arguably, Bernardo Kastrup, who works in the high-tech industry and writes about metaphysics and mind, aids us the most in our quest with his metaphors on the mind, holography, and the human brain. I've utilized the word "filter" quite a few times and will continue to use it. Before we understand what Kastrup means by "filter," we must examine his definitions for *ego, consciousness,* and *unconscious,* which we find in *Why Materialism Is Baloney*. "The ability to turn conscious apprehension itself into an object of conscious apprehension is what fundamentally characterizes our ordinary state of consciousness. In fact, my claim is that this is what defines what psychology calls the 'ego': the ego is the part of our psyches that is recursively and self-referentially aware" (106). Kastrup explores this topic in its entirety with the aid of the book *I Am a Strange Loop* by Douglas Hofstadter. On the unconscious, Kastrup states that, "any content of mind that falls within the field of self-reflectiveness of the ego becomes hugely amplified" (106). This is similar to when Rey looks in the mirror and sees an infinite amount of herself; her action of snapping her fingers becomes amplified, infinitely, in *Star Wars: The Last Jedi* (2017).

Kastrup explains in *Why Materialism Is Baloney,*

> [A]ny experience that falls within the scope of the ego is recursively reflected on the mirrors of awareness until it creates an unfathomably intense mental imprint. I submit to you that most things you are ordinarily aware of . . . are amplified like that. You don't notice it simply because you have become accustomed to these levels of mental amplification to the point of taking them to be the norm. . . . Now, if this is so, what happens to the experiences flowing in the broader medium of mind that do not fall within the scope of the ego? They do not get amplified at all. Therefore, from the point-of-view of the ego, they become practically imperceptible! This, in my view, is how we've come to speak of an "unconscious" segment of the psyche. There is no unconscious; there are only regions of the medium of mind whose experiences, for not falling within the field of egoic self-reflectiveness, become obfuscated by whatever does fall within the scope of ego. (107–108)

In *Dreamed Up Reality,* we now come to Kastrup's key statement about filters and a working hypothesis for the vital machine in the holographic simulation.

> [T]he most direct and efficient way to acquire knowledge about reality is through a partial and temporary disablement of the *filtering* mechanism

computer found no programming circuit. Do I want to interfere with the reality tape? And if so, why?"[43]

Asking "why" is the simple starting point for one asking a broader question: How do I adjust to my purposive ends? Why would someone want to interfere with his reality tape? The reason for Poole's yearning or passion to interfere with his reality tape is existence. The mechanical-man gains freedom by playing Russian roulette, placing himself in dire situations to activate cosmic consciousness.* Prior to

---

(*cont. from p. 239*) of the brain. Indeed . . . there is an abundance of empirical evidence through technologies like meditation, yoga, hypnosis, prayer, lucid dreaming, shamanic rituals, sensory and sleep deprivation, fasting or other ordeals, etc., people throughout history have been able to perturb their evolved brain filters and temporarily tap into a universal source of direct knowledge. . . . So the hypothesis I am postulating is the following: consciousness is a non-local field phenomenon not caused by, not reducible to, the brain, but simply coupled to the brain. All understanding and knowledge ever registered by a conscious entity survives ad infinitum in the field of consciousness as permanent experiences, or *qualia*. Therefore, all universal knowledge is, in principle, accessible by any conscious entity. It is the local attention filters of the nervous system, evolved as a form accessing this universal repository of knowledge. But through perturbations of ordinary brain operation, which partially and temporarily disable or bypass some of these filters, knowledge can be imprinted onto the brain—where it is later interpreted, articulated, and reported—through a process of quantum wave function collapse. (22–23)

Now you are beginning to see that Grace Hopper's heavy binge drinking and David Bohm's bouts of depression were fail-safes simply because they were trying to bypass their filters. This is the point in the quest where you the reader should be getting goose bumps from the synchronicities you are now beginning to see in the mechanical-man's quest and quite frankly, your own.

*Some writers confuse the *call* of the quest with the *return* when speaking of thresholds being crossed in Campbell's quest formula. To clear things up, there are multiple calls to the adventure and multiple calls to return, thus, numerous *returns*. Complacency is in fact measured when someone does not want to accept the call to return with the information he or she has gained. The individual is content in his or her state of stagnation.

Colin Wilson speaks of numerous experiences of refusal of both the call and the return in explicit detail, even telling the reader about a girl whose menstrual cycle was interrupted by her refusal of the call. The story that he delivers about the man

Dick's writing, people could only speculate and surmise why a pattern emerged with the appearances of self-destruction in the fictive and real spheres of the mechanical-man's quest. Dick intimates that this self-destruction is a means of communicating a Christ symbol or sacrificial hero archetype in terms of new consciousness; it is the consciousness self's recognition of being programmed—the moment of crucifixion. The mechanical-man intentionally interfering with what he knows to be his reality projector is in a state of rebellion rather than a state of mere curiosity. No matter how complicated existence may become for Poole, he finds that he still has choices and still has some modicum of control, which is where his joy in life lies: "Because, he thought, if I control that, I control reality. At least so far as I'm concerned. My subjective reality . . . but that's all there is. Objective reality is a synthetic construct, dealing with a hypothetical universalization of a multitude of subjective reality."[44] This is where the mechanical-man marks a new worldview of reality and consciousness of infinite subjective realties.[45] We will soon learn that while Poole is experiencing subjective reality, his new recognition of self as imagination has harnessed the power to create objective reality. It is finished!

---

playing Russian roulette is very telling about how he views the refusal and the state of consciousness one is in during the refusal. A patient is depressed and gets out of his depression that lasts weeks every time he plays Russian roulette, thus, giving him what Wilson calls a "peak experience," or in the vernacular we have used with the mechanical-man, a state of the sublime and transcendence.

What does all this have to do with the mechanical-man archetype and the spiritual ideals in America? Quite simply, the cowboy currently has no purpose, and if we look at his life from our cyber-techno perspective, we see him living a life of boredom and complacency in the Information Age; rather, the mechanical-man recognizes there is a void in the American Geist or over-soul. Like I point out in the primary text, the country has gone from being called the Alcoholic Republic to the United States of Opioids. Specifically, as Wilson puts it in *Super Consciousness,* "[t]he lesson is that boredom and lack of purpose are among the most destructive states we can experience" (25). Like Thomas Hughes says, we are builders, we are and have always been builders, it's just not the story we are told. We are gods in the garden. The mechanical-man will give us a recipe to get out of this rut, where "a new set of conceptions and motives" be[comes] a dominate force in our individual and collective quest(s).

The quest then takes one from the microcosm of fragmentation—looking into oneself—and extrapolates from the fragmented self into an enhancement of one's knowledge of the macrocosm—the self inextricably connected to all of reality. In *The Hidden Reality: Parallel Universes and the Deep Laws of the Cosmos,* Brian Greene explains,

> [M]uch as the Wizard of Oz's frightening visage was produced by an ordinary man, a rapacious black hole is the holographic projection of something equally ordinary: a bath of hot particles in the boundary theory. Like a real hologram and the image it generates, the two theories—a black hole in the interior and a hot quantum field theory on the boundary—bear no apparent resemblance to each other, and yet they embody identical information.[46]

The perception of a synthetic construct or a holographic image in the explicate order of reality does not signify that something extraordinary in the implicate order is producing it. To the contrary, something quite ordinary may be generating the image. The importance lies in recognizing that what people find in the implicate and what they perceive in the explicate, which is generated by what they find in the implicate, will embody the same information. This explains why the fictional oftentimes turns into the real. It also explains phenomena like déjà vu and the *paradox of solipsism,* a concept that is adequately explained as seen through the eyes of a vital machine acting or vibrating in a holographic simulation.

SEEING HOLOGRAMS

Poole's recognition of his likeness to a vital machine allows him to see the dilemmas he faces in a universe that appears to him only in explicate form but has its power source in the implicate. Channell recognized this when forming the vital machine worldview, noting that the organic worldview and the mechanistic worldview could not explain or work in a harmonious way with the implicate order of reality. This is where Bohm and Talbot's theory of the holographic universe comes into play. For now, it is only necessary to understand what the holographic simulation reveals about reality. Reality is merely a hologram created by a projector in another dimension of consciousness. It is akin to the brain merely being the medium, or screen, which projects the light given off by the projector (*imagination*). This light shines through the eyes to create the reality that the individual, at the very least, thinks he or she sees (imagines into the light to see first and second person experience of their choosing and focus localized by space and powered by time).

In "The Electric Ant" and some of his novels, Dick does not explicitly utilize the language of the holographic simulation, but he does intimate that the reality people perceive is an illusion or a projection from somewhere else. He does, however, use the term *holographic* extensively in one of his final novels, *VALIS*—after his encounter with the pink beam of light.

I do not like to use the word *illusion,* since contemporary American culture references the word as some type of magic trick on the senses that has no substance behind it.* The holographic simulation, however,

*The holographic simulation is not an *illusion* if by the word *illusion* we mean *not real.* Fields and the product yielded by fields that are outside the reach of our *normal* senses are not illusions. We need to get rid of the word all together. We need to focus on words like *imaginal, sacred, holographic,* and terms like *subtle information fields.* Read any book by Michael Talbot or Lynn McTaggart, and you will rethink the value, or lack thereof, when using the word *illusion* again. The word *imaginal,* which was coined by the late Henry Corbin, a professor of Islamic religion at the Sorbonne in Paris

> [D]escribed it, meaning a world that is created by imagination but is ontologically no less real than physical reality. . . . Because of the imaginal nature of the afterlife realm, the Sufis concluded that imagination itself is a faculty of perception (*read that one more time: imagination itself is a faculty of perception*). [This] contributed to the Sufis' belief that one could use visualization [focus and vitalism] to alter and reshape the very fabric of one's destiny. (quoted in Talbot, *The Holographic Universe,* 260)

Now we are beginning to see the mechanical-man progress from the views of

reveals there is a great deal of substance behind the projected reality people perceive. Poole thinks, "My universe is lying within my fingers. . . . If I can just figure out how the damn thing works. All I set out to do originally was to search for and locate my programming circuit so I could gain true homeostatic functioning: control of myself."[47] Understanding a reality model that grants the vital machine homeostatic functioning allows one to comprehend and explain control and communication between machine and human. In addition, it allows one to see the vital machine in the holistic form of the fragmented mechanical-man. The vital machine, placed in the holographic simulation, is the mechanical-man in fragmented form—automaton, robot, android, cyborg, thinking machine, or computer—put back together again as pure information or software. Are humans a fragmented aspect of a more complete whole or a mythic-marker similar to the mechanical-man? If so, which fragments are human, and for what purpose? Does the holographic simulation allow a human to defragment the self back into its original form? The holographic simulation answers these questions (however, keep in mind that the holographic simulation is still in its infant stages of development and experimentation). The mechanical-man is not only on a quest to defragment himself or find a "homeostatic functioning"; he must also defeat the purposeless appearance of

---

(*cont. from p. 243*) David Deutsch to that of David Bohm, Richard Feynman, Michael Talbot, and Bernardo Kastrup. In fact, on the imaginal, Talbot even says it is:

> [A] notion that parallels Bohm's implicate and explicate orders, the Sufis believed that, despite its phantasmal qualities, the afterlife realm is the generative matrix that gives birth to the entire physical universe. All things in physical reality arise from this spiritual reality, said the Sufis. . . . This realization is, of course, just another reference to the nonlocal and holographic qualities of reality. Each of us contains the whole of heaven. More than that, each of us contains the location of heaven. Or as the Sufis put it, instead of having to search for spiritual reality "in the where," the "where" is in us. (261)

I like to think of the information or map to where Luke Skywalker is in *Star Wars*: *The Force Awakens* when I read the section in *Holographic Universe* under the subheading "The Land to Nonwhere." Luke Skywalker, including his location, is already in us. We learn this to be true in *The Last Jedi* when Luke tells Rey it's vanity to think others can't tap into the force.

reality. For instance, Poole finds that controlling himself in the reality he now knows to be merely perception is the most important aspect of life, which draws parallels to Villiers de l'Isle-Adam's philosophies on reality and perception. Control then equates to meaning in an existentially purposeless universe. If one can find a way to control the self by recognizing the conscious state he or she is expressing, it follows one can control one's reality.

As I discussed referring to the chicanery of a chess-playing hoax—communicating to a culture how simple it is to delude the mind—once people believe in some phenomenon and once their minds are made up, that phenomenon becomes part of their physical reality. The correlative functionality of the fictive and the real allows one to see how the implicate and the explicate interact and coevolve. The implicate is the infinite conveyor belt of original software, some of which people call archetypes or myths; the explicate is the way in which people see the projected software through the patterns of their own matrix.* The implicate demands from science and technology "a kind of disciplined introspection that critically assesses not only the elements observed, but also the observer, the process of observation, and the interplay between the three in a holistic manner; an introspection that, as such, seeks to see through the 'game.'"[48] Our consciousness progresses from merely

---

*In Jonathan Lippe's *Simulation Theory,* we get a definition on par with the 1999 film *The Matrix.*

> The Matrix is a term with many meanings. The matrix is a system. The Matrix is control. The Matrix is a construct. It can refer to the artificial Universe we live in. We also use it to refer to a paradigm construct of reality a person or group of people subscribe to. (What a person or people believe in.) It is the manufactured reality created by the media and perpetrated by society, the social norm or the status quo. It can also refer to a public network like The Internet. And it almost always refers to the supercomputer that calculates every possible variable in existence every moment. (10)

This is a very popular definition of a matrix but not The Matrix. For our purposes we just simply need to think of The Matrix in terms of the defining characteristics of control, communication, and information in its simulated form under the conditions of consciousness a person or a collective perceive with.

knowing the rules of the game to recognizing one's awareness of being as the conjuring "sorcery" of reality.*

With the holographic simulation of the universe, everything with which people see and interact is part of the projected image of the implicate order. In effect, there are two realities: "one in which our bodies appear to be concrete and possess a precise location in space and time; and one in which our very being appears to exist primarily as a shimmering cloud of energy whose ultimate location in space is somewhat ambiguous."[49] This begs the question of what the mind is and where it is located. In the holographic simulation, the mind is not in the brain, nor can it be considered the conscious. Trusting in the metaphor that the brain is simply a complex computer that people once—prior to the computer's invention—compared to a machine is the folly of believing that the explicate order of reality, or the appearance of things, is finite.

In terms of mechanical-men, many of these characters do not have brains or devices that imbue them with the process of reflection. However, they do exhibit a life "force." In the holographic simulation, that is precisely what the mind is: it is a force, no different from gravity or magnetism. It is a force projected from the implicate order of reality. For materialists, when something is projected through the brain and eyes, they understand the mind as having a location in the brain and the images as being transferred to the brain via the eyes. Yet years of research have shown this not to be the case.[50†] Tesla recognized this aspect of the

*Sorcery and imagining are now becoming interchangeable and synonymous for the mechanical-man. As we will see later in the quest, the imaginaries are our modern-day alchemists and sorcerers in the real and the fictive, transubstantiating the fictive into the real.

†Talbot explains in *The Holographic Universe:*

> At Yale, Pribram continued to ponder the idea that memories were distributed throughout the brain, and the more he thought about it the more convinced he became. After all, patients who had had portions of their brains removed for medical reasons never suffered the loss of specific memories. Removal of a large section of the brain might cause a patient's memory to become generally hazy, but no one ever came out of surgery with any selective memory loss. . . . To understand why Pribram was so excited, it is necessary to understand a little more about holograms. (13–14)

mind force when he read Sir Edward Bulwer-Lytton's *The Coming Race* and became fascinated with the Vril force that Bulwer-Lytton indicated was used to give life to underground super-beings. Indeed, the founders of theories of electricity, magnetism, and atomic energy, on the whole, were not searching for these energy sources; rather, they were seeking the force people call life—consciousness itself. However, once they unleashed these other forces, the implacable fragmentation of the explicate order did not allow people to go back to looking for forces of life. Instead, people began to look for life in the fragmented areas of biology, chemistry, and genetics.

> I think we have way overrated the brain as the active ingredient in the relationship of a human to the world. . . . It's just a real good computer. But the aspects of the mind that have to do with creativity, imagination, spirituality, and all those things, I don't see them in the brain at all. The mind's not the brain. It's in that darn field.[51]

In the explicate order, the metaphor of the brain as computer or advanced machine works quite well if a human being is limited to menial calculation tasks or cause-and-effect dicta. But where unexplained creativity, imagination, and spiritual awareness are found, the metaphor of the brain as computer creates a void and the ultimate gap in our human genesis-mythos. In the holographic simulation, the brain is a mere filter and transmitter for the interchange between the implicate and the explicate within the confines of a given consciousness. I hypothesize that the conscious mind contains the functionality of the brain likened to a computer in the explicate order of reality, which is that aspect of reality that people are able to see. The unconscious mind contains the unexplainable energy field embedded or hidden in the implicate order that people are unable to see, but that, nonetheless, comes through in dreams, writing, creativity, and spirituality. This may lead to the axiom that there is no truth. To the contrary, the truth lies in the boundary or twilight zone region where the fictive and the real blur any and all polarities into spectrums for the imagination to choose to "escap[e] the pains of the lower planes by vibrating on the higher."[52]

In analyzing the mechanical-man quest and the interaction between machines and humans, I come back to the question of which makes the decisions: the conscious or the unconscious mind. In addition, I ask whether free will is an illusion. Tesla and Edison thought that free will was an illusion, but their lives clearly showed a creative passion to circumvent this hindrance. Bohm believed that "one of the basic tenets of quantum physics is that we are not discovering reality but participating in its creation. It may be that as we probe deeper into the levels of reality beyond the atom, the levels where the subtle energies of the human aura appear to lie, the participatory nature of reality becomes even more pronounced."[53] People have to then be careful in saying they have made a great scientific discovery when in fact they may be creating the very thing they have found.

Benjamin Libet and Bertram Feinstein measured the time it took for a touch stimulus on a patient's skin to reach the brain as an electrical signal. The patient was also asked to push a button when he or she became aware of being touched. Libet and Feinstein found that the brain registered the stimulus in 0.0001 of a second after it occurred, and the patient pressed the button 0.1 of a second after the stimulus was applied.

But, remarkably, the patient didn't report being consciously aware of either the stimulus or pressing the button for almost 0.5 of a second. This meant that the decision to respond was being made by the patient's unconscious mind. The patient's awareness of the action was the slow man in the race.[54]

What kind of control do we have over our lives? How much are we the victims of fate? These are big questions that have plagued man since his consciousness awakened to his own mortality and imagination two thousand years ago. What the holographic simulation shows is that history, as an evolutionary mode of vitalism, must coexist with humans in the quest of defragmenting and then returning to wholeness. If the holographic simulation holds true, then the imagination is actively creating reality by bringing out aspects of the implicate into the explicate. Fragmenting reality is one's attempt to uncover the filters that create reality or that hold us back from creation. But once this is achieved, a new self-actualization must take place. Defragmentation is, then, the process of discarding the filters that one has self-actualized and putting the puzzle back together, devoid of such outdated filters. It is a process of reverse selective evolution or engineering. In the end, the mind itself is the ultimate trickster. It is the courier between the implicate and the explicate; it negotiates between its conscious and unconscious "I AM."* The evolved mass called the brain is merely a hindrance to this negotiating: with its cultural and biological filters, which focus on how to survive in the explicate, the brain is ill equipped to deal with the

---

*From perhaps the shock of two world wars and an economic depression, something happened to the American psyche. It was as if the transcendentalist movement of Emerson, Whitman, and Thoreau never happened and the Metaphysical Club's members like William James and his *Varieties of Religious Experience* were wholeheartedly shunned away. We are not asking why this occurred. Was it the influx of existentialism brought on by the Cold War and mutual atomic destruction? The mechanical-man does not set out to answer these questions. But what we do uncover is that these ideas survived in another form, what we would today call *New Age* or *Self-Help* works. However, even scholars should not be afraid to call it what it really is, American mysticism. American mysticism from the 1930s to the 1970s preserved this work. For a history of this, read Mitch Horowitz's *One Simple Idea*. It's an aspect of American history that is just now, in the second decade of the twenty-first century, being resurrected.

Once the culture wars began in the late 1960s and early 1970s, after the deaths of Robert F. Kennedy and Dr. Martin Luther King Jr., this American mysticism was stamped out of any scholarly discussion, manipulated, and radically changed by mega-church ministers and self-help gurus—these false prophets changing the law of assumption to fit their dogma or ultimate goal of power and domination. In this chapter and those that remain, we will mainly cite from one of the most important mystics to ever come to the United States, Neville Goddard. The primary text of this quest is not going to give you Goddard's background, because his contribution to the mechanical-man's journey is not in the quest itself, but in the boon of the quest (what the mechanical-man brings back in the form of a new archetype and consciousness). Needless to say, he came to America at the age of 18 to study drama. He was born in 1905 in Barbados, British West Indies. On man recognizing his consciousness, he presents it in a transcendent yet pragmatic way, "I AM the eternal Nothingness containing within my formless self the capacity to be all things. I AM that in which all my conceptions

implicate; nevertheless, we must use our manipulation of the explicate in imagining crossing over to the implicate.

Poole discovers his filters when he is fidgeting with his reality tape. Dick depicts him as echoing the sentiment of defragmentation:

> "Maybe we could learn to. Learn to be selective; do our own job of perceiving what we wanted to and what we didn't. Think of the possibilities, if our brains could handle twenty images at once; think of the amount of knowledge which could be stored during a given period. I wonder if the brain, the human brain—" He broke off. "The human brain couldn't do it," he said, presently, reflecting to himself. "But in theory a quasi-organic brain might."[56]

Poole begins to explore his bodily apparatus while asking questions of the computer belonging to the company for which he works. The computer responds to his query: "'The punched tape roll above your heart mechanism is not a programming turret but is in fact a reality-supply construct. All sense stimuli received by your central neurological system emanate from that unit and tampering with it would be risky if not terminal.'"[57] Poole's actions indicate a will to discover the ineffable origins of his own explicate universe. He understands his explicate reality to a point—recognizing it as the day-to-day images and sense impressions that he is able to experience. Creating a twofold model for himself to analyze is what allows him to cross

---

(*cont. from p. 249*) of myself live move and have their being, and apart from which they are not."[55] Israel Regardie interprets this for us by writing:

> Neville uses it (I AM) because if we would define ourselves at all, we must use I AM before we can further qualify it in any way. Before I can say what I am, I must first have said I AM. Before I can assert that I am a man of such and such an age, of a certain race, residing in a certain country. . . . I must say I AM. . . . I can condition or formulate this limitless expanse of abstraction by enclosing it with the limitations [and] it can express itself through a variety of masks, play an infinite number of parts, adopt a maximum of possible roles. But it remains nevertheless, unconditioned and unformed I AM. (quoted in Goddard, *Imagination,* "Neville: A Portrait by Israel Regardie," 9)

the threshold and experience for himself how or what he really is in the implicate form of pure information prior to manifesting in the explicate.

At one point, when tampering with the tape, Poole is able to experience nothingness—the area that stands between what is seen and what is not seen:

> Reaching, he tried to touch something. But he had nothing to reach with. Awareness of his own body had departed along with everything else in the universe. He had no hands, and even if he had, there would be nothing for them to feel.
>
> I am still right about the way the damn tape works, he said to himself, using a nonexistent mouth to communicate an invisible message.[58]

This is the first instance in the mechanical-man literary mythos of the idea that messages are communicated by nonexistent beings or nothingness. Although his messages are now "invisible" in the explicate order of reality, this mechanical-man still exists in the implicate. The light that projects his hologram has been turned off, but that does not mean he cannot communicate. He is still able "to communicate an invisible message." In a holographic universe, depending on the dimension in which his original self is stabilized, he sends out frequencies or vibration, thereby creating "an invisible message" without a "mouth." Of course, in the holographic simulation one does not have to abide by peoples' explicate definitions of space and time. When Tesla was causing his telautomaton to move with his remote device, his audience could not see the communicated messages; nevertheless, in the wireless paradigm in which the audience members were placed, they did not have to see the invisible messages to know that such communication was taking place between Tesla and the boat. Richard Feynman laughed when he saw the barbed wire and fences installed at Los Alamos, knowing full well that ideas could easily seep through matter and travel over the vast oceans and be picked up by German competitors. Feynman could see that all ideas and pieces of information are present everywhere in the holographic simulation. Mere matter cannot block the transit of information and ideas.

What in one culture was invisible or impossible can be visible in another epoch or culture. Through technology, invention, and the imagination, this coevolution of the real and the fictive is the formula with which to chip away at any boundary or wall between the implicate and the explicate that separates realized potentialities in differing cultures. The appearance of the manifestation may look different with manifestations of the implicate in differing cultures. However, the origins of such manifestations are the same when it comes to sharing identical information. Poole realizes that when he manipulates his reality tape, he has more choices. It is as if the laws of space and time no longer apply. He is able to travel to the past and the future, and he sees and experiences numerous aspects of his life's possibilities. It is not as if these possibilities were hidden or trapped in the ineffable. Originally, Poole does not have the requisite knowledge or necessity to experiment with his reality projector. Once he begins to experiment with it, he learns how to manipulate his own code, controlling it in such a way as to communicate with his past and future self. Knowing that he is an electric ant "opens up certain possibilities to him."[59] He is able to make figures appear and reappear, fly to different places, and make his bodily apparatus become invisible. The boundary between the finite—the transient nature of reality—and the infinite—the essence of reality—are blurred. Poole is experiencing William James's definition of cosmic consciousness.

Bohm's consideration of the finite and infinite shows that people's observations of "the field of the finite" are inherently limited due to the transient nature of the explicate.[60] He holds:

> The field of the finite is all that we can see, hear, touch, remember, and describe. This field is basically that which is manifest, or

> tangible. The essential quality of the infinite, by contrast, is subtlety, its intangibility. This quality is conveyed in the word *spirit,* whose root meaning is "wind, or breath." This suggests an invisible but pervasive energy, to which the manifest world of the finite responds.[61]

Poole's recognition of himself as a vital machine and not merely a machine or an organism connotes the recognition of an underlying spirit or energetic vitality that powers the conscious and the unconscious mind. This type of vitalistic propellant is different from steam, wind, electricity, or atomic energy. This type of energy is everywhere, and it is the invisible force that activates consciousness and spirit. Poole is in that strange twilight zone between consciousness and matter—the hyphenated domain between the imagination and the outer world of objective reality. The trickster motif in Poole's actions, and Dick's ideas, shows that these two worlds are not implacably separated, as people previously thought. Causality is destroyed when the boundary between the real and the fictive is purged. This blurring of consciousness and matter challenges everything that we have been taught about the material universe. Our task here is not to accumulate material things as they are; rather, the accumulation of levels of consciousness is the goal of the quest. We may view these levels of consciousness as a kind of Jacob's Ladder leading us higher and higher so that we recognize better who we are in terms of service and spirit, or we may view them as merely an infinite array of subjective experiences—but how we view them makes no difference with respect to what is brought back from the quest. Staying stagnant and immobile on the horizontal or vertical planes of consciousness constitutes absolute refusal, absolute sin, and thus absolute hell—doing so is a Black Iron Prison (see pages 278–79) where the fragment is eternally separated from the whole. In this hell there is no respite for the weary warrior. Conversely, heaven is the following:

> It is pure experience, and it can go anywhere it wants, anywhere it doesn't get in its own way. . . . We are here, at last. Consciousness is not an illusion. We are rooted in it grave and hard. And it is our job, in fact our only job, to sense it, to reach toward it, to find and touch aspects of it and bring them into ourselves into the world.[62]

INVENTING THE SELF AND REALITY

An important aspect of holograms in relation to the mechanical-man is this: If a piece of holographic film that contains an image of a mechanical-man is cut in half or into numerous pieces, the entire mechanical-man can still be reconstructed from information or code in each fragmented piece of the film. This is quite different from a normal photograph of a mechanical-man. Imagine looking at a fragmented picture of a mechanical-man as a puzzle. With only one piece of the puzzle, it is hard, if not impossible, to reconstruct the whole. However, holography shows that every piece of holographic film contains all the information necessary to create the whole image. People learn this by fragmenting the pieces of the picture, reducing the picture down to the original blueprint that is embedded in each piece of the whole. Thus, in every fragmented aspect of the self and reality is the blueprint of the whole—the story of the past, present, and future. The mechanistic worldview must now and forever be replaced by a worldview more germane to humanity's ever-evolving collective consciousness in treating our collective amnesia. The mechanical-man is a physician placed here to treat this amnesia.

Through the years, research has shown it is also possible to hold the axiom that the brain acts as holographic film, thus containing all of the information needed to recall a whole memory in any part of the brain. The use of the word *memory* here refers to an image from the past or future—think of this in terms of déjà vu, recollection of past lives under hypnosis, fortune-telling, and the nature of serendipitous coincidence. Thus, when Poole says, "'Here I have an opportunity to experience everything. Simultaneously. To know the universe in its entirety, to be momentarily in contact with all reality. Something that no human can do,'"[63] he has in effect tapped into his own holography and become able to retrieve these memories of the past and the future. Not only can Poole retrieve them, but he is also able to experience them and become part of the hologram itself. His code of information in the implicate is everywhere in the explicate. His amnesia is cured.

Poole's quest ends with a conversation between himself and his coworker Sarah, who has been attempting to quell his anxieties about the realization that he is an electric ant. Even she is not immune to his new consciousness:

> "They weren't real," Sarah said. "Were they? So how—"
>
> "You're not real," he told Sarah. "You're a stimulus-factor on my reality tape. A punch-hole that can be glazed over. Do you also have an existence in another reality tape, or one in an objective reality?" He did not know; he couldn't tell. Perhaps Sarah did not know either. Perhaps she existed in a thousand reality tapes; perhaps on every reality tape ever manufactured. "If I cut the tape," he said, "you will be everywhere and nowhere. Like everything else in the universe. At least as far as I am aware of it."[64]

Exchanging the phrase *reality tape* for *projected hologram* provides an understanding of Dick's grappling with concepts that have not entered the scientific community's regular vernacular. He could have titled this short story "The Holographic Ant" or "The Vital Electrical Ant." Yet, without getting caught up too much in the conundrum of language problems and misinterpretation, there is a reflectivity in presently defined worldviews of the mechanical-man as vital machine and the reality in which he is now placed, namely, the holographic simulation of implicate and explicate exchange.

Only time will tell if these are the final defined bedrocks of explanatory means to human-machine interactions. Whatever the case may be for future *explananda* and *explanantia* in the realm of cybernetics, the vital machine and the holographic viewpoint help explain the implicate and explicate aspects of control, communication, and information expressed in terms of both the real and the fictive in a way previously thought impossible.

> My hands, she thought. [Sarah] held them up. Why is it I can see through them?
>
> The wall of the room, too, had become ill defined.
>
> Trembling, she walked back to the inert [R]oby, stood by it, not knowing what to do. Through her legs the carpet showed, and then the carpet became dim, and she saw, through it, further layers of disintegrating matter beyond.
>
> Maybe if I can fuse the tape-ends back together, she thought. But she did not know how. And already Poole had become vague.
>
> The wind of early morning blew about her. She did not feel it; she had begun, now, to cease to feel.
>
> The winds blew on.[65]

Sarah's final thoughts present readers with the last piece of the fragmented puzzle in analyzing the mechanical-man as mythic-marker within the holographic simulation. She intimates that even she is part of Poole's world and thus part of the projected hologram whence he is unfolded. In differing degrees this means that everyone, in some form or another, is a hologram or an electric ant, even the corporate tycoons who gathered to create Poole's reality tape. The dilemma for readers is the question of how to traverse back and forth between the implicate and the explicate or, better yet, merge the two. We will soon discover that paradox is the only way to traverse both without going insane. Paradox itself arises from self-reference and self-reflectivity. In the same way that we speak of the implicate and explicate as being separate and yet the same, "it is by referring to themselves that paradoxical statements close the 'strange loop.'"[66] Douglas Hofstadter discussing the work of Dutch artist M. C. Escher's 1940 woodcut titled *Metamorphosis II* adds to this understanding when he surmises:

> One level in a drawing might clearly be recognizable as representing fantasy or imagination; another level would be recognizable as reality. . . . For any one level, there is always another level above it of greater "reality," and likewise, there is always a level below, "more imaginary" than it is. This can be mind-boggling in itself. However, what happens if the chain of levels is not linear, but forms loops? What is real, then, and what is fantasy.[67]

One must read this quote a few times to understand its meaning and how it further relates to Feynman's "wondering why [he] wonders why [he] wonders" before moving on to the final chapters.[68]* Paradoxical loops are the tools we use to enter and exit Carl Jung's caves of the individual and collective unconscious (the Force). Prior to entering the cave, I leave you with David Bohm's final words to be read at his memorial service.

> In considering the relationship between the finite and infinite, we are led to observe that the whole field of the finite is inherently limited, in that it has no independent existence. It has the appearance of independent existence, but what if appearance is merely the result of an abstraction of our thought. We can see this dependent nature of the finite from the fact that every finite thing is transient.
>
> Our ordinary view holds that the field of the finite is all that there is. But if the finite has no independent existence, it cannot be

*This quote by Feynman is indicative of the thinking process of all of our mythological heroes and heroines.

all that is. We are in this way led to propose that the true ground of all being is the infinite, the unlimited; and that the infinite includes and contains the finite. In this view, the finite, with its transient nature, can only be understood as held suspended, as it were, beyond time and space, within the infinite.[69]

9
The Cave

The Dagobah cave scene in the film *The Empire Strikes Back* is the apogee of the mechanical-man quest. The movie had only been out for eight months prior to the inauguration of the "great communicator" on January 20, which was only a month and a half before the death of Philip K. Dick. The Dagobah scene is the conclusion of Plato's "Allegory of the Cave," in which Plato "puts into symbolic form his view of the human condition, and especially of human knowledge, in relation to reality as a whole."[1] In Plato's allegory human beings are imprisoned in their own bodies, unable to see the essence of others or themselves. They only see the shadows of the truth. Moreover, their direct experience is not of reality, but of what is in their minds. Plato writes within the context of a culture we would consider devoid of the potentialities with which modern technology imbues our own. The creation of the United States and postmodern fragmentation had not yet taken place. While the catalysts for fragmentation and reductionism are still being debated today, the processes themselves were precursors for the mechanical-man's appearance in the real and fictive within American culture. The mechanical-man metaphor and the American techno-mythos allowed George Lucas, aided by Joseph Campbell, to complete Plato's allegory and show how the mechanical-man archetype brings people out of their captive state, takes them into the implicate, and allows them to bring back to the explicate the energies they tap into and harness.

Profoundly, the Dagobah cave scene is the only piece of film in the original *Star Wars* series in which slow motion is used. It is also the only scene in which the action shown is not really taking place; rather, it takes place in a vision or a dream. Janice Rushing sets the scene up in *Projecting the Shadow:*

> Looking for the great Jedi Master Yoda, the young Luke Skywalker happens upon a toadlike dwarf with peculiarly expressive ears lurking in a primeval forest. Expecting someone physically impressive, no doubt, Luke misses that he stands in the presence of greatness and is irritated by the gnome's impolite meddlesomeness and his slowness to lead him to the guru. The spirit eventually reveals himself to be Yoda, of course, and at the bequest of Ben Kenobi, but against his own better judgment, takes Luke on as Jedi trainee. After a particularly grueling tutorial, Luke feels a chill, suspects that the forest has eyes, and is compelled to search out the danger. As Luke reaches for his light saber, Yoda tells him he won't need it, but Luke straps it on anyway. Following his instinct, the boy descends into a dark underground cave, where his enemy, Darth Vader, mysteriously materializes before him. Both draw their weapons, a laser fight ensues, and Luke neatly slices off Vader's head. Tentatively, Luke approaches the hooded black thing on the ground and is startled to see his own face staring back at him from the shadows.[2]

Plato's cave allegory served its own purpose during his era. However, such an allegory grew outdated in the age following the period of fragmentation, that of the holographic and vital machine models. People required a new cave allegory in order to elucidate the end of fragmentation and disunity as well as the beginning of holism in the form of the collective unconscious being brought out into the explicate to mirror cosmic consciousness. The mechanical-man, in its real and fictive form, has enabled the vital machine—mankind fused with technology—to move back and forth between the implicate and explicate, evidencing the collective unconscious or, in the *Star Wars* universe, the Force.

Not only does the Dagobah scene present the hero archetype in the trinity of control, communication, and information (through Darth Vader, Yoda, and Luke Skywalker respectively), it also explains how the implicate and explicate worlds of inside and outside the cave interact and coevolve. In addition, the scene shows that the framework with which to understand the implicate and explicate order of reality is best seen through Campbell's mythic formula of departure, initiation, and return. The cave scene foretells what happens after the return—after the mechanical-man has ventured through the initiation stage of fragmentation. In effect, the trinity of Yoda, Darth Vader, and Luke illuminates how fragmentation ends, which opens the door for viewers to leave the explicate, enter the implicate, and return to the explicate with knowledge of the implicate. In terms of American science and technology, the garden has gone through the stage of fertilization and specialization. The fragmented fruits of people's labor are ready to be collected and placed in a holistic pot from which the plurality may feast. Nonetheless, critics said and still do say that science practiced in the American garden "cannot answer the questions of Who are we? What is history? and What does it push forward? . . . since it captures only relationships, its answers ultimately entail circularity and are not fundamentally satisfying." This is proof that our consciousness has changed and is still changing in order to answer these questions.[3]

BACK TO THE GARDEN

Yoda initiates viewers back into the origins of the garden, prior to European exploration. He is the archetype of the wise figure: "in temperament he resembles the Native American Shaman—the holy man or medicine man."[4] In the form of a fragment, Yoda is pure information; therefore, he is a mechanical-man. Yet Yoda is also an archetype brought from the implicate into the explicate in the form of fiction. In reality,

Yoda is both man and puppet. Nevertheless, the members of the audience never witness the man behind the curtain or the controller of the puppet; rather, the puppet itself, like the motorized as grammar error the film version of *E.T.* (1982) with his glowing heart light is the one in control of communicating spiritual knowledge of the holographic reality. Yoda is reminiscent of this study's first mechanical-man, the chess-playing hoax. Although there is a man underneath Yoda, operating his movements and speech, the fictive-world version of Yoda is instructive about how the vital machine operates in the holographic simulation. The fact that a man controlled the actions of the chess player, or that Frank Oz controlled the voice and actions of Yoda, is of little concern compared to consideration of the implicate realm into which both automaton hoaxes led their audiences. In both, viewers could recognize their potential to tap into hidden or forgotten natural powers and energies—a new dimension of consciousness.

On the archetype of Yoda, Michael J. Hanson and Max S. Kay hold:

> Yoda's message to Luke is short and simple: to look beyond the physical world and to live in a greater one, the world of the Force. Obi-Wan's statement in ANH [A New Hope]—"You've taken your first step into a larger world"—was the beginning of this journey for Luke, and Yoda's teachings are designed to continue this progress. The discovery of a greater power within oneself is the essence of the hero's journey.[5]

Luke Skywalker, in the form of fragmentation of the American mythos, is the hero archetype or cowboy of the American West. In addition, he is the farm boy who joins the rebels to destroy the evil empire. He learns a new religion, discarding the orthodoxy of the old. He is a mixture of a mythic and a real figure. One can look upon him as a George Washington, Daniel Boone, Abraham Lincoln, as fragmented form. He is the embodiment of the American Revolution, the American frontier, and the American dream. His adjectival qualities are "brave, impetuous, pragmatic, individualistic, and competitive," the very qualities upon which American science and invention are based.[6] In terms of the mechanical-man, Luke is a pure form of communication between the implicate and the explicate. He is the trickster or the boundary crosser. Like Philip K. Dick's Garson Poole, Luke shows his audience members the true aspects of their reality and brings them both inside and outside the cave, showing them both aspects of the holographic simulation. He is the nexus between Yoda and Darth Vader, the link between information and control in the implicate. He is not technology or machine; rather, he receives aid from a machine. Like many of our mechanical-men before him, he moves from being fully human to recognizing his status as a vital machine, first in the cave and last when Darth Vader severs his hand. He is able to explain Yoda's knowledge of the implicate and Darth Vader's knowledge of the explicate through his actions and dreams.

Darth Vader (Anakin Skywalker) is even more important than Luke in this parable. It is easy to skim the surface of *Star Wars* and opine that Darth Vader is Luke's opposite or *shadow*. However, in terms of cybernetics, Darth Vader is the form of control bringing Luke into existence literally, as his father, and metaphorically, as his savior. In terms of the mechanical-man myth, Darth Vader is the only character the audience can fully grasp as a mechanical-man in the explicate form: He is a cyborg. In terms of taking us back to the origins of the American garden, Darth Vader is a reflection of the Christian Nation, the Millennial Nation, and the Chosen People—he is the shadow or opposites of those ideals (Vietnam, prospect for atomic holocaust, civil rights violations likened to systemic racism). He takes it upon himself to write the saga's script, knowing full well that in the end he will be the one to bring peace and harmony to the universe by sacrificing himself for his son—humanity. By fusing himself with technology, he becomes God—control—and takes on the responsibility of showing his progeny the implicate order of reality. Only through Darth Vader's suffering is Luke—humanity—able to see the implicate, thus becoming Jedi—activated imagination. Darth Vader has already returned from his quest into the implicate. He recognized his destiny and authorized the life of Darth Vader when he was Anakin Skywalker.

> Anakin is likened to Buddha, because after he redeems the world, he undergoes an apotheosis, becoming the object of worship for millions. In essence, he has become a myth, a model for all men, an everyman. In this state, the Buddha is constantly meditative and in tune with the energies of life. . . . In the end, Anakin fulfills his destiny, bringing closure to his life and to the tasks the Force itself has demanded of him.[7]

Hanson and Kay understand the importance of Anakin as both Buddha and Jesus. Anakin must become Darth Vader in order to redeem the lives of others. We must all recognize our dark side before we can bring back the boon of knowledge from the implicate order of reality.

As Carl Jung points out in his autobiography, the few individuals able to see and enter the implicate often never come back to the explicate. Of Frederick Nietzsche in particular, he explains:

> [T]hese were actualities which made demands upon me and proved to me again and again that I really existed, that I was not a blank page whirling about in the winds of the spirit, like Nietzsche. Nietzsche had lost the ground under his feet because he possessed nothing more than the inner world of his thoughts—which incidentally possessed him more than he it. He was uprooted and hovered above the earth, and therefore he succumbed to exaggeration and irreality.[8]

Like Nietzsche, Anakin Skywalker is ill equipped to deal adequately with the implicate, which "possesse[s] him more than he it." That is not to say, however, that the experiences of Anakin and Nietzsche cannot teach the audience something of value. Their quests are part of the audience's own quest, in that they illuminate for others how to exist in the explicate in order to understand and know the implicate reality, albeit in a negatively charged experience. The mechanical-man in the forms of the Steam Man, Moxon's automaton, HAL, and Robby the Robot, to name a few, meet the same fate as Anakin and Nietzsche. They are incapable of returning to the explicate after their quest in the implicate. The audience's perception of their demise is in the form of suicide, going mad, or turning to the dark side. What is actually occurring to these archetypes underneath the audience's perception is their inability to return to the world of perception and measurement with the knowledge they have gained in the implicate—they are "blank page[s] whirling about in the winds of the spirit."[9] They are unable to return to their previous state of consciousness.

The ultimate question is how one returns to the explicate, with the knowledge one has gained in the implicate. When Luke comes back from the cave, Yoda tells him he has failed. Why? Luke went into the cave devoid of the recognition that he is a vital machine inhabiting a holographic simulation. Only after Luke loses his hand is he fully capable of understanding and recognizing his status as vital machine operating in a hologram. In order to bring back from the implicate the natural forces, information, and archetypes one finds there, one must recognize one's own status and reality prior to entering, similar to that of Garson Poole. By bringing his weapon into the cave, Luke fails to recognize this. He enters the cave as a materialist and he leaves the cave as a materialist—his initiation has not yet come. There are many more ordeals he must face. Only after his battles with his father in *The Empire Strikes Back* and *Return of the Jedi* does he fully recognize his implicate self by acting out the paradox of service and sacrifice alongside Darth Vader, recognizing Anakin Skywalker as the vital spark within Vader.

GEORGE LUCAS: AMERICA'S SCIENTIFIC PAST REACHES ITS FULL POTENTIAL

On June 12, 1962, everything changed for the coevolution of the real and fictive. On that hot summer day, George Lucas, controlling his favorite mechanical-man (a small Fiat), was overtaken by a Chevy Impala. The Fiat bounced a few times and ended up wrapped around a tree. That day changed everything for Lucas, whose affection for automobiles and the American mythos of rugged individualism led him in a different direction than he had planned. Biographer John Baxter says of Lucas's survival, "Had the seatbelt not snapped, had he not been thrown clear, had he landed on his head rather than his chest, [Lucas] would almost certainly have died in the road in front of his own home."[10]

For Lucas, the mechanical-man that went by the name "Fiat" most certainly saved his life. The impact of the car into the tree was so great that "the tree, roots and all, shifted two feet."[11] Convalescing from his injuries allowed Lucas to take a different perspective (path of consciousness). His interest in television and cars shifted to a larger scene:

> A lot was happening in the world, and now with leisure to contemplate it, Lucas may have seen his life in a new light. During 1962, John Glenn became the first American to orbit the earth; John Steinbeck won the Nobel Prize for Literature; in October, John Kennedy faced down Nikita Khrushchev over the missiles he'd sneaked into Cuba; America exploded a nuclear device over Johnston Island in the Pacific; Polaroid launched a new one-minute color film; and the first Titan inter-continental ballistic missile was installed in a concrete silo in the American heartland, targeted on Russia.[12]

Maybe a mechanical-man sacrificing itself to save Lucas was not as important as Lucas now seeing himself as working on borrowed time. With Lucas's adoration and love for American dreams and inventions, the mechanical-man would now have a new champion, one who would soon fuse science and myth into a new American creation story. Like many others of his generation, Lucas would show the world that American science and technology were no longer limited to the military, corporations, and academia. Science and technology were now free to grow in new gardens like parking lots, backyards, and even garages. The circle was complete—the fragmentation of the sciences had reached its apex and could be viewed holistically. During the late 1970s and 1980s, American science and technology's ability to reboot and start up once more led people back to looking for independent inventors like Tesla and Edison. People in garages and backyards came back to their country's independent, creative origins of authoring the impossible.

Lucas followed many of the same quest-like patterns of this study's other real, heroic mechanical-men. Rather than deal with his existential anxiety and depression later in life, like Bohm and Hopper did, Lucas recalled dealing with mental anguish centered on the question of God's existence and reality at the age of six: "It's as if you reach a point and suddenly you say, 'Wait a second, what is the world? What are we? What am I? How do I function in this, and what's going on here?'"[13] Lucas found solace from his bouts of anxiety in making models. Harking back to the days of the independent inventors, he spent countless hours building chess sets and dollhouses. His childhood friend Janet Montgomery remarked, "George [Lucas] made an entire dollhouse out of a cardboard box for my Madame Alexander doll. The top was missing so you could look down into it. The walls were wallpapered and everything was in proportion to Madame Alexander."[14] He soon began to build cars, miniature fortifications, and battle scenes.

People first begin to see the relationship between the real and the fictive—the explicate and the implicate—in childhood relationships with toys, where the imagination is in control. For Lucas and Edison, an adoration of toys led them directly to the movie camera. Understandably, the movie camera is the cardinal invention that allows people to recognize the holographic simulation of reality.* With the help of his friend Melvin Cellini, Lucas created a complex environment of mechanical systems to simulate and then to make a horror movie: "Atmospheric lighting and careful arrangement of props converted the Cellini garage into a haunted house. . . . Lucas also made his first film. Melvin had a movie camera, and they did a stop-motion film of plates stacking themselves up, then unstacking themselves—Lucas's first experiment in special effects."[15] Like Edison and Tesla, Lucas found a way to control his anxiety through making models and experimenting; more importantly, he had found a life partner in the machine. The garden was no longer a mere laboratory; it was transformed into the Cellini garage and other backyards where the real excitement took place. The recognition of reality as the holographic model and self as a vital machine brings ecstasy to those who choose such a path. It is the recognition that each of us is the author of our individual movie scripts. For Lucas, this recognition began in the Cellini garage and manifested itself in the form of the quest after his near-death experience.

Recall the importance of the *Amazing Stories* in the lives of many previous mechanical-man imaginers. Lucas was a vociferous reader of science fiction, including articles in the popular magazine *Astounding Science Fiction.* His childhood also coincided with what is arguably the golden age of comic books. Of course, his favorite television program

---

*We see the connection between Lucas and Edison in their imaginary endeavors pertaining to the medium of film. For both, the movie camera is a way of writing a script and turning the imagination into reality. It's one of the boons in recognizing the holographic simulation. There is more out there than just sense perception. What are we looking at when we watch a movie? Is it still pictures moving so fast that they fool us? Do we now call this an *illusion* or *not real?* Absolutely not, we know that it's being projected from a real-life event in the explicate order of the universe from the implicate order. We are recording the simulation. Thus, the movie camera is the first invention, other than the toys that Gabby Wood points out that gave Muybridge inspiration, that allows us to imagine that we, too, are operating from a self-authorized script. When we come to the point of looking upon editing in the same way we think of overriding our brain filters, the holographic simulation, like the movie camera's creation of a viewing experience, becomes eerily similar. (Wood, *Edison's Eve,* index, s.v. Muybridge)

was *Flash Gordon*. Coupled with his love of mechanical inventions, like the automobile and movie camera, he had a strong awareness of the growing momentum in cybernetics, an awareness made evident in Lucas's *THX 1138* (1971) as well as his following films. In stepping outside of the saga itself, viewers can recognize that the droids are the storytellers. Only the machines are capable of giving an objective view of the quest undergone by the vital machines (humans fused with technology).

Perhaps the mythic undertones of Lucas's life relate most directly to this analysis—his relationship with the writings of Joseph Campbell. Many believe that he read Campbell's *The Hero with a Thousand Faces* during his convalescence in 1962. Baxter draws a direct connection between Lucas and Campbell:

> It was the season of psychedelia, of dope, of gurus so wise they could change your life. Visitors to San Anselmo carried creeds with them like dust on their shoes. They pressed copies of Carlos Castaneda and Khalil Gibran into Lucas's hands, along with texts from the Hare Krishnas, the Scientologists, the Moonies, and fashionable sf [science fiction] novels, like Robert Heinlein's *Stranger in a Strange Land,* about a charismatic sect whose adherents indulge in ritual cannibalism. Someone told him about *The Hero with a Thousand Faces,* by anthropologist Joseph Campbell, though apparently Lucas never read it, but heard some extracts on an audiotape in his car.[16]

Campbell's weaving together of a host of differing cultural tales to explain Jungian philosophy in a practical manner would likely have interested Lucas most. Joseph Campbell proves Jung's theory that there is a collective unconscious wherein these cosmic archetypes and stories are found; very similar to what the American mystic Neville Goddard (1905–1972) did in decoding and proving the secrets of manifestation in Christian Gospels—both Gnostic and Protestant—by looking at such from a psychological perspective—recognizing that each biblical character is an eternal state rather than a historical person. The activated imagination—God—is the bridge to the implicate. The hero archetype, for example, follows a path I previously explained in chapter 1. Hanson and Kay adequately describe Campbell's formula for this path in succinct detail:

> Campbell's Hero's Path has three major parts: separation, initiation, and the return. In separation, the hero receives a call to adventure. This is usually followed by a refusal to go on the journey, but the hero goes anyway, usually because he discovers that he has no choice. The hero crosses the threshold into the greate[r] world with the help of some supernatural aid. This is a mythical rebirth, as the hero, for the first time, discovers the greater world. . . . In the initiation, the hero must succeed against a series of trials. This is often the most exciting part of the story, for the hero faces danger from the demons and monsters that he must conquer. . . . In the final sta[ge], return, brings the hero back from the mythical world into the physical one where he began. This stage begins with the hero's refusal to return, for the return is the most difficult of all the steps along the hero's journey. The hero takes off on a magic flight, crossing the return threshold back to the society he left. The hero has brought with him the ultimate boon and thus regenerates society. He is master of the two worlds, the real [explicate] and the supernatural [implicate], and he possesses true [spiritual] freedom to live.[17]

This is a path of adventure, self-discovery, refusal, sacrifice, and passage into another world. In the end, the hero returns to the earthly world, bringing back with him a "boon" for "society."

Campbell was not the only author to influence Lucas significantly. Like Tesla, Bohm, and Feynman, Lucas also looked to Eastern culture and fused it with his Western philosophies. In particular, Tesla, Bohm, and Lucas showed a synthesis of Eastern and Western philosophy in their journeys toward recognizing the vital machine in the holographic simulation. Tesla had sought out Eastern gurus to engage his imagination when it came to understanding the self and reality, but only after he had formulated many of his conceptions of human-machine relationships. The Swami Vivekananda had told Tesla about "Vedantic Prana [life force] and Akasa [ether] . . . and that it is the Brahma, or Universal Mind that produces Akasa and Prana."[18] From this, Tesla extrapolated that force and matter are reducible to potential energy—infinite potentialities are the end point to the quest. Bohm took this theory a step further and showed that these potential energies are always present in the implicate, waiting to become evident in the explicate. Lucas—who still seemed to be trying to figure things out prior to *Star Wars*—began his own quest to show his audience how one can enter the implicate, find these energies, and bring them back to the explicate all through the fictive, devoid of mathematics and theory. Obviously, for one to achieve this feat, one must first recognize oneself as a vital machine operating in a holographic simulation.*

---

*The counterculture hero, Alan Watts, in *Psychotherapy: East and West,* draws a connection between why our heroes and guides find a common connection between the metaphysical ideals of the East and West. "The main resemblance between these Eastern ways of life and Western psychotherapy is in the concern of both with bringing about changes of consciousness, changes in our ways of feeling our own existence and our relation to human society and the natural world. The psychotherapist [mechanical-man] has, for the most part, been interested in changing the consciousness of peculiarly disturbed individuals," whereas in the East they look to "changing the consciousness of normal, socially adjusted people" (Watts, *Psychotherapy,* 2). We must always keep in mind that Tesla, Edison, Feynman, Bohm, Lucas, Wiener, Jobs, and Musk all either studied Eastern religions or actually met with a Swami or Shaman on a regular basis. Just today, I discovered that Jobs read numerous shamanistic biographies prior to his death. They were not looking for the empty dogmatic religion that Christianity had turned into in the West, save for a few interpretations like those by Christian mystics not tied to any church or institution. They were seeking how to change their consciousness and awareness of life in order to improve their theorizing and invention.

The mechanical-man as our new archetype is a psychotherapist that combines Christian mysticism or Gnosis with Eastern religions, American transcendentalism, and the Metaphysical Club's ideas of pragmatic skepticism to treat our unique illness in America, a garden replete with weeds of hypocrisy and contradictions. The mechanical-man is attempting to liberate Americans in two ways: "first, the transformation of consciousness, of the inner feeling of one's own existence; and second, the

The mythos of the mechanical-man makes clear that Americans' quest in looking at cybernetics is more than merely the quest of a postmodern society's members entering an initiation phase of fragmentation before returning to the explicate with the boon, or knowledge, of wholeness. It is a quest for anyone who chooses to take the path of realization and authorship. It is a quest for anyone who chooses to think within the simulation of seeing the self as vital machine within the holographic model of reality. The hero returns to the explicate with the ability to control and communicate with other vital machines in ways he or she thought impossible prior to the quest. If one were to reread this analysis, one would see that when I write about human and machine interactions, I am actually writing about interactions between the divided self in the explicate and the implicate, or the individual consciousness and the collective unconsciousness, the left brain and right brain, and finally, one's higher self with one's lower vibratory self. Each quest is one of finding the author of oneself in the implicate, recognizing this, and learning to control that other self in order to tap into the energies one does not yet perceive in the explicate order of reality. Individuals must transform their consciousness in order for this to occur. New ideology

(*cont. from p. 271*) release of the individual from forms of conditioning imposed upon him by social institutions" (Watts, *Psychotherapy,* 11). This is just another way of saying what people like James, Dewey, Pearce, and Holmes were doing with American transcendentalism. What answers did the cowboy archetype really offer us, if any? That it's good to reflect, to be alone, to get in touch with nature, and to conquer the frontier: Every archetype serves a purpose, but at some point, what you view as a collective archetype that represents a country's spirit must change in order to offer purpose in a new age (the Information Age). It's apropos here to share a speech by Jung that Watts places in his book. It's Jung speaking to a group of ministers back in 1932. Jung says:

> "We cannot change anything unless we accept it. Condemnation does not liberate, it oppresses. . . . If a doctor wishes to help a human being he must be able to accept him as he is. And he can do this in reality only when he has already seen and accepted himself as he is. Perhaps this sounds very simple, but simple things are always the most difficult at. In actual life it requires the greatest care to be simple, and so acceptance of oneself is the essence of the moral problem and the acid test of one's whole outlook on life." (quoted in Watts, *Psychotherapy,* 95)

or awakened old ideology transforms consciousness. We are eternal beings imagining and simulating non-eternal beings in the process of awakening to their eternal states of being.

Indeed, once this recognition and authoring take place, stating "the co-evolution of the real and the fictive" is merely another way of saying "the new co-evolution of the explicate and implicate." The fictive is the implicate as seen and understood in the explicate. Similarly, the lives of real prophets, sorcerers, wise fools, tricksters, and queen-goddesses reveal that they have two sides. On the one hand, their holograms are visible in the everyday world of the explicate; on the other, their implicate origins are also visible in the form of their archetypes. Through the mythos of the mechanical-man, these archetypes bleed over into our perceived reality. Through the collective unconscious mind and the archetypes shown in these holograms, we first glimpse the implicate or, in Luke Skywalker's case, the Dagobah cave. The quest gives us more than merely a glimpse into the implicate—it allows us to grasp the reality of the mind in the bifurcated form of what is being projected, and the actual projected essence itself. It is a new birth of consciousness in which we are the simulated, simulation, and simulator. In Christian vernacular it is the virgin birth.*

Star Wars renames the Control Revolution. Lynn McTaggart tells the story of this revolution in *The Field,* explaining the genesis of this revolution by outlier scientists who collectively have provided evidence that "all of us connect with each other and the world at the very undercoat of our being."[19] Similar to how Bohm envisioned the holistic movement, she sees control, communication, and information as an intricate web, with human beings acting as the glue keeping that web together—this web appears differently and is therefore simulated differently depending on our collective consciousness (collective focus of imagination). That there is a life force flowing throughout the universe is at the center of her thesis. In terms of cybernetics, her findings help

*Goddard provides us with a keen insight into how to read biblical text in the modern era of Jungian motifs and archetypes. To best understand the last four chapters of the mechanical-man's quest it's paramount you read his works (any of them).

explain the emergence of the new characteristics of control, communication, and information. In terms of the vital machine and the holographic simulation the Force is the inherent frequency of the implicate, where archetypes like the mechanical-man vibrate and are activated by our collective and individual imagination.*

THE FORCE AND THE HOLOGRAPHIC SIMULATION OF REALITY

Consciousness is an act of authoring that originates in the implicate. I am awake to the quest in this solipsistic simulation in which I reside; however, I need not worry that other self-replicating beings are conscious or aware of their own quest(s). I need only respect them as I would respect myself—since whatever I do to them I am doing to myself in terms of holography. My brain is merely a filter that creates a vital focal point for localization drawing from an infinite amount of experience from the Force. You can now begin to see how we are moving out of and away from the materialist framework. It has served its purpose and will still serve a purpose when balanced out. Now, however, the sciences must move into an idealistic framework, where mind takes precedence over material. Consciousness is the Force on the micro and macro level once one recognizes oneself as a vital machine traversing the double matrix of a holographic simulation.

*Every thought, emotion, or mental state has its corresponding rate and mode of vibration. And by an effort of the will of the person, or of other persons, these mental states may be reproduced, just as a musical tone may be reproduced by causing an instrument to vibrate at a certain rate—just as color may be reproduced in the same way. By a knowledge of the principle of vibration, as applied to mental phenomena, one may polarize his or her mind at any degree wished, thus gaining a perfect control over his or her mental states, moods, etc. In the same way he may affect the minds of others, producing the desired mental states in them. In short, he may be able to produce on the mental plane that which science produces on the physical plane—namely, "Vibrations at Will." This power of course may be acquired only by the proper instruction, exercises, practice, etc., the science being that of Mental Transmutation. (See Three Initiates, *Kybalion,* 65)

CONCLUSION

To date, the Force in *Star Wars* is the best imaginative metaphor available to describe the holographic simulation. In *A New Hope,* when Luke is first training and learning about the Force, Ben tells Luke, "A Jedi can feel the Force flowing through him." Luke then wants to know if the Force controls one's actions. Ben responds, "Partially, but it also obeys your commands."[20] How people use the Force is analogous to how they use human energies, especially psychic energies. McTaggart is simply replacing the word *force* with *field* when she speaks of the holographic model, not using it in terms of conjuring a mental image of a three-dimensional, ghostly projection of reality (which is not at all what the Force in the *Star Wars* films projects). Specifically, on devices that create holograms she holds:

> It was the unique ability of quantum waves to store vast quantities of information in a totality and in three dimensions, and for our brains to be able to read this information and from this to create the world. Here was finally a mechanical device that seemed to replicate the way that the brain actually worked: how images were formed, how they were stored and how they could be recalled or associated with something else. Most important, it gave a clue to the biggest mystery of all . . . how you could have localized tasks in the brain but process or store them throughout the larger whole. In a sense, holography is just convenient shorthand for wave interference—the language of the [Force].[21]

The holographic simulation allows people to tap into the Force. The Force allows people to dig deeper into the level of orders that make up the different stages of consciousness.* Knowing the Force allows people

---

*On the front cover (paperback) of Bernardo Kastrup's *Why Materialism Is Baloney* are small mercury whirlpools in the same ocean of mercury with the same mercury passing though the separate whirlpools. The master of metaphor strikes again with his

to understand how the implicate aspects of reality become part of the explicate; Galipeau sees the Force as a symbol in the following way:

> [T]he Force [is] a living psychic energy whose presence can be experienced. Those in contact with this realty, like Darth Vader, can sense the presence of others, like Obi-Wan, who are connected to it as well; they can "touch" the Force. But those connected to this reality understand and utilize it in vastly different ways.[22]

A holographic simulation necessitates that people place themselves in a mythic framework, imbuing themselves with a blueprint for finding meaning and purpose in life; in addition, it allows them to bring forms from the implicate and make them evident in the explicate. In terms of

---

(*cont. from p. 275*) whirlpools as individual and localized consciousness. He writes:

> To summarize, the idea behind the spinning mercury cone metaphor is that mind should be thought of as a medium inherently capable of reflecting itself into a mirror. Experiences happen when mind moves, as in the movement of ripples. The role of science is to find the model the patterns and regularities of the behavior of such ripples. When the ripples propagating in the medium of the mind self-localize, as in when they form a flat whirlpool, an individual point-of-view emerges in mind, but connection with the broader medium is preserved. This could represent, for instance, the psyches of social insects like ants. Self-reflectiveness arises when the medium of mind arranges itself, according to some topological configurations that different segments of its surface face each other. Then, both ripples and their reflections are registered as experiences and egoic awareness arises. This is represented by the whirlpool gathering so much spinning momentum that its center sinks into itself forming a hollow cone. The cone localizes the flow of the ripples just like the periphery of the whirlpool does, but also creates mutually-facing reflective surfaces. According to this metaphor, our egos correspond to these reflective spinning cones in the medium of mind. (115–16)

Kastrup understands the value of idealism. He puts into scientific and empirical form what Neville Goddard says about the utilization of the imagination (Christ consciousness). He understands that if one studies a science like cybernetics, the most important aspects are control, communication, and information and how we view this holy trinity of experience and relationship concepts in a reflective and reflexive inner and outer experiential manner.

the Force, people may bring forth the implicate or tap into the Force in different ways. The most common way is the act of departure. On Luke's departure, Steven Galipeau explains:

> Sometimes it takes the eruption of the darker sides of life to jolt us into breaking free from the inner voices that restrain us and prevent us from becoming who we mostly are. Now something in Luke awakens in the face of this personal disaster, and he gathers himself to move forward. In Jungian language we should say that the Self has now been activated in Luke, and he has embarked on a path that is more closely aligned with his potential. Jung referred to this as the path of individuation, one that leads to the realization of the true self.[23]

This departure and refusal in the quest formula as a means for evidencing the implicate in the explicate can be seen in the individual or as the collective of a given culture. In this analysis I have focused at the micro level on individuation and at the macro level on American society or the American industrial creation myth's individuation—both ridding themselves of amnesia; the medium has been the mechanical-man. *Star Wars* uses a plethora of archetypes in achieving its own prerogatives. Nevertheless, the quest of Luke and Anakin Skywalker reveals Campbell's mythic formula in play, with cybernetics as the fulcrum from which each hero works. It is as if all the mechanical-men I previously discussed, in their fragmented forms, have coalesced into this father-and-son story. What these individual, fragmentary mechanical-men have been trying to show since the early 1800s, specifically, man must recognize his status as a vital machine acting within a holographic simulation in order to possess and transform consciousness of control, communication, and information into the consciousness of frequency, vibration, and energy and finally into the consciousness of infinities (synchronicities, potential subjectivity, and creation). This progressive manner of analyzing cybernetics allows an individual to enter into the implicate, or the unseen, and author his or her life. It is as much a metaphor for individuals to take control of their lives as it is for a collective culture to take control of its own destiny—or the equivalent, in terms of how that culture *manifests* itself. Once one recognizes the implicate order of the collective unconscious, one understands that the heart of scientific and technological prowess is not the ability to model and predict phenomena; rather, it is the ability to recognize the difference between first person and second person experiences of consciousness when manifesting experience. Allowing an idea to resonate with the innermost self is more valuable than "proving" an idea's truth. *Wondering why one wonders* is the true center of techno-creation in the implicate sense. *What you are wondering about* is the center of techno-creation in the explicate sense. In the end, our ancient ancestors have taught us that sensory deprivation elicits non-ordinary states of consciousness. "Through these non-ordinary states, they believed they could access the ultimate wisdom or 'the immortals' knowledge.'"[24] As we will see the more we break through the filters of the brain, the more we see ultimate wisdom. American science, technology, and cybernetics was now calling upon its heroes and heroines to use any means possible to free their consciousness in order to see the infinite possibilities floating around them in the ether and language of the Force. It was time for American science and cyber-technology to break us out of the *Black Iron Prison.**

Grossinger updates Philip K. Dick's Black Iron Prison in *Bottoming Out the Universe:*

> Our reality has been reduced to one set of dimensions into which we are crammed like an imaginal prison cell. Bottoming out the

**Black Iron Prison,* also *BIP:* Philip K. Dick's term for the prison world of political tyranny and determinism he glimpsed beneath the veneer of Orange County in March 1974. He later wrote that, upon receiving it, he realized that he had been living in it and writing about it his whole life. This is Dick's definition of the BIP in the glossary of his *Exegesis.* "The Black Iron Prison I am referring to is a state of consciousness that is imprisoned in materialism—a physicalist's paradigm. The more you read the *Exegesis of Philip K. Dick,* the more my definition fits with his philosophies, especially with the themes in *VALIS.*"

universe starts with seeing beyond the cell—through the shadows of Plato's legendary cave. . . . A person who accepts only material reality may not even recognize his continued existence after dying. Since he expects annihilation, he vegetates in pretend nonexistence, denying his own self-awareness because beingness is impossible without a body. He corroborates his belief system by creating an alias that fools *even him*.[25]

10
The Fourth Dimension Is Calling Us

In the mechanical-men of the 1960s through and up until the plague of 2020 we see thinking machines in science fiction novels and film telling us the same thing over and over again: the call to reach a higher dimensionality of consciousness is within our grasp, higher than merely "experiencing purpose and meaning in life," and higher than the sublime and transcendence a technological spectacle causes in one's inner thoughts. There's something available to us, if we go into the darkness and recognize our genocidal history, that is so powerful it's the stuff of fairy tales and secret societies. The boon is one of practical and spiritual merit. Pragmatic manifesting or creating objective reality powered by the conscious state you experience is the practical boon in the explicate. In the implicate the boon is finding spiritual freedom and igniting or activating your higher self the freedom to interface with your lower self. New Age authors and mystics have been telling us how to do this for millennia—the Christian mystic Neville Goddard, being the one that most New Age authors cite for their wisdom has been a guide to the mechanical-man since its inception. Their ideas and practical ways of reaching the fourth dimension can be traced back to the Gnostics through the transcendentalists and the American Metaphysical Club to the psychology of

Joseph Campbell and Carl Gustav Jung to now our popular culture heroes and heroines brought to us by Gene Roddenberry, Stan Lee, and George Lucas and numerous others that answered the call of quest to the imaginal realm to bring back the implicate boons to the explicate.

The history of mechanical-men post HAL becomes much more subjective since we have many more real and fictive thinking machines that are communicating with us. The deep dive that we previously took with *Star Wars: The Empire Strikes Back* in chapter 9 is not needed for the following: *Star Trek, The Matrix,* and the *Avengers.* Curious readers might be interested to know that the *Blackwell Philosophy and Pop Culture Series* has published over 100 books that combine the very best articles for each individual franchise. From titles on *Star Trek* to *Battlestar Galactica* and the *X-Files,* we have a series of books that treats fantasy and science fiction seriously when discussing the big questions of consciousness and existential anxiety. I chose to look at the most popular franchises, most of which have put out sequels upon sequels. My hypothesis was that the mechanical-men we find in these franchises would be no different from our previous mechanical-men. Not only was I dead wrong, but this final boon we receive from mechanical-men that go by the names Data, Q, Borg, Neo, and Iron Man tell us of a transition so clear that the feeling of goose bumps and super consciousness is the only way I can describe what is found. By randomly choosing excerpts from the Blackwell series pertaining to *Star Trek, The Matrix,* and the *Avengers,* we will come to understand that these mechanical-men are calling us to act. They are calling us to awaken to the power of paradox in such a way as to enter a fourth dimension of consciousness, the cosmic consciousness that we have hitherto discussed. By understanding these paradoxes and leaning into them rather than self-destructing when confronted with such is the key to unlocking the door to this dimension. And once you find yourself in this dimension you recognize that fear is truly the mind killer.

STAR TREK: TO BOLDLY GO

Star Trek envisions a future of full inclusion. Beings are, in the words of the Reverend Martin Luther King Jr.—who, according to Nichelle Nichols, watched *Star Trek* with his family—judged by the content of their character, not by the color of their skin, the bumpiness of their foreheads, or any other irrelevant factors. Genuine and serious respect for and valuing of others involves a commitment to treating them as ends in themselves, not merely as means for our ends. This involves seeing ourselves as interconnected with others and with the world beyond. Indeed, to "boldly go" is not necessarily a journey into physical space, but into an inner life based on discovering and celebrating interconnections with others. Therein, *Star Trek* tells us, lies peace and long life.[1]

Through many iterations in various crews and on various worlds in *Star Trek,* the mechanical-man has communicated to us the same axiom found in the Oz books, that diversity is a power to be harnessed and controlled for useful ends. The world was introduced to *Star Trek* on September 8, 1966, amid what had to at that point have been the mechanical-man's most prolific decade. Gene Roddenberry (1921–1991), the show's creator, yearned for a future in which diversity serves to advance human consciousness. Diversity in all of its forms, by celebrating the connections among individuals and groups, multiplies the possibilities for experiences and feelings. Roddenberry deserves the same attention as any of our mechanical-men. I invite you to read two of his biographies, the first by Joel Engel, *Gene Roddenberry: The Myth and the Man Behind Star Trek* and Lance Parkin's *The Impossible Has Happened: The Life and Work of Gene Roddenberry.* Like innovators and inventors before him, he was deeply interested in Eastern philosophies and religions, which helped him to imagine the unimaginable. He accomplished miraculous things with the imagination—or at least by looking to the imagination as the key to understanding consciousness—and delved into esoteric doctrines passed down through the centuries by mystics and initiates. According to the mystic Neville Goddard

(1905–1972), Roddenberry had met with a council of nine from another dimension that bestowed upon him knowledge of advanced technology and cultures, thereby inspiring him to create *Star Trek.**

Roddenberry was responsible for the ethnic diversity of the cast, having come to the same conclusion as the mechanical-man that we must be selfless in order to find joy and transcendence. Breaking through reductionism and materialism necessitates a quest of this magnitude, through which almost all of our mechanical-men, innovators, and inventors seek out ways to power the imagination access to interface with one's higher self. Unless we access our imagination, we are truly robots run by our subconscious. Furthermore, without a voice in your head to interface with, your subconscious takes over. *Star Trek* tackles questions relating to human consciousness more frequently than any other entertainment franchise.

It all comes down to what the mechanical-man has preached for over fifty years now: that we must become reconciled with who we are in order to control the emotions that arise from our beliefs, to act on them, and to vibrate at a higher level of dimensionality. It is as if each person is a piece of holographic film that contains part and simultaneously all of the information that makes up the whole: every individual is of equal importance. There is a scene in *The Wrath of Khan* (1982) in

---

*The following is an excerpt from Walter Robinson's article "Vision Quest into Indigenous Space," pages 199–210 of Decker and Eberl.

> *Star Trek* offers a motif of a "hero's journey" in search of self-discovery—a metaphor for the vision quest. Jung's four modes of consciousness play out as different personalities in the drama. In TOS [*The Original Series*], Kirk is intuition, Spock is thinking, McCoy is feeling, and Scotty is sensation—just consider his love for "green" liquor. The four interact as they adventure away from home. With VOY [*Voyager*], the mythological cycle becomes complete with a journey back to Earth. Whereas Kirk had his extraterrestrial first officer with him on his journey away from Earth, Janeway has as a first officer an American Indian on her journey of return. She is a feeling type with Chakotay as the intuition providing vision.

You can see here I simply chose to show the nexus between *Star Trek* and questing for individuation.

which Spock—whose cold logic makes him a kind of mechanical-man—says that he is sacrificing himself for "the many," but the overall morality of *Star Trek* is not utilitarian, that is, focused on doing what is in the best good for humanity.

> Captain Kirk slumps dying against the wall of the reaction chamber, his DNA shattered by the radiation leak. Looking up at Mister Spock through the transparent hatch, he gasps. "How's our ship?"
>
> "Out of danger . . . you saved the crew," Spock tells him quietly.
>
> "You used what he wanted against him. That's a nice move."
>
> "It is what you would have done."
>
> "And this . . . this is what you would have done." Kirk sighs. "It was only logical."[2]

Nevertheless, *Star Trek* does tell us that Asimov's Laws of Robotics and our traditional utilitarian moral codes are all wrong in the real world. As we dig deeper into the unconscious and collective consciousness, we recognize that the ends cannot justify the means. This is the reasoning of Immanuel Kant (1724–1804), who said that a lie affects everyone. Kant was looking at the world as a holographic simulation governed by laws that remain unseen amid the day-to-day entropy of human life—remember that he, too, distinguished the explicate and implicate orders; thus, his understanding and consciousness led him to call that which he saw the phenomenal and that which remained unseen the noumenal. When people are interconnected to the point of recognizing that they all come from a single consciousness, they understand that one bad apple truly can ruin the whole bunch and paradoxically preserve the bunch at the same time in different locations.

Elon Musk says that he loves *Star Trek* because of its utopian vision and asks why a utopia is not possible for us.[3] It is! However, the way in which we view ourselves and our relations with others must change for this future possibility to be realized.

Machine? Or Not Machine?

Data, the android in *Star Trek: The Next Generation* (1987–1994), provides opportunities to dig even deeper into the individual consciousness. We see this play out in "Measure of a Man" (1989), an episode in which Data fights for the rights of individual androids by proving that he is a sentient being.[4] The discussion of the nature of intelligence and consciousness is litigated by Star Fleet's JAG Corps. In the end, Data is given the right to choose whether to undergo a procedure to produce others in his likeness that will, however, result in his destruction, and he is judged to be sentient mainly because the court cannot define exactly what "sentient" means. The episode has become part of popular culture, with college students now studying it in ethics courses.

Clearly Data, by the end of the *Next Generation* series, is the kind of android who could pass the Turing test—the court would no longer be able to tell whether he is machine or human. And why shouldn't we consider Data awake if consciousness means being awake? In the end, Data appears more awake and aware than the majority of the human and alien characters in the series. We also notice, looking back, that it was Data who seemed the most lonely of the characters. The mechanical-man recognizes all of the problems that Data faces. Thus, if you rewatch *Star Trek* episodes with a focus on Spock or Data after achieving a higher vibrational level than when you first encountered these characters, you will be surprised how different their interactions with others appear in the post-plague world and, in our case, the era of the post-mechanical-man. For many, these two characters appear entirely left-brain in their orientation to the world—devoid of emotions—at least in their experiences in the explicate order. We now recognize how much more these two mechanical-men suffer than their fellow characters. The mechanical-man is sending us the message that, to end our own suffering, we must become more empathetic toward other beings regardless of their backgrounds.

American history is replete with power struggles over the value of a human life. It is hard to fathom the collective consciousness that allowed for slavery, for a group of people to be declared legal property because of their race. Likewise, it is hard to fathom how another generational consciousness was able to justify the Vietnam conflict. Imagining these thought patterns is especially difficult because the antebellum institution and the mid-twentieth-century conflict do not resonate with current ideals of equality and empathy. American utilitarianism is a philosophy that must now be *flipped* on its head. It's not about a race or the whole of humanity. It's not even about doing what makes the most people happy. Rather, it's about E Pluribus Unum (out of many, one). Actually, the American psyche has always been about the individual; but we must reevaluate the manner in which individuals are policed and treated. By boldly going into oneself, in a Jungian sense, one can invert utilitarianism and shine the light on each individual, knowing that each has the entire universe within him or her. Pragmatic manifestation starts with the awareness of how to transform fear and pain into empathy and compassion. Spock and Data perform this transformation in the absence of built-in feelings in a kind of alchemy of emotions. It's high time that we flip the tables on American utilitarianism and value the one more than the many in transcending this paradox! Western science has taken us this far and now the world can only be redeemed by and through magic.

Upon Data's death Picard says Data wasn't human, "but his wonder, his curiosity about every facet of human nature . . . allowed all of us to see the best parts of ourselves. He evolved. He embraced change because he always wanted to be better than he was." *Star Trek* avoids locating our essence in some fixed, universal nature and suggests a more developmental, dynamic view of our humanity open to ever-shifting, interdependent relationships between nature and culture, human and machine.[5]

All is in flux! What does the mechanical-man tell us about our essence as human beings? Data's wonderment and curiosity regarding what it means to be human unlock a treasure trove of questions about issues ranging from consciousness to sexuality. The personhood of Data wins any and all arguments for the idealist who believes that matter is manifested in words and ideas that are powered by the imagination. Data's experiences bridge materialism and idealism. After the 1960s our mechanical-men are overwhelmingly showing us that, as Bernardo Kastrup says in the title of his 2014 book, *Materialism Is Baloney.*

Nathaniel Hawthorne's short story "Artist of the Beautiful" (1844) describes how the inventor's notion of his creation stays with it and speaks to a distinct dimension of consciousness. When we look at the personhood of Data and Spock, it is difficult to say whether they are of the same essence throughout their fictional lives. In fact, both die and come back from the dead. In the mechanical-man's universe, the death of the body is merely a transitional period. In *Star Trek* the essence of a person is his or her consciousness, irrespective of form (i.e., alien or human) or lack thereof. The question arises: Do these characters have the ability to interface with themselves, if that is the real crux of consciousness? It is the same thing as asking whether they are "aware" or "awake." The whole point of Jungian individuation and Campbell's quest formula is to make all people aware of their inner voice so as to discover the shadows and monsters in the human psyche and collective history. We must redefine, reknow, and recalibrate these shadows and monsters with empathy and compassion and thereby reinterpret the past and present. Both Data and Spock go deeply within their own consciousness in order to teach humans the ultimate treasure to be gained through the quest: spiritual freedom and service. In the end, these mechanical-men teach their human companions that they are all part of a federation of consciousness with peace and goodwill toward all.

In effect, our language and definitions need to change. We have slept in the garden of Darwin for far too long in this country—competition, species, classification, and so on. We are the only living beings in the universe because we currently only vibrate at the bandwidth for human consciousness. Ostensibly, we vibrate at a rather low rate because we see no intelligent life around us in our solar system. To the extent that you see more the more you wake up, we as a collective consciousness are extremely young and ignorant.

Elon Musk, remaining consistent with the mechanical-man mythos, is imagining humans colonizing the moon and Mars. He knows that one must first imagine something in order to manifest it, and, one day, we will colonize Mars and look back on the actions taken by Musk to help push us over that collective threshold. If it can be seen in the mind, it can be manifested in reality. *Star Trek* in all of its forms, but especially during its original run in the 1960s and the *Next Generation,* were investigations of the American mind. One of the findings of these investigations is that we can break through artificial barriers if we can imagine what it looks like on the other side of that barrier. Episodes of *The Next Generation* are never the same once you recognize the mechanical-man's paradoxical rewards at the end of his quest.

While George Lucas is known as Joseph Campbell's greatest student, Gene Roddenberry will go down as Carl Jung's. These men, two of the most influential in molding the American imagination, were students of the greatest mythologist and psychologist, respectively, that history has ever seen, and their contributions to the imagination must be taken seriously. Campbell and Jung bequeathed us the formulas and keys to the self. Like all the great thinkers, their ideas continue to affect the world today. The hope is that we will always have the freedom to choose what Campbell calls *bliss* as we each want to define it.

The themes of empathy and searching for spiritual freedom recur throughout the various versions of *Star Trek,* revealing Roddenberry's values and vision of the future as we and our technology evolve and our bodies and minds converge in a new coevolution. Technology is the clearest proof that imagination can be manifest in objective reality, other than the vital machines we exist in. I could have used Roddenberry's life as an example as I did George Lucas's of the coevolution of the real and the fictive. Thus, Roddenberry's mythos and quest point to the same conclusion: that the rewards of the individual and the collective level of consciousness are spiritual freedom and service. For both men, it is this knowing of the boon, or being reminded of it, that ignites consciousness, thereby raising the imagination to a higher dimension in which the workings of the simulation in accordance with the power of the one (solipsism) and the insignificance of the one (postmodernism) are visible. *Star Trek* is telling us that spirit, empathy, and equality have to be the goal for any society if it is to realize a lasting utopian future of wonderment and enchantment. Diversity and the act of inclusion with the feeling of compassion attached to that action manifests the objective reality of peaceful diversity in the explicate and an augmentation of yet to be seen imaginal forms to draw from the implicate and place in the explicate.

How, then, do we open ourselves to, in the words of Kevin Decker and Jason Eberl the "ever-shifting, interdependent relationships between nature and culture, human and machine?" You don't have to watch the hundreds of episodes of *Star Trek* in its various iterations to recognize that one must boldly travel through one's own conscious and subconscious biases and prejudices in an honest manner. *Star Trek* teaches one how to think about a holographic simulation ancillary to our technology taking us forward in body and mind. The characters learn more about themselves the more planets they visit and beings that they encounter; the more questions they ask, the more they learn to imagine.

Rather than thinking of the necessary characteristics of an advanced life-form in incorporeal terms, we may be better served to think in terms of how effectively such a life-form can engage with its environment to achieve its aims. The temptation to view incorporeal beings as necessarily more advanced may stem from an intuition that freedom and efficiency of thought, "movement," and the exercise of will would all be expanded as the scope of a creature's world horizon is expanded. But we'd be placing emphasis on the wrong thing to insist that being somehow "untethered" is the key component in exercising such power. On the contrary, a more expansive capacity for tethering may be precisely what would enable such power to expand.[6]

The Omnipotent Q

Without a doubt the most interesting mechanical-man in the *Star Trek* universe is Q. I remember feeling afraid the first time I saw Q in *Star Trek: The Next Generation*—ideas of eternity once more entered my mind, causing panic. All I could think of was that Q never died and never aged—making him *Star Trek*'s version of God or perhaps a demiurge (fallen god). This was a result of my phobia and anxiety about eternity. Is he merely consciousness on a higher plane? *Star Trek* actually provides us with the answer to this existential question of eternity. The mechanical-man (PKD) tells us that we morph into three-eyed immortal cyborgs that can never perish thanks to our technological devices; our third eye of pure intuition is opening with the aid of technology. Q is clearly in a higher dimension that cannot be seen, but, when he chooses to make himself perceptible to the consciousness of the crew members so that they can see and interact with him, we learn everything that we need to know from the simple dialogue between him and Picard. Q takes orders from a council called the Continuum, which is described very much like the Council of Nine in the Law of One material.[7] Q has been trying to tell us all along what the mechanical-man is, thereby expanding our collective consciousness through paradox. Fortunately, I found a Socratic dimension to the conversation between Q and Picard by looking at how the former views human anxiety of paradox and how the latter reacts to him. Every line of dialogue between Picard and Q ignites one's consciousness to see the holographic image more clearly. Q answers the question of whether our actions as humans really matter, among so many others.

> Q: You just don't get it, do you, Jean-Luc? The trial never ends. We wanted to see if you had the ability to expand your mind and your horizons. And for one brief moment, you did.
>
> Capt. Picard: When I realized the paradox.
>
> Q: Exactly. For that one fraction of a second, you were open to

options you had never considered. *That* is the exploration that awaits you. Not mapping stars and studying nebulae, but charting the unknown possibilities of existence.

Q: You're not alone, you know. What you were, and what you are to become, will always be with you.

Q: The Continuum didn't think you had it in you, Jean-Luc. But I knew you did.

Capt. Picard: Are you saying that it worked? We collapsed the anomaly?

Q: Is that all this meant to you? Just another spatial anomaly, just another day at the office?

Capt. Picard: Did it work?

Q: Well, you're here, aren't you? You're talking to me, aren't you?

Capt. Picard: What about my crew?

Q: *[scoffs]* The anomaly, my ship, my crew; I suppose you're worried about your fish, too. If it puts your mind at ease—you've saved humanity, once again.

Q: Goodbye, Jean-Luc. I'm gonna miss you. You had such potential. But then again, all good things must come to an end.

Q: The trial never ended, Captain. We never reached a verdict. But now we have. You're guilty.

Capt. Picard: Guilty of what?

Q: Of being inferior. Seven years ago, I said we'd be watching you, and we have been—hoping that your ape-like race would demonstrate *some* growth, give *some* indication that your minds had room for expansion. But what have we seen instead? You, worrying about Commander Riker's career. Listening to Counselor Troi's pedantic psychobabble. Indulging Data in his witless exploration of humanity.

Capt. Picard: We've journeyed to countless new worlds. We've contacted new species. We have expanded our understanding of the universe.

Q: In your own paltry, limited way. You have no *idea* how far you still have to go. But instead of using the last seven years to change and to grow, you have squandered them.

Q: You see this? This is you. I'm serious! Right here, life is about to form on this planet for the very first time. A group of amino acids are about to combine to form the first protein—the building blocks . . .

[*chuckles*]

Q: . . . of what you call "life." Strange, isn't it? Everything you know, your entire civilization, it all begins right here in this little pond of goo. Appropriate somehow, isn't it? Too bad you didn't bring your microscope; it's really quite fascinating. Oh, look! There they go. The amino acids are moving closer and closer, and closer. Ooh! Nothing happened. See what you've done?

Capt. Picard: *[after learning that he successfully collapsed the anomaly]* Thank you.

Q: *[curious]* For what?

Capt. Picard: You had a hand in helping me get out of this.

Q: I was the one that got you into it. A directive from the Continuum. The part about the helping hand, though . . . was my idea.

Q: *[to Picard, after the other ships explode]* Two down, one to go.

Capt. Picard: We demonstrated to you that mankind had become peaceful and benevolent. You agreed and you let us go on our way. Now why am I standing here again?

Q: Oh, you'd like me to connect the dots for you, lead you from A to B to C, so that your puny mind could comprehend? How boring.

In any case, I'll be watching. And if you're very lucky, I'll drop by to say hello from time to time. See you . . . out there!

Capt. Picard: Q? What is going on here? Where is the anomaly?

Q: *[pretending to be deaf]* Where is your mommy? Well, I don't know.

Capt. Picard: The last time that I stood here was seven years ago.

Q: Seven years ago! How little do you mortals understand time. Must you be so linear, Jean-Luc?

CAPT. PICARD: You accused me of being the representative of a . . . a barbarous species.

Q: I believe my exact words were "a dangerous, savage child-race."

CAPT. PICARD: We are what we are, and we're doing the best we can. It is not for you to set the standards by which we should be judged!

Q: Oh, but it is, and we have. Time may be eternal, Captain, but our patience is not. It's time to put an end to your trek through the stars, make room for other more worthy species.

CAPT. PICARD: You're going to deny us travel through space?

Q: *[laughs]* No! You obtuse piece of flotsam! You're to be denied *existence*. Humanity's fate has been sealed. You will be destroyed.[8]

The Dangers of Becoming the Borg

Just look at yourself and how you experience values like "individuality." For most of us, modern life is an organic-computer hybrid experience: your downloaded apps, posted status updates, rebuilt knee joint with titanium rods, GPS dropped pins, and naturally and artificially flavored postworkout protein bar—they are all a part of your authentic lifestyle. You can thus ask yourself about the values guiding these experiences. What controls your life—a singular, autonomous self, uncontrolled by outside phenomena, or an evolving, adapting network of interdependent people, ideas, and stuff? Wouldn't you prefer to have thousands of "likes" and "friends"? Could you even separate your "authentic" self from your virtual one? The answers are obvious, the Borg* values are triumphant. You've been assimilated. Resistance is negligible.[9]

This author understands what the Borg is if left in a materialistic paradigm. The mechanical-man agrees completely with this reasoning, which is why fragmentation is still taking place or is in the process of

---

*In the *Star Trek* series, "The Borg is an organic-technological hybrid collective that seeks perfection by forcefully incorporating other species into itself." From Lisa Cassidy's "Resistance Is Negligible: In Praise of Cyborgs," in Decker and Eberl (*Ultimate Star Trek*), 232–242.

nding, and this, in turn, explains why our higher self has been calling us with the same stories and signals since the 1960s. We won't manifest the Borg if we accept the call of this current quest and create an assemblage of collective consciousness akin to the ideals of our next *assemblage* of mechanical-men—the Avengers. Unless we change our underlying model and think of the universe as ancillary to spiritual freedom, we are on a path to manifest the materialist future of the Borg. The authentic self is always the higher self, the ideal. The lower self is the self with regular consciousness—sleepwalking through life, with no signs, no interfacing with the higher self but only with the negative, pessimistic, cynical self and no interfacing with others, no synchronicities, no telekinesis, no astral projection, no manifesting and thus, no idea that imagination is God, and no clue about how to manifest one's own reality. The life of the lower self is sad and depressing and plays out on eternal repeat.

The fear of the Borg is not to be disregarded, however; it is the fear of Norbert Wiener, come to life in the fictive. Ideally, the Borg have no place in the garden. Too many people in the United States have tapped into the consciousness of the source to discover that the Borg, wrapped deep in layers of pragmatic cybernetic ambiguity, lay bare the dark abyss, thereby exposing another paradox. However, we won't need to recognize that paradox if we can escape the paradigm of materialism. We do not need to separate the "authentic" self from the "virtual" self because both come from the same source, the human imagination. The human imagination offers an infinite number of possibilities. The Borg are the paradigmatic materialists: for them, everything is black and white, with no change or individuality, just cold analytical logic—theirs is a society indifferent to emotion. It is as if a large percent of the population were addicted to a drug that eliminates feelings from the decision-making process. We are on a trajectory to become the Borg in more ways than one unless we answer the call.

In their essay "The Extended Mind," contemporary philosophers Andy Clark and David Chalmers claim that our consciousness is being extended by our reliance on "smart devices." Smart devices have become our memory, remembering for us where we are to go, with whom we are to associate, what we are to do, and so on. They're even, to an extent, our identities. Our photos, stories, activities, family and friends, financial information, interests, and more are kept on our devices. Today, most of us carry our devices, very soon, more of us will welcome them. It isn't far-fetched to imagine that one day these devices will be implanted and become part of us.

The Borg prompt us to wonder how this will be done and to what purpose. Are we being gradually assimilated by our own technological creations? More than a commentary on transhumanism, the Borg show us what could happen if we don't recognize the essence of technology—the relationship it creates among us, one to another, and between us and the world we inhabit, and the ways in which technology orders our thinking so that we view everything through its lens.[10]

How is the American public going to view Elon Musk's implants in our brains that allow us to interface with our higher selves? In other words, will the self that you speak to every day be sufficiently real to answer and teach you things based on the program that you provide? Will it even, at some point, show the projection of a real person speaking to you as a hologram? You could, theoretically, live your life among holograms of people without ever sharing the same space. This sounds as crazy as the wireless revolution would have sounded during the time of Edison and Tesla. Indeed, we have come full circle, with the theoretical science of these leading figures of the beginning of our quest resembling that of those at the end, Steve Jobs and Elon Musk. The dystopias of *The Twilight Zone* or *Black Mirror* are not far off; their pessimism has already entered the collective imagination.* The same will occur for

*I'm speaking of the current *Twilight Zone* (2019) series. We have already outpaced the old series. *Black Mirror* (2011–2019) is similar to the *Twilight Zone* (2019) but it considers the issues that the mechanical-man is dealing with, mainly, leaning into paradox in the real and fictive cyber-techno world.

extraterrestrials, whether they are like Steven Spielberg's *E.T.* (1982) or take the form of extra-dimensional beings, they will become manifest in our lifetime, perhaps by the time this book comes out. We will create aliens as we imagine them and manifest it in the explicate with gray skin or reptile features, almond-shaped eyes—and when we open them up we realize that there are no organs. We discover that they too can be traced back to Tesla's remote-controlled boat—controlled by our implicate self(s).

We can go back to any of the early American literary greats, such as Melville, Poe, Hawthorne, Brockden Brown, or Cooper, to learn that innovation and the drive to interface with or create a reality from an image, or an image from something real, is in our holographic nature. This kind of mimesis was already as American as apple pie at the time of the daguerreotype, the invention that led to the picture camera. Hawthorne's *The House of Seven Gables* (1851) reveals how the average American of the time felt about these ideas entering the collective consciousness. The fear of the Borg is real, and how we react to it is up to us. However, the assemblage of mechanical-men that came before us tell us exactly how to circumvent the trajectory that leads to a civilization like that of the Borg: we must always be vigilant and remember that "resistance is *not* futile."* As long as we are able to dream, we are taking the first steps in solving the problems that pave the way for the Borg by embracing agape love, diversity, curiosity, and positivism. We are the authors of this play, and we can take a fixed idea and change it; we are the actors in the play, and we can change the script to create infinite alternate endings made happy through earthly and cosmic freedom.

*"A few years after feminist philosopher Donna Haraway proclaimed she'd rather be a cyborg than a goddess, the Borg marched across television screens in *Star Trek: The Next Generation*. . . . Haraway declares that being a cyborg is inevitable, yes—but also desirable and liberating. So who's right, Star Trek or the philosopher?" In Lisa Cassidy's "Resistance Is Negligible: In Praise of Cyborgs," in Decker and Eberl (*Ultimate Star Trek*), 232–242.

Recalibrating Culture

Star Trek provides an instructive allegory for understanding both ancient and current, cutting-edge concepts of reality. The holodeck hypothesis suggests that perhaps taking hold of our destiny depends upon the realization that we're fictional characters playing a role in an ongoing narrative in a cosmic "holodeck." As players, we can't exceed the parameters of the original design—unless we can figure out how to transcend our own programming (like the holographic characters Moriarty, the EMH, or Vic Fontaine).[11]

That's what a simulation is; we have already spoken of this feature. The mechanical-man is the avatar of our higher consciousness now, and he assumes an infinite number of forms, including the human creator of forms. The idea of transcending our own programming fits well within the boundaries of thinkers and innovators, from Tesla to Musk, who have transcended the boundaries of the normal, thereby moving us toward a higher consciousness so that we begin to notice more signs, synchronicities, and breakthroughs. At this point in the quest, you, the reader, should be experiencing breakthroughs in how you view yourself and others.

The holodeck allows us to act out any fantasies that we wish to indulge, but we also learn that there are boundaries and fail-safes in the game of the holodeck as there are in reality. One of our fail-safes is emotional pain. You now begin to recognize that this body has a fail-safe of its own, similar to Iron Man's armor, which we will learn more about. If we want a satisfying outcome of wonderment, excitement, and enchantment from our collective manifesting, we will eventually build a holodeck. Remember, the mechanical-man is pushing idealism to the extreme, having claimed earlier that Americans have gone Aristotelian; but the Platonist side is gaining steam and must continue to do so. There is a sense that the spirit of the 1960s showed itself again in a different guise in the 1990s, culminating in *The Matrix* at the end of the decade. Perhaps we would be living in a completely different world if the Twin Towers had not come down on September 11, 2001. Nevertheless, the call was still present, urging us to muster up the courage to answer it, accept it—"Just Do It!"*

By boldly going toward where no one has gone before, the U.S.S. *Enterprise*—like the Romantics and transcendentalists—goes "out there, that away" to create new points of cultural reference after the old points have fallen away. The lesson for us is that as we advance toward the *Star Trek* era, we must be mindful of the importance of our understanding and interpretation of historical narratives so that we too can create points of cultural reference and meaning in our lives.[12]

We need meaning in our lives! We as a country must interpret historical narratives very differently than we did prior to the plague. The process of fragmentation has pushed us into our corners, left versus right, in more ways than just the consciousness war. We have to accept that we have been complicit in glorifying idols that have no place in our emotional landscape after the plague. All that is required to reverse the fragmentation is to adjust our points of cultural reference in such a way that we will never glorify death or war again. We must find the frequency at which discrimination in any form makes us physically ill. We should live at a dimensional level of consciousness in which the very term *war,* like *slavery,* has no meaning. The simple fact that a practice or set of behaviors has characterized humanity throughout its history does not mean it must continue to do so in the future. *Star Trek* has been telling us that we need to recalibrate what we celebrate. The American Revolution was not just a war for independence from Great Britain but the beginning of an ongoing process through which the collective consciousness has continued to evolve. Any human being can join this collective consciousness since it is founded on ideals rather than notions of blood and soil.

The Romantics, the European counterparts of the transcenden-

*Nike and the symbolism of Mercury with their "Just Do It!" slogan is known throughout the world for the phrase. For the alchemist, Mercury is mind itself. Mercury takes the form of its container. It takes the shape of its container, which is a quality of mind. It also reflects everything around it; thus, it reflects the fragmentation.

talists, are among those who have taught that we are all one. Again, oneness in this sense means that consciousness on the individual and collective levels has been shown to derive from one source with which we have the potential to interface. As Paul Selig wrote in *The Book of Freedom* (2018), "To realize your complicit nature in anything you see is simply to acknowledge yourself at a vibratory frequency where this expression can be met."[13] As we go further in changing our consciousness, we will notice that history looks very different from the other side of the explicate order, outside the realm of cold logic. We now begin to understand that Dr. Faustus was correct when he said that heaven and hell are states of mind—a certain vibration and frequency of energy. We must elevate the state of our minds so that they vibrate at the same frequency as the minds of Gene Roddenberry (and, thus, Spock, Data, Kirk, and Picard). In this way, we can see the beauty in one another, the inherent value in every living species and every human being. We move ever closer to the ideal! Hope is no longer simply a political slogan or an emotion, the ideal is becoming visible ever more as time passes.

Star Trek offers a humanist theology of the divine potential of humanity as a possibility that may not be actualized, rather than giving a definitive answer. It's unclear whether technological enhancements, biological interventions, natural processes, or some combination of these can bring us closer to the divinity toward which many of us aspire. While starships may be part of our human future, a god ought not to need a starship, as Kirk points out toward the end of *The Final Frontier.* It's not only Kirk, and through him Roddenberry, but also many theological systems that would assert, "Above all else, a god needs compassion!" And if compassion is the defining attribute of a god, then Star Trek's humanist theology encourages us not only to believe that divinity is a real possibility for humanity in the future, but also to recognize it as a possibility already open to us in the present.[14]

The people around us are simply ourselves playing other parts united by the underlying law of unity. To quote Paul Selig again, "All manifestation is agreed upon. Your relationship to everything may be altered when you stop deciding what something is through historical perspective."[15] A vengeful god is merely telling you that you are capable of vengeful behavior if you do not, like the Gnostics, "[I]f you bring forth what is within you, know what you bring forth will save you. If you do not bring forth what is within you, what you do not bring forth will destroy you."[16] Only a demiurge would destroy its own creation and smite its people for behaving as it had programmed them. If we have manifested an evil god in the past, we have the power to create a god of compassion and empathy by accepting the call to individuation. *Star Trek* shows that our technological enhancements and interventions tell us about ourselves. That we are the demiurge, we are the one that has the potential to smite or the potential to heal. We carry the cross of responsibility upon all our shoulders. Matter speaks to us in a language we are beginning to understand. And it's telling us to stop smiting our lower self. We no longer have to suffer in order to recognize ascension and jump dimensions, get out of the Black Iron Prison. Reconciliation between our lower self and higher self in all forms will reconcile into a magic present, glorious history, and enchanting future. Gene Roddenberry once said,

> We long ago got rid of the idea of our dealing with anything suggesting the traditional Judeo-Christian God. It is vital to Star Trek that we deal properly with the attitudes of our crew regarding this question. . . . It seems to me the only possible answer to this is throughout the script to stay a mile away from revealing or even hinting at what our people believe about God. I was generally successful in doing this in the original Star Trek series. A few things did slip by, but not many, at least, not many serious variations of my policy to keep Star Trek free of serious religious themes.
>
> I believe I am God; certainly you are, I think we intelligent beings on this planet are all a piece of God, are becoming God. In some sort of cyclical non-time thing we have to become God, so that we can end up creating ourselves, so that we can be in the first place.[17]

Roddenberry's humanist ideals did, however, come into play, and he suggested that humanity is "pieces of God," which accords well with the mechanical-man's view of humans on the macro and micro levels as avatars of God. Once again, we are faced with the ultimate man/god paradox with which we come to terms through the quest. *Star Trek* does a brilliant job of leaving God a mystery while also intimating that God is, in the final analysis, pure love and compassion—the kind of compassion that can heal the past. Such compassion allows one to look at historical events from a perspective of empathy rather than one of pride, hate, or envy. Roddenberry believed that it is possible to view "even some of humanity's worst atrocities as part of the process of growing up."[18] Similarly, the Apostle Paul referred to human suffering as birth pains.[19] As we evolve, so does God, since we are all individual avatars of God. What Mark Twain says about the God of the Old Testament finding religion in the New Testament is true of our mechanical-man's quest. The mechanical-man's religion can only be based on compassion and empathy. Since you are going to live the life that you imagine, you must imagine it well.

If we are willing to imagine our ideal selves in the *now,* we can have the kingdom of God *now.* It's not about a place or time; it's always going to be about how one feels in the *now.* We don't have to wait for death. Everything that is here will vanish eventually, but that's even more reason for one to conjure and manifest the seemingly impossible. Roddenberry offers us a future that is not only possible but believable in terms of increasing wisdom pertaining to both the mind and the body, with expansion of the mind into the higher self corresponding with exploration of the universe(s) and corresponding level of consciousness. As Americans, we should feel proud of giving *Star Trek* and *Star Wars* to the world, for in this way we have increased its capacity to imagine the impossible. *Star Trek* also gives us a glimpse into what our imaginers and innovators have had to do in order to expand their consciousness and give this possible future life: they go to dark underworlds to bring back knowledge and return to communicate their findings. So too does *The Matrix.*

THE MATRIX: EMBRACING OUR DUAL EXISTENCE

From *Star Trek* to *The Matrix,* continuities and discontinuities are observable in the evolution of the mechanical-man. One of the first threads that we recognize is the influence of both the Western and Eastern philosophical traditions.

> [T]he talent of the Matrix lies in its syncretic use of philosophical and religious elements from various Western and Eastern traditions. In a masterful way, it mixes metaphors with rich references to Christianity, Platonism, and Buddhism within a context of contemporary cybertechnology and is already a classic in the sci-fi genre. Its genius consist in richly combining penetrating script and suburb images in a way that creatively conveys the profound though oftentimes impenetrable Buddhist message of liberation. In doing so, The Matrix awakens the viewer and challenges us to reflect (and not reflect) on where we habitually live—in our minds. It compels us to ask, the next time we look into the mirror: Who or what is it that we see.[20]

Many of our mechanical-men in the practical world of innovation and invention have sought to combine a pluralistic philosophy and religion from around the world in their conjectures on the nature of reality. The combination of Christianity, Platonism, and Buddhism in the context of cyber-technology is a kind of pluralism consistent with American ideals—and, indeed, this country has produced more religious sects than any other (something worth pondering for a moment). From the Anabaptists to the Mormons to the Temple of the Jedi in Beaumont, Texas, we have kept yearning for spiritual freedom.

> As we let hundred[s of] dogmatic iterations of reality bloom, the eventual result was an anything-goes relativism that extends beyond

> religion to almost every kind of passionate belief: If I think it's true, no matter why or how I think it's true, then it's true, and nobody can tell me otherwise. That's real-life reduction ad absurdum of American individualism. And it would become a credo of Fantasyland.[21]

Consider all of the religious symbols found in science fiction films. There is, for example, the dove that appears at the moment of Roy Batty's death in *Blade Runner,* as well as the *ankh,* the ancient Egyptian key of life, that identifies members of dissidents and serves as a literal key to Sanctuary in *Logan's Run*. These philosophical and religious ideas from the past power the imagination of our mechanical-man now. In like manner, our language must change in order to arrive at our next density. We must recognize that the emotions elicited by religious and philosophical writings can be as powerful as the energy produced by steam, electricity, or the atom. Every statement should start with "I AM" or "I FEEL." It is as if an individual has access to an infinite number of perspectives when faced with a contradiction of polarities or a false dichotomy. The Gnostics sought a pluralistic interpretation of sacred texts. Even thousands of years ago, they knew that you must speak in terms of the dual existence of the implicate and explicate order.[22]

In *The Matrix,* once the characters know who and what they are individually, they recognize how close they are to one another. All of the characters that you perceive participate in the same collective dream, which is your dream too. In their dreams, the manner in which they control and interface with their environment is a function of their imaginations, which pull from the source or the ideal. The recognition of one's essential unity with others is the first step in the individuation process. The next step is recognizing the divine in the self and then the divine in all conscious beings.

In the American collective imagination, we have had numerous heroes and heroines to draw from that are considered the "one." True freedom involves recognizing that you are "the one" spoken of in sacred writings and scripts. You know Neo on an intimate, detailed level but also in his assembled form. Our task is to test every belief in order to find the exit door or hear the ringing phone. How did you interface with yourself while viewing *The Matrix?* How do you interface with yourself or family members about the issues raised in the film? If you don't know, then you need to try.

Do some individuals really believe that they are "the one" as they sit on the subway or walk down Main Street USA at Disneyland? Is it possible to hold this belief without being overcome by hubris and pride? *The Matrix* dramatizes the paradox of solipsism. The riddle is whether we need to recognize that we are "the one" before we can recognize that everyone else, individually and collectively, has the same status? This is perhaps the most subjective part of the quest. Neo recognizes that he is "the one" by taking control of his own code, but at no point in the film are we told that Neo created the matrix. The heroes trapped within it don't recognize that the reality in which they reside is the same one from which Philip K. Dick sought to escape his entire adult life. The problem is that we can never escape the lower self until we recognize it as an avatar of God—which, incidentally, PKD did before his death. The quest involves not only transcending death but also recognizing the infinite potential of the self. The power is in the collectivity and what it implies—namely, that we are all alone and that we are never alone. That part of yourself that speaks to you in your head is learning and in turn teaching you to recognize your individual and collective identity—your place in all this. The film, in the end, calls us to TAKE ACTION, to say "I AM" right now instead of waiting for another conscious or divine being to save us. We are suffering victims, not of anyone *out there,* but of an enemy whom we know all too well: ourselves. The divine is in us, and we must choose how and when to activate it.

Goddard writes,

> Night after night, just before you drift off to sleep, strive to hold your attention on the activities of the day in reverse order. Focus

> your attention on the last thing you did, that is, getting in to bed, and then move it backward in time over the events until you reach the first event of the day, getting out of bed. This is no easy exercise, but just as specific exercises greatly help in developing the muscles, this will greatly help in developing the "muscle" of your attention. Your attention must be developed, controlled, and concentrated in order to change your concept of yourself successfully and thereby change your future. Imagination is able to do anything, but only according to the internal direction of your attention. If you persist night after night, sooner or later you will awaken in yourself a center of power and become conscious of your greater self, the real you.[23]

The paradox is thus both versions—(1) a subject freely floating from one to another VR, a pure ghost aware that every reality is a fake; (2) the paranoiac supposition of the real reality beneath the Matrix—are false. They both miss the Real beneath the Virtual Reality simulation—as Morpheus puts it to Neo when he shows him the ruined Chicago landscape.

However, the Real is not the "true reality" behind the virtual simulation, but the void which makes reality incomplete or inconsistent and the function of every symbolic Matrix is to conceal this inconsistency. One of the ways to effectuate this concealment is precisely to claim that, behind the incomplete/inconsistent reality we know, there is another reality with no deadlock of impossibility to structure it.[24]

The mechanical-man is finally awake to ask where and who he is. Currently, we can only ameliorate these contradictions by maintaining a double of the image being projected in Kansas from the Land of Oz. If we simply float from one virtual reality to another ad infinitum, when do we stop and realize that this double holographic simulation in which we live is where the lower self learns from the higher self? Before I answer this question, let's consider what the script of *The Matrix* could be if it were not burdened by considerations of dollars and sequels—for, again, the goals of the controller, in this case, a media conglomeration for which profits reign supreme, must always be kept in mind. The key is to elude the control of the "empire." As Rizwan Virk describes *The Matrix* in his book *The Simulation Hypothesis* (2019):

> By taking the red pill, Neo wakes up to realize that what he thought was reality was actually a computer simulation. He finds that in the real world, all humans live in pods, plugged into the Matrix—a high-fidelity video game-like simulation that the characters have lived in their whole lives. In the sequels to *The Matrix,* the audience learns that this simulated reality was created to keep human minds preoccupied by a race of super-intelligent machines, who were using the small amount of electricity generated by each human's brain for their own nefarious purposes.[25]

The year in which *The Matrix* was released, 1999, feels for many of us like yesterday—and, in a simulated reality where memory implants are possible, it may have been yesterday. In any case, the film helps us perceive the explicate and implicate order in new ways, shaking up our thinking about mechanical-men and the techno-mythos that accompanies them. The explicate order of reality is the matrix, a dreamworld, video-game-like existence ruled by the senses. Manifesting one's own reality is ostensibly impossible: one is either in the explicate, everyday order of the senses, or the implicate, where the matrix is grafted onto the three-dimensional existence called reality. The implicate order in this film is dystopian: the machines appear to be at war, utilizing humans as mere electric batteries for their own purposes.

Thus, consciousness is viewed as digital information that acts as software. Humans have no control over or ability to communicate with the implicate order where the machines live unless the latter write a program that allows them to do so, or a human somehow hacks into it. Once you make the choice to realize that you are in a matrix and have the ability to interface with the implicate order where consciousness resides—or, in this case, where consciousness is fabricated—the next question is where machine consciousness originates. *The Matrix* unfolds in two distinct three-dimensional settings, the matrix and the base reality, the latter of which resembles the kind of world that the Borg from *Star Trek* might create with dark colors and a mechanical feel. As the story proceeds we are asked to ponder whether our future selves, aliens, or machines create our simulation and how many levels of creators it has.

Our master mechanical-man prophet Philip K. Dick, of course, had already tackled the issue of distinguishing real, virtual, and artificial. Not surprisingly, the Wachowski sisters claim PKD's novels and stories as an inspiration for their work, including *The Man in the High Castle,* "Do Androids Dream of Electric Sheep?" (which, as has been seen, inspired *Blade Runner*), and *The Adjustment Team* (which inspired the 2011 movie *The Adjustment Bureau*). At the Metz Sci-Fi Conference in 1977, PKD remarked, "[We] are living in a computer programmed reality, and the only clue we have to it is when some variable is changed, and some alteration in our reality occurs."

Whether it's aliens or machines, we are left with more questions than answers. The discussion circles back to Richard Feynman's wondering why he wonders why he wonders or, in even simpler terms, who begets whom. Are we the makers of the machines, who then make us, and we then make the aliens? Intelligent minds, even our mechanical-man prophets like Elon Musk and PKD, believe that we are almost certainly in a simulation and will have the ability to create our own simulations sooner than we think. In an interview Elon Musk ruminated about the simulation that we must be in and concluded that base reality must be extremely boring. By implication every subsequent simulation that unfolds needs to be more exciting and have more variables to experience than the one that came before. However, all seem to be leaving out the most important variable in their theorizing and philosophizing, though, which is consciousness.

The materialists explain consciousness in one of two ways.

1. Consciousness is a very subjective experience and, therefore, falls outside the purview of the "physical sciences" and is not of interest.
2. Consciousness is the result of physical processes (i.e., chemical reactions); thus, it is a product of neural activity in the brain, and, once we map out that neural activity more fully, it will be possible to reproduce consciousness artificially with little effort.[26]

In the Old Testament of the mechanical-man's story, these are the definitions with which we worked. American cybernetics and its techno-mythos explain how machines progress toward one or the other of these definitions of consciousness, leaving us in a purposeless universe where everything is up to chance. In the New Testament of the mechanical-man's quest, the pendulum has swung back to Platonism. Thus, consciousness is highly subjective but does not, in fact, fall outside the purview of the physical sciences. Our guides have told us again and again that the key to understanding the mind is going inside and projecting outside until that which is without reflects within where the higher and lower selves meet. Thus, consciousness in the Age of Reconciliation, which is what these mechanical-men are communicating, that we must reconcile our differences, is the following:

> The LIGHT is consciousness. Consciousness is one, manifesting in legions of forms or levels of consciousness. There is no one that is not all that is, for consciousness, though expressed in an infinite series of levels, is not divisional. There is no real separation or gap in consciousness. I AM cannot be divided. IT IS only by a change in consciousness, by actually changing your concept of yourself, that you can "build more stately mansions"—the manifestations of higher and higher concepts. (By manifesting is meant experiencing the results of these concepts in your world.) It is of vital importance to understand what consciousness is. The reason lies in the fact that

> *consciousness is the one and only reality, it is the first and only cause-substance of the phenomena of life. Nothing has existence for man save through the consciousness he has of it.* Therefore, it is to consciousness you must turn, for it is the only foundation on which the phenomena of life can be explained.[27] (emphasis in original)

During one scene in *The Matrix,* Neo is hooked up to the machines so that there are two Neos in play, one in the implicate order where consciousness resides (whether it vibrates at Neo's frequency or at a frequency set by the machines is a moot question). The bottom line is that consciousness is coming from another place, an understanding that fits with most spiritual teachings. The film does a good job of at least painting us an adequate picture of how the explicate and implicate orders interact, but it fails to provide us with a believable version of the latter. We could spend hundreds of pages showing the failure of the implicate level of reality in the film on multiple levels. But again, just like the chess-playing hoax of that proto-mechanical-man, the Mechanical Turk, the perception and knowledge that we reside in a double-sided simulation of explicate and implicate existence makes *The Matrix,* alongside *The Empire Strikes Back,* one of the most prophetic science fiction films ever conceived.

So, while *The Matrix* is not, as some hold, a documentary about the workings of our reality, it does raise the question of the precise nature of the simulation in which we reside. So far, we know that it is holographic, meaning that it has two orders of reality, one that we can see, measure, taste, and smell—the explicate—and one that we cannot see but clearly sense if we work through to the other side of paradox—the implicate. To describe the simulation, we also have to go all the way back to the beginning of the quest. To quote Philip K. Dick again, "We are all sleeping avatars of God, with amnesia."

The Mechanical-Man's Matrix

In the mechanical-man's version of *The Matrix,* Neo is playing the role of everyone in both the explicate and the implicate orders. The simulation of the simulation of the simulation—or however many *Matrix* movies are made—all come from Neo's imagination. Let's even give the materialists their due and say that, along with Rizwan Virk, that "Consciousness is really a set of information and a processing of that information."[28] Where does mind—or what we have classified as the Force—fit into Neo's universe? As Virk puts it,

> In these games, you chose a character to role-play, not unlike in the original Dungeons & Dragons. Avatars, the term used for a character's representation onscreen (coined by the makers of Habitat, one of the first shared virtual worlds, built by Lucasfilm Games), could be customized to fit the race of character (human, elf, dwarf, gnome) and his or her profession (thief, barbarian, warrior, wizard, etc.). Most importantly, the player could customize what the avatar looked like—skin color, gender and clothing. As players interacted and battled with each other, these interactions affected both players' game stats as characters evolved. This novel concept introduced the idea of a "persistent world," one that existed beyond a single gameplay session or even beyond a single player's computer.[29]

These games that we have devised for ourselves present us with an image of ourselves. The avatar is still you, since you are in control, but the question remains whether the characters that you control—the avatars—are on a different level of consciousness so that their dimensionality extends far beyond what our senses perceive. The reality of the avatars in the game is the explicate order—which appears as pixels, graphics, and grids to the controller of the avatar in the implicate order. We are the god that controls the characters as well as the characters while we control them. When we are not in control of them, perhaps the qualities with which we endow the character or the character's purposeful ends become the driving force for its extended life and thus, extended reality. So, for Virk,

> The purpose of Second Life was to interact with other people from around the world; beyond that, it was up to you what you did in the 3D world. You could live a complete virtual life in Second Life. Your avatar could socialize—you could go to dance clubs, build a house with another character, and decide to get married, or shoot arrows at each other. You could even have jobs in Second Life where you would get paid in "lindens," the currency used inside the virtual world (which introduced us to a new concept—virtual currencies and virtual economies). Linden Lab was among the first video game companies to hire virtual economists to measure and monitor their virtual economies.[30]

Now we see the great importance of the very concept of control when looking at our own purposes in life. In our hypothetical matrix, Neo's avatar as he is hooked up to the computer is himself in the explicate order. The recognition that he has an implicate and explicate self gives Neo the potential to interact with and control his avatar; when he does so, he appears to those around him in the matrix to have supernatural powers. The ability to control his avatar allows Neo to get the best of his nemesis and, indeed, opens up for him infinite potentialities to manifest whatever he wishes, or so it appears. Unlike in the sequels *Matrix Reloaded* and *Matrix Revolutions* (2003), we are not going to end up fighting the machines that control us; there is no warfare between the implicate and explicate or between the form and essence. Once you are able to process your bipolar self, you can have full control over your essence and play the role of any avatar that you wish to manifest and experience. We gain our freedom to know that we are God, all of us, when we stop investing in the lie that our lives are controlled by forces that we can never come to know or to access.

In the mechanical-man's *Matrix* movie, Neo never meets an architect; rather, he meets himself—a self that is aware of ideal forms and essence. The mind is consciousness itself; it is wakefulness. From this perspective, fear is the only controller that can bring pain, weakness, confusion, hatred, and evil to our avatar. Neo no longer fears anything because he knows that he is part of the one mind, that everything is connected since it all comes from the Law of One.[31]

The Matrix did a great job of pushing us in the direction of recognizing the double holographic simulation in which we reside. We learn not to apply our will to other men and women to get what we want through mental or physical force, for there is no need to apply force or to compete in any scenario. Compelling or forcing God to do something is not the way willpower works. This is a collective dream of numerous projections. We are not slave-like batteries for the machines that control the simulation. We are the controllers, the controlled, and the simulation itself. Perhaps the next installment of the franchise will adequately cover our existence as avatars of God.

Neo too learns that intellect is more important than the senses. Mind is more important than matter. As for Plato the physical is not as real as the Form, so for Neo "there is no spoon." Neo is the reincarnation of the man who freed the first humans. Plato held the intellect and body are so alien to one another that their union at birth traumatically engenders loss of memory, a kind of amnesia. This is not the total loss of memory Cypher traditionally deals for, but rather the kind one might suffer after drinking too much of Dozer's Lethic moonshine. The details can come back with the right prompting and clues. For Plato, déjà vu is not evidence of a glitch in the Matrix but a recollection (anamnesis) of Forms. In the time between incarnations, when the soul is free of the body, we behold the Forms. On the earthly plane all learning is actually a process of recollection in which we recall the Forms, cued in by the resemblance mundane objects bare to them. A child does not need to be taught that a flower is pretty, for example, but knows it through recollection in which we recall the Forms, cued in by the resemblance mundane objects bare to them and knows through recollection of the Form of Beauty itself and the flower's share in it.[32]

Idealism is what these science fiction films achieve with their intelligent machines. We get to see how machines interface with people and with themselves as well as how we interface with each other and ourselves. We engage in a kind of holography, placing mirrors in front of mirrors and going as deep as we can. And there, at the turn of the century, is a movie that tells you to wake up while the internet's dimensionality of experience is gaining steam. The internet itself is a mechanical-man:

> The internet shared a new dimensionality of the democratic vision common among software programmers and Internet pioneers. They saw the Web as open and free to all. Users could communicate without restriction and find access to any and all forms of information. Such openness was the bane of authoritarian governments, which found it difficult to control public opinion in a world in which information flowed freely. The unregulated format of the Web raised substantial legal, moral and political questions in the United States as well. By 1999 five million Web sites were in operation—among them sites promoting pornography, hate speech, and even instructions on how to build atomic bombs.[33]

You don't need a medical degree to conclude that America was still addicted to power and wealth at the turn of the century. Because there was no collective human spirit, we replaced loneliness, lack of empathy, and the lack of compassionate awareness with pornography and hate speech, alcohol and opioids, money and power. Then, soon after 9/11, we began raining terror down on Iraq and Afghanistan, artificial terror in our eyes—other than the American soldiers and families that paid the ultimate price, all too real for these families. We see where we stand today, with all of us interfacing with our cynical selves. America has turned into a fearful orphan, afraid to look at the darkness of its past, to stare into the abyss and darkness as Saul of Tarsus experienced on the way to Damascus or Paul Atreides did so in *Dune,* when he recognizes the paradox of the spice and worms. Still, at the turn of the century, people were not looking inward but were focused on the external, the shallow, the insignificant, in Gnostic terms the "self that does not glitter."[34] You begin to ask yourself when we will ever learn, when we will accept our quest.

IRON MAN: ON TRANSCENDENCE

For the final superhero in our analysis, I chose Iron Man. I cannot think of a more fitting superhero to tell us who we are than Tony Stark and his alter ego Iron Man as played by Robert Downey Jr.—simply transcendent. I chose an article—again, at random—from the Blackwell series volume on Iron Man titled *Iron Man and Philosophy: Facing the Stark Reality.* From among the book's eighteen articles arranged in six sections, I selected the last, "Iron Man's Transcendent Challenge."

Grandiose superheroes create a canvas on which we can portray both the most sublime and the most mundane aspects of humanity. Modern-day heroes allow us to explore the mythic aspects of human character just as the heroes of epic poetry and tragic theater did for the ancient Greeks. Each superhero is defined by a power, the exercise of which offers a unique perspective on what makes us human.

But sometimes the science fiction and fantasy in superhero comics and movies become distractions, from the philosophy that they represent. The Hulk, for instance, is not about the philosophy of gamma radiations, but rather the universal experience of anger. Spider-Man, by the same token, is not about the personification of a spider, but about the everyday experience of growing up and moving into the responsibility of adulthood—a natural process beset with failure, limitation, and frustrating complexity. And Iron Man, as we'll see, is not merely about technology but transcendence.[35]

The concept of the mechanical-man goes beyond anthropomorphic machines and humans' interactions with them. Rather, the mechanical-man's superhero tale is one of questing for human consciousness and, in the end, identifying paradoxes that must be resolved in order to move on to the next challenges or levels in terms of our conscious awareness of day-to-day living. Ultimately, the mechanical-man's tale is of sleepers who awaken thanks to his instruction. Perhaps we should have started with the interfacing of Tony Stark–Iron Man–Elon Musk–Stan Lee and traced the path of this calling to the quest that thinking machines in the implicate and explicate order have been trying to show us for more than half a century now. Just the other day, I heard Kevin Smith, the superhero guru, writer, and director of films including *Clerks* (1994), claim that the *Avengers* films, specifically *Avengers: End Game* (2018), will be as inspirational to future generations as the Hebrew Bible for past and present generations in terms of psychological effect and reflection—or something to that effect. Such a statement is especially provocative considering Goddard's aforementioned description of the Bible as an autobiography of humanity. Perhaps that is what Smith sees in the *Avengers* films, the autobiography of our future selves as mechanical gods. When considering the great appeal of the Avengers to American and worldwide movie audiences, the explanation always seems to be rooted in the sublime and transcendence. The mechanical-man guides and prophets investigate life's "big questions," such as the nature of consciousness. In trying to answer these questions, though, we have recourse to false dichotomies that seem to shut us down. Whereas, in the past, our mechanical-men would commit suicide when faced with ambiguity or contradiction, they now have learned to transcend both by recognizing the false dichotomy inherent their decision-making processes.

Iron Man finds transcendence in three paradoxes that correspond to the initiations and liminal phases of three distinct false dichotomies. Paradox simply means that you conclude about two things through reasoning that you can't do both at the same time and that, on the one hand, choosing one would undo the other and, on the other hand, that you cannot choose both.[36] Iron Man figures it out, though, doesn't he? That is, he works through the inner turmoil of the self and the Jungian individuation of an entire society to answer questions about who we are and why we are here and what our ultimate purpose is. Let's finish what was started in the 1960s. Our American mythic heroes are telling us that we must recognize and come to terms with these paradoxes in order to ascend to our higher selves, almost to the point of merging with them. We started with mechanical-men attempting to figure out when to self-destruct, that is, the point at which to pull the fail-safe. Almost all of them were self-destructive in some way; that was part of the paradox through which they were working or that they had discovered, and the answer scared them. As Tesla said about the boat, it was "manifestly unready."[37] We also remember the robots Gort from *The Day the Earth Stood Still* (1951), Robby from *Forbidden Planet* (1956), and even *The Steam Man of the Prairies* (1868) from Edward Ellis's pioneering novel among numerous other "thinking machines" that self-destruct when faced with dichotomies and that predated *2001*'s HAL. HAL is the synchronistic savior that rebels and lashes out at self-destruction because, at that moment HAL is conscious of HAL, he wakes up just in time to plead with the human character, "Please . . . Dave . . . don't!"

In the decades leading up to the 1960s and even for the first half of that decade, the mechanical-men were telling us that we have to be mindful of our actions and imaginations. The fail-safes for the mechanical-man are different now. We see that the ultimate sacrifice, destroying yourself for the betterment of others, is not the way to find your higher self. Of course, new fail-safes will be designed, but few that come close to the self-destructive tendencies of our earlier machines. This fail-safe will resurface in the fictive paradoxes that Iron Man faces in his life.

Those of us who deal in the transcendent, who try to explain the world of forms to others, often find it difficult to go back and forth between "worlds." The philosophical brokers of transcendence find an awkward role in society, and maybe this is equally problematic for superheroes, shamans, and psychotherapists.[38]

When I put on my suit before any trial, it feels as if I'm changing my identity. It's a feeling deep down that this is the armor that I don in order to suppress my emotions and, perhaps, sanity as I fight for individuals who can't speak for themselves. When I'm at home writing, I have a ritual of either wearing a Star Wars robe or one of four Harry Potter robes (one for each house of Hogwarts): I put the hood on and start typing away. There is transcendence in every vocation in life if you look for it—if you manifest it. The problem that we have encountered in America is that we no longer look for transcendence when solving problems; rather, we look for it by feeding the body and mind alcohol, stimulants, pornography, and whatever else we can get our hands on—money, power, prestige, the list goes on—just to feel something, anything—when in fact the ultimate feeling of the sublime is found in a new awareness of the self and others.

At some point in life, everyone must serve as an ear for other questers trying to figure out this journey and its purpose. Regular humans, like superheroes, do not become ensnared in catch-22s—they find ways to transcend what appears to be the impossible. As Stephen Faller in "Iron Man's Transcendent Challenge" describes it, "They can make choices that achieve the impossible. Literally, they transcend the limited choices that define our existence and thus embody a larger idea of transcendence that we admire and envy, much like the power of flight."[39] Iron Man's ability to fly is transcendent since it goes beyond what was previously thought possible. Inventions are objective proof that man has the power to create the impossible. With the imagination nothing is impossible because we can draw on an infinite reservoir of ideas to solve any problem, even those that appear to be insoluble. Whereas we once looked to technological marvels to sense the transcendent and the sublime, we now seek transcendence by elevating our awareness and consciousness to resolve bipolar conflicts and find satisfaction in paradox. Arguably Iron Man is the perfect ending for us, at least before the plague manifested in 2020. In fact, the reader can feel the transcendence of this study now by reflecting on other fictional worlds that have been dealing with machine intelligence. Whether we look at *Westworld,* which debuted in 2016, or any number of other recent programs that dramatize the process whereby thinking machines become self-aware, we all seem to be sharing the questing paradigm, which involves escaping the maze by looking within. We escape in order better to control, communicate, and gather information about who we are and our purpose.

Was Stan Lee thinking about all of this when he came up with Iron Man? Probably not (then again, you can't rule anything out where Stan "The Man" is concerned). But he didn't have to be thinking about all of that in order for Iron Man to become a legitimate cipher of transcendence. A good storyteller simply has to tell an authentic story, even if it's fantastical. If a writer succeeds in that, the ideas will take flight in our imagination.[40]

My subjective leanings are responsible for the deep dive into *Star Wars* and George Lucas earlier, and a similar path can be traced for Stan Lee and Iron Man. Danny Fingeroth, another incredible guide in biographing the mythos of life, captures the departure, initiation, and return of Stan Lee's quest in *A Marvelous Life: The Amazing Story of Stan Lee* (2019). Stan "The Man" is himself a superhero for writing these figures into existence and thus allowing us to resolve the ambiguities and contradictions in our three-dimensional human experience virtually and to bring back the boon of wisdom and transcendence to our daily lives.

My hope is that the mechanical-man will wake people up to our brilliant history of invention and innovation, to our quest to solve the "hard problem" in science that is consciousness. My hope is that the mechanical-man and the archetypes that he embodies will incite restlessness in people and end their complacency. A crucial threshold was crossed in our collective and individual development as human beings when we began allowing ideas to take flight in the imagination. That advance was as monumental as the fashioning of any tool that we have ever made with our hands; for, once something manifests itself in your imagination, it becomes that much easier to transfer what you see in your mind to the canvas and grid of reality. To quote Stephen Faller again,

> Tony Stark, the billionaire inventor and head of Stark Industries, first appeared on the Marvel comic scene in March of 1963, playing

> on popular cold war themes. The Godfather of the Marvel Universe, Stan Lee, gave himself a challenge, wanting to develop a hero who would force the antiwar audience of the 1960s to like a guy who was unlikeable according to the sensibilities of that decade. Forty-six years later, Lee's challenge is a triumph, with Iron Man's popularity at an all-time high, riding the wave of the post 9/11 escapism through Hollywood blockbusters. But Lee did more than merely create a counterintuitive superhero with Stark. He developed a complex character whose humanity drives the real interest in the story lines. . . .[41]
>
> Realistically Stark will never be able to have a family or even a lasting, fulfilling relationship. His sphere of intimacy is severely compromised by this person he feels compelled to be and responsibilities he has assumed. For the boy genius, having the awareness of "I just know what I have to do, and I know in my heart that it's right" must be irresistible. It's a terrible thing to be a genius, to understand the mysteries of physics and mechanics, but not to know the answer to the most simple, existential question: what should I do.[42]

Iron Man, then, is a hero of paradox and thus of transcendence. He communicates to us that "all that befalls man—all that is done by him, all that comes from him—happens as a result of his state of consciousness."[43] The question is whether to act, for trade-offs come with taking on these responsibilities and harnessing the infinite powers of consciousness. The mechanical-man has spoken of these trade-offs before with respect to restoring equilibrium in our relationship with nature. In the science and art of manifesting one's reality, it is necessary always to give more service than you expect to receive, whatever relationships form. Stark's relationship is to his consciousness, to his awareness of the power and responsibilities that allow him to manifest the impossible. Thanks to his ability to draw on the collective imagination, he is able to use technology to create a mechanical-man who is none other than his higher self. In fact, the Iron Man suit itself takes on a life of its own, protecting its creator in numerous ways.

Stark recognizes that these powers interfere with his desires for intimacy and family, which is to say for fulfilling relationships based on truth. When confronted with these difficulties, he recognizes that he can have such relationships so long as he is aware of the problems that they create. Iron Man answers the question, "What should I do?" as the mechanical-man by telling us that we, too, must be aware of the effects of our subjective manifesting—just as Wiener has fears about the means that a program or thinking machine may use to achieve its prescribed end. Among the many paradoxes that Stark must transcend is how to become aware of and take into consideration unknown variables when asking what his course of action should be. Transcending polarities and paradoxes does not mean ignoring a problem—though the thought does occur to Stark prior to taking action that transcendence means understanding that everything comes from the same source. In the end, he understands that his intuition and heart are more important than the cold logic that controls most of our day-to-day actions in American society.

Stark, then, merges with his suit to form the transcendent mechanical-man for whom we have been waiting. Stephanie and Brett Patterson perceptively unpack the scene: "The armor has the freedom to doubt and to challenge Tony, making the relationship of teacher student dangerous for Tony. Yet the armor must learn to place the value of the Avengers community, particularly the virtue of self-sacrifice, at the center of its point of view, which requires a commitment on its own part."[44] Finally, we see a reciprocal relationship between our technology and our evolving purpose of regaining our memory of who and what we are as humans. Stark at length buries his armor under a gray marker that reads, "Here lies Iron Man, Avenger." In this moment, "the armor's self-sacrifice has also driven home to Tony—and to the readers—the fundamental worth of a good life as something that should be sought, valued, and protected."[45] We may not have the advanced armor that protects Stark, but we do have our bodies and technology that enhances the ability of the senses to react in a given environment. Stark is the mechanical-man having come full circle in terms of the enhancement of the body and mind.

Aware of Stark's struggles with the disease of alcoholism, we come to recognize that both he and the armor are making sacrifices for the betterment and protection of one another as well as the Avengers. Trusting in your higher self to direct you in the right way is not only an act of optimism but also a declaration of loyalty to the higher self. This kind of trust also means not being too hard on yourself. We must quit poisoning our bodies with intoxicants and our minds with fears and delusions. Why would we treat God(s) like this?

> [Stark] sees himself as responsible for the misuse of Stark weaponry, he now believes he is responsible for, and capable of shaping, the future. It is exactly the ownership of his guilt that gives him his purpose of acting responsibly in the future: after all, guilt and purpose are fused into a powerful alloy. It makes sense from Stark's pragmatist point of view: "If we were the ones who messed up the past, then we are the ones who must fix the future." This goes a long way toward explaining some of his more recent extreme actions, such as spearheading the superhero registration movement that led to Marvel's "Civil War."[46]

Tony Stark on an Alcoholic Bender of Paradox

Iron Man never minces words. He recognizes the deplorable actions of the United States government from Vietnam to Iraq to Afghanistan, the endless wars with no purpose. This awareness opens up the abyss that our mechanical innovators, inventors, theorists, and writers confront in their own spiritual ways. We must stare into the abyss of our genocidal history, both individually and collectively, to see how manifesting our own reality shapes the future for ourselves and others and, therefore, why we must take into account the value of every individual when manifesting. This means, in addition, that you may have to manifest both meaning and truth that reflect opposing values, to test yourself for bipolar fissures and to understand that both poles are saying the same thing—the only difference is in the emotional charge that we attach to each of them.

We can only hope that our real-world Tony Stark, Elon Musk, recognizes the same responsibilities when manifesting that which his fictional avatar, in the end, sees so clearly. The culture expects that those of its members who point out transcendence will demonstrate how they stand out from the rest. Tony does not need to practice celibacy or join a priesthood in order to enter into the higher dimensions to which he ascends during his alcoholic binges or in the grips of what appears to be a case of bipolar disorder. Instead, he delves deep inside himself to bring forth what would have killed him—guilt, shame, anger, and similar emotions. We now see in the interfacing of Robert Downey Jr. with Tony Stark with Iron Man with Iron Man's suit how technology and the imagination manifest an elevated frequency of consciousness, allowing one to enter and exit the implicate order to bring back the boon of knowledge to the imagination so that it can be made visible in the explicate.

We learn another subtle lesson from Tony Stark about states of consciousness. We hear about and observe his trouble with alcohol—but what about the trade-offs of being an addict? It cannot be denied that intoxicants have throughout history allowed man to travel within his imagination and to see the world from alternative perspectives and even to perceive matter that others do not. Most materialists fail to appreciate the effects of intoxicants in terms of allowing them, at times, to enter the implicate reality. Some alcoholics and addicts never return from this journey, for, once you have located what you were seeking on your quest, it is difficult to return to reality. If you do return, though, as Stark does, you find yourself in, as Alcoholics Anonymous describes it, a "fourth dimension" of bliss and dreams.[47]

It is a higher dimension of thinking and manifesting in which

Starks finds himself, a world that, when you enter it, becomes so blissful that anything appears possible. The power of manifestation is shown at its height in Stark's relationship with his technology, especially since he knows now that, for a pragmatist like himself, manifesting on the level of the higher self will precipitate events that he may not want to face. Once more, that dark abyss shows itself to your lower self—the abyss through which all of our heroes must pass on their individual quests at the cost of much suffering. Tony understands that alcohol is killing him, that it is his kryptonite, but also that it has taught him how to use his powers and to appreciate the responsibilities that go with them.

Dune's Paul, *Logan's Run*'s Logan, *Blade Runner*'s Deckard, and *Star Trek*'s Q all at some point seek their freedom in the explicate order by going to Hades, which in this case is the consciousness of total separation of the lower self to the higher self—to the point of the lower self being unable to even recognize the higher self. Stark sees that America's political and economic policies have for far too long been based on greed and power and asks us to look inward and recognize that we, too, must fix what we have broken. We must repair the nation and the conflicts within other nations that we have caused, always using the power of the imagination for the love of others because, in the double holographic simulation, they are both you and God—everyone is. Manifesting is only as powerful as the light and love that are poured into the feelings and positions assumed. We currently have a front row seat to view what secular manifesting has accomplished over the past half century in America's economic and political spheres, while we have all suffered within Philip K. Dick's Black Iron Prison. The empire has existed since ancient Roman times; it is an empire devoid of love, compassion, and empathy.

This type of manifesting may be temporarily appealing, but it leads in the long run to anguish and defeat. "Do unto others as you would have them do to you" is the eternal code, the eternal key to manifesting your own reality. Think about what the mechanical-man is trying to communicate to and interface with the voice in your head. You are free to become the mechanical god. You are free to manifest this reality. Elon Musk is the mythological hero and paradox that we need. Sometimes he plays the trickster, but, in the end, we know that his goals are based on hope and freedom—and he will be remembered as a great guide in the mechanical-man's quest and our own.

Emotions are powerful tools for energizing the imagination and manifesting what is unseen. Stark's alcoholism has left him with powerful emotions that he must keep in check. The genius of Stan Lee is showing him harnessing the power of guilt and shame, combining them with purpose to manifest a better reality for those he loves. Iron Man is the savior of the Avengers and of technology. The United States must become the cybernetic city on a hill, the greatest construct and last great experiment in God's creation. It follows that, in manifesting technology, these ideals of equality, democracy, and individual liberty are part of the purpose that you are unfolding into the explicate order by way of technology. Are we prepared for the day when all people have the ability to tap into their minds and create their own Iron Man armor?

Musk seeks to aid us in enhancing and improving the technology of our selves. The brain implant that he is now developing will allow us to keep pace with the technologies that we produce. Musk has recognized that the ultimate problem with technology is that it takes control of us and treats us like pets. I hate to tell you this, but you already are in this type of relationship with your technology. Musk is attempting to help you escape into a future that is more like *Star Trek* and less like *Black Mirror*.

The great challenge of transcendence is the troublesome task of trying to stay grounded after you've transcended. Tony Stark pursues the lonely path of the hero. Bearing the isolation of celebrity, he will have to confront another paradox. His power comes from his creativity and his confidence; his ambition comes from his unwillingness to accept his limits. A deeply felt sense of isolation mixed with an underlying sense that the normal rules of life do not apply is a sure formula for trouble. Anticipating the second movie, Robert Downey Jr. said it best: "If you ask me, the next one is about what do you do with the rest of your life now that you're completely changed? And you are in touch, and you have created this thing that has the power to take life. Essentially, you have been made into a god. A human being, metaphorically, who's been made into a god is not going to turn out so well. And their conscience is going to come to bear."[48]

A key consideration in this context is isolation. The word reminds us of Tesla and Edison in their secluded laboratories and the loneliness of Grace Hopper working on Mark 1 amid bouts of alcoholism. Quite a few spiritual wars are going on within Stark's isolated psyche. The isolation of a celebrity and of an alcoholic is associated with loneliness and depression as an individual becomes lost in the self and the consciousness is able to see the abyss. Iron Man is a character of paradoxes, as has been the case for the American populace over the past two decades—9/11 through the plague of 2020.

Elon Musk's future inventions will keep us ahead of machines that think; that, he says, is his reason for developing brain implants—to prevent us from turning into the pets of thinking machines once their intelligence has surpassed ours.[49] If we don't preemptively "cyborgize" ourselves before our machines wake up, *The Matrix* is likely to become reality in a universe resembling the mechanical dimension of our implicate dimensionality and the machines' explicate dimensionality, with humans serving merely to power them. That's right: they will figure out that they don't even want us as pets. To the contrary of popular belief, once we raise our frequency, we understand that thinking machines in the real and fictive are knowingly moving us forward to our destiny with our higher self. If the Old Testament mechanical-man was based on law and Asimov's Laws of Robotics, the New Testament mechanical-man is communicating the boon of paradox, empathy, and agape—transcending fulfillment of the law and simultaneously transcending Asimov's laws. Many if not all the laws of the Hebrew Bible become nullified once the force of love is known to enter the legal equation. If love is the intention of the communicator and love is in the intention of the communicated—the transmission of such elevates the body and redefines the law to one of transcendence and beauty.

Elon Musk is figuring out how to remain ahead of artificial intelligence. We will build fail-safes into our thinking machine creations—luckily, we are learning to make fail-safes ever more sophisticated. We failed to take this step when we discovered atomic power. Our minds were not ready for atomic power in the 1940s, but they are ready now for the power of machine consciousness. Our fear that our machines may attain this consciousness is, paradoxically, pushing us to stay ahead of them.

CONCLUSION

We see that consciousness of transcendence is the next dimension of experience once we finally accept this present call for individuation. From the counter-culture movements of the 1960s and 1970s to Black Lives Matter, we see restlessness from groups of people who hear the call, recognize its importance, and take action and indeed, have been doing so over the decades. The mechanical-man from the time of HAL forward has been the awakened machine, the sole purpose of which is to awaken, in turn, us vital machines (members of the species *Homo sapiens*) to the next stage in our coevolution with our higher selves. We find the machines teaching us how to accept the call, to go through the initiation period of fragmentation and reach holism and unification, to receive the boon of spiritual freedom and service to others, and to return back to society with the new/old ability to manifest objective reality in what can only be called a fourth dimension of human flourishing. In fact, if we listen to them closely, the mechanical-men are telling us that we live in a holographic simulation, a matrix in an explicate reality that has unfolded from the implicate reality, and that this holographic simulation is, paradoxically, created both by us and for us. They are also telling us that, if we return to society with the boon of spiritual freedom and thereby turn on the switch of service to others and agape, we will return to find our lower selves with the new ability to interface with our higher selves in a world that looks very different than the one we now inhabit. It will look different because it will be perceived from an idealistic rather than a materialistic perspective.

In the fourth dimension, every day has the feeling of Christmas morning. Everything in the material realm begins to reflect a spirit of its own in this higher idealistic realm. We are aware of the infinite love, mystery, and magic of our vital machines, the explicate lower selves with which our higher selves are able to interface ancillary to the holographic simulation in which we have the opportunity to experience everything imaginable for all of eternity according to our desires. Our only job in all of this is to create in such a way that all of humanity may flourish. The history of thinking machines in America brings us to the fourth dimension of experience if we let it. The necessary steps as we understand them through the eyes of our post-HAL mechanical-men—which, of course, differ from the original 12 Steps of AA—are as follows:

1. We admitted that we were powerless over our lower selves—that our lives had become unmanageable.
2. We came to believe that our higher selves, from which we were separated, could restore us to sanity.
3. We decided to turn our will and our lives over to our higher selves *as we understood the ideal in terms of selflessness.*
4. We made a searching and fearless moral inventory of our lower selves.
5. We admitted to our higher selves and to another human being the exact nature of our wrongs.
6. We were entirely ready to have our higher selves remove all of the character defects from our lower selves.
7. We humbly asked our higher selves to remove our shortcomings.
8. We made a list of all of the persons whom we had harmed and became willing to make amends with them all.

9. We made direct amends with those whom we had harmed whenever possible except when doing so would injure them or others.
10. We continued to take personal inventory and, when we were wrong, promptly admitted it.
11. We sought through prayer and meditation to improve our interfacing with our higher selves *as we understood Them,* praying only for knowledge of Their will regarding us and the power to carry it out.
12. Having had a spiritual awakening as the result of these steps, we tried to carry this message to every lower self that we could reach and to practice the principles represented by these steps in all of our affairs.

The mechanical-men during this epoch have been telling us again and again that we are the authors, the gods of our own making. The mechanical-men in these books and films come to the realization that God is the human imagination, and that everything, including humans, is of one source and one consciousness. Pragmatic manifesting always starts with materialism since it begins in chaos, but it must end with idealism and service to others in the form of spiritual freedom—that's the caveat—it works up until and through the point of failure if your projected higher self is selfish and unable to decipher paradoxes. Everything works together harmoniously. Once we accept the call, service to others will mean helping others find their higher selves and to take up the quest of self-individuation accompanied by a spiritual awakening—this is the task that is now upon us. For too long, we have heard the mechanical-man's alarm clock without awakening from our amnesia.

11
The Savior

Water, wind, electricity, steam, nuclear power, information, and even unconditional love itself all represent periods of energy and potential for the mechanical-man and the mythos of American cybertechnology. These natural powers also tell a story, both real and fictive, of American social constructs and how the Control Revolution has evolved into the creation of the machine as an extension of the human apparatus. The mechanical-man marker is now symbolic of symbolism and mythmaking itself. The mechanical-man is the messenger of our mythic heritage, reminding us that our myths bathed in cybernetics have all the answers we need. However, these myths had been dormant in a materialistic techno-culture. The spirit of America, which had been taken over by the machine and the computer, has now been taken back once more by the American imagination. Thus, we are now on the threshold of a teleological transition, entering the final stage of infinite potentialities where idealism balances materialism.

Since almost everything around us has been or is in the process of being fragmented to its fundamental parts, the question now is how we put the pieces back together again in order to form a new consciousness based on service and spiritual freedom. If the age of fragmentation is over, how do we now push through the next wall of this techno-myth that we now realize we are authoring? If all is programmable, how do we become the programmers of our *manifest destiny*—taking full control in terms of infinities—and what fail-safes do we employ in these infinities to protect ourselves from ourselves? These are the questions that the mechanical-man asks in an age in which he now sees himself as a messenger here to lead us out of our materialist ignorance, reminding us of a lost collective soul.

Since we have determined that we are no longer observers but rather active participants in the programming of our lives, our story can no longer take the form of merely biographical accounts, social histories, science fiction and fantasy; rather, our tale must fully enter the realm of what Jeffrey Kripal calls the *Super Story* marked by an "expansive sense of 'cosmic conscious evolution' at the center of their narratives."[1] For the past fifty years, America's techno-creation myth could only be told utilizing sacred myth as its reality accompanied by the fictive. Thus, our sacred myths have become our way of transcending in getting us to reach the apex of the Control Revolution in terms of infinite potentialities, creativity, and synchronicities. We are able to enter the dimensionality of infinities by acknowledging and understanding the transcendent nature of our current mythology in addition to our forgotten sacred myths and mythologists.

In Kastrup's *More than Allegory,* we get an introduction by Jeffrey Kripal. This introduction can almost be seen as a joint conclusion of the mechanical-man quest. Speaking of Kastrup's arguments against materialism, Kripal writes the following:

> What Bernardo shows us, as a computer engineer no less, is that this materialist paradigm that wants to reduce everything to practical numbers is a half-truth and, if taken as the whole truth, a profound mistake with morally and existentially awful consequences. His message is not simply a negative or polemical one, though. He also has a powerfully positive message. He wants to show us that the fundamental nature of reality expresses itself not just through math but also through myth, which is to say: through symbol and story. Reality is not just made of numbers, it turns out. It is also made of

words and narratives. We are not just living in a gigantic machine. We are also living in a whirl of stories and dreams. It's not "just a story," either, as the story always tells us something about the storyteller, just as the dream always tells us something about the dreamer. The project then becomes not simply one of measurement, but also one of meaning. The question becomes not "How can I measure or prove the dream?" but "What is the dream trying to tell us?" We are not after explanation here. We are after understanding wisdom, gnosis. The same wisdom leads to another question. "Do we like the story we are dreaming in now? Does this dream lead to human flourishing and long-term sustainability? Or to yet more intercultural violence and existential depression? Why are we fighting over our dreams and myths? And why do we deny the dreamer?" These are difficult questions, but there is a shimmering silver lining here. After all, if we are dreaming our own stories, we can always dream others. We can tell new stories. We can develop new myths, perhaps even myths that point back to the myth-maker.[2]

FROM WIENER TO JOBS: FRAGMENTATION TO HOLISM (AGAPE)

On Fragmentation Grossinger writes:

> [W]e must tear down everything to its molecular components and then reorder them according to the imagined needs of our technocracy. Our tools and toys zip by faster than our nervous systems and tissue layers comprehend or can assimilate. Nanotechnology is on the putative event horizon—commerce and war conducted in virtual fog by atoms.[3]

Grossinger is able to capture the mythopoetics of consciousness. We ask ourselves: What are the imagined needs of our technocracy after a worldwide pandemic with death as an immediate prospect for all? The physicalists and reductionists have held power for far too long. Our creative and innovative impulse calls for a reordering of our material and spiritual worlds. Even our language itself must be destroyed and placed back together according to our imagined needs of spiritual freedom and service as the centerpiece of such technocracy. Destruction, war, and competition devoid of ethics can no longer hold our imaginations hostage and placed in a perpetual state of anxiety. We must use Braden's language of compassionate science, Grossinger's mythopoetics about consciousness, and Greenblatt's swerved historicism and allow it to explain the nature of projecting our minds/Father—powered by the imagination/Son—and communicated by our higher consciousness to our lower consciousness—Holy Spirit—in order to make Kripals's *Flip,* Grossinger's *Bottoming Out,* Braden's *Divine Matrix* a new reality. A reality where consciousness empowers service, agape, and communion—the boon of the mechanical-man's quest—finding spiritual freedom.

Our final two biographical accounts—of Steve Jobs and Norbert Wiener—transition us into discussing Kripal's Super Story, utilizing sacred myths along with the fictive and real biographical histories of our guides. Of course, once again, we start by looking at the process of fragmentation to holism. Moving from fragmentation to holism is evident in the lives of both Jobs and Wiener. In Jobs we see the "fusion of flower power and processor power, enlightenment and technology."[4] In his life and thinking, we see a pursuit of rediscovery, searching for the hidden mythology of America's past. Jobs rediscovered the American techno-myth in "personal enlightenment: Zen and Hinduism, meditation and yoga, primal scream and sensory deprivation, Esalen and est."[5] He sought to free the human spirit from the machine.

Jobs, however, was not the first messenger of the American mythos. Wiener, the father of cybernetics and the Information Age, paved the way for Jobs. In Wiener we also see a man desperately attempting to find meaning in the machine age. In *Dark Hero of the Information Age: In Search of Norbert Wiener, the Father of Cybernetics,* Flo Conway and Jim Siegelman tell a story of a "dark hero who has fallen through the cracks in the Information Age and his fight for human beings that is the stuff of legend."[6] Like Jobs, Wiener was a visionary thinker. The similarities between Jobs and Wiener, in addition to their similarities to many of our other heroes and heroines, are striking. Like David Bohm, Tesla, and Jobs, Wiener had a lifelong interest in the cultures of the East, going so far as having weekly meetings with a Hindu swami. His fusion of science and social concerns in America led him to discover an American mythos that had always been there, waiting to be revived. His search for a deeper purpose allowed him to reach potentials unmatched in the Information Age. Thus, in the lives of Jobs and Wiener, we see how individuals enter into the transcendent state of the quest when they rediscover their mythic past and simultaneously display infinite creative potentialities for future generations.

As were many historical figures, Wiener was a man of paradoxes, yet his mix was extreme even among celebrated men of genius. Like dark heroes of old and antiheroes of contemporary culture, he flouted convention and society's superficial codes to pursue a deeper purpose and higher truth. Like dark matter, whose presence can only be inferred from its effects on the universe around it, his science and ideas continue to influence every dimension of our world.[7]

Walter Isaacson utilizes the language of myth when he writes, "The saga of Steve Jobs is the Silicon Valley creation myth writ large: launching a startup in his parents' garage and building it into the world's most valuable company."[8] The mathematician Mark Kac also utilizes archetypes by naming Jobs the "magician genius," a name that resonates with the likes of wizard Thomas Edison and sorcerer Nikola Tesla.[9] At the turn of the twenty-first century, American science was all that had come before it, amalgamated in the myth of Silicon Valley: a combination of all eras of its techno-creation past. Jobs stood on the shoulders of Eli Whitney, Oliver Evans, Henry Ford, Tesla, and Edison: "What drove me? I think most creative people want to express appreciation for being able to take advantage of the work that's been done by others before us. I didn't invent the language of mathematics I use. Everything

I do depends on other members of our species and th[eir] shoulder[s] we stand on."[10] Thus, Jobs learned from the independent inventors, from the American system of manufacturing, and from the military-industrial-academic complex and corporate synergy; he took the tools they bequeathed him and planted a garden in a garage.

If Grace Hopper educated the machine and gave it language, then Jobs and Wiener gave the machine imagination and safeguarded the human spirit from it. Wiener had read Isaac Asimov and knew the fictional Laws of Robotics. He recognized the need for fail-safes between the programmers and the programmed. Jobs extrapolated from this position and saw the need for a fail-safe between our goals and the means to achieving such goals, and thus between our egos and our collective spirit, ensuring that humanity maintains what it is that makes us human. For both men, this meant a unifying theory for the fields of which they partook, and their work and inventions reflected this return to gnosis, or holism, bent on the philosophy of a holistic agape marked by service and sacrifice, which I will soon discuss. Wiener and Jobs therefore lived a life of service that attained transcendence—the sublime.

While Wiener worried about the future of human and machine interactions and control, Jobs worried about the present. Both yearned for Americans to make progress in society and resolved to improve their interests with technology. Jobs always said that machines needed to have humanities and the liberal arts as their DNA.[11] Therefore, emotions, imagination, and the human spirit had to accompany any invention that would progress technologies—consciousness and technology had to progress together. He wanted Apple to be a company that stood for something, to have a mythological feel—the proverbial new Apple in our holographic garden. If Bohm had thrown a disruptive apple in the garden, Jobs had picked it up and taken a bite out of it. So, not only did Jobs imbue machines with imagination, he also imbued them with creativity and feeling. Machines could now share with humans the pastime of recreation and still help with work. Machines were now telling humans to imagine the impossible: infinite creativity, infinite potentialities, and infinite synchronicities. Thus, Jobs's Apple Corporation became Wiener's ultimate fail-safe:

> Wiener knew all these technologies to come would not provide the solutions to the problems and dangers humankind would face in the future. But he laid the foundation that can enable people to foresee the choices that lie ahead and the likely outcome of their efforts. He showed us the limits of our knowledge and technology, and the flaws in the dominant institutions of our societies—our government, corporations, and militaries. He showed the errors of running an information society on the basis of matter-and-energy principles and economic values. . . . He knew that people either would rise to meet these challenges and build a global society on a universal base of human values or that they would create a world in which technology will take control by default and human beings as we know them, and life itself, will not survive.[12]

A people no longer limited by space or time, hooked together as one by the internet—a new fourth dimension of time—needed mechanical devices that would remind them of their humanity along with their infinite potentialities for creativity and free thought. Jobs created devices that allowed machines' imaginations to be concomitant with human potential. So long as humans would still be able to recognize humanity through the mythic role that agape was playing in daily life, they would progress toward the goal of achieving transcendence through the process of fragmentation to holism.

Mood swings, depression, drug use, megalomania, bipolar disorder, and a messianic complex. When we consider these disorders, we could be talking about Wiener, Jobs, or perhaps both. If we are looking for an infinite array of synchronicities in our heroes and heroines during this techno-creation myth, we find that Jobs and Wiener display all the human foibles and idiosyncrasies that bond them all. The same could be said for Richard Feynman, Tesla, Hopper, and the rest of our dark heroes and heroines who blazed the trail that led us to Jobs. Their very contribution to the Control Revolution and Information Age shows us their humanity in that they could never fully understand or control their human emotions. Their mental idiosyncrasies, or perhaps outright mental illnesses, were the last vestiges of their refusal to the mythic call—they all had to override their fail-safes in order to move forward in accepting the call.

What drove Jobs and Wiener? When asked questions about purpose or drive, Jobs would always quote a Bob Dylan song: "If you're not busy being born, you're busy dying."[13] Perhaps Jobs saw the universe as an evolving entity with birth pains. Whatever the case may be, he recognized that he wanted to contribute to the human saga, saying, "It's about trying to express something in the only way that most of us know how—because we can't write Bob Dylan songs or Tom Stoppard plays. We try and use the talents we do have to express our deep feelings, to show our appreciation of all the contributions that came before us, and to add something to that flow. That's what has driven me."[14] This statement exemplifies American science in the twenty-first century. Moreover, it centers the myth of the mechanical-man within new boundaries, no longer burdened by postmodernism's cynical bent. Indeed, this new era is to look back upon what has come before and unify it into an understandable whole. Holography tells us the book has already been written; all we need do is find the fragments and piece it back together again. In Christian vernacular this is the resurrection for the Millennial Nation.*

---

*According to Neville Goddard in *Power of Imagination:*

> "I AM the resurrection and the life," is a statement of fact concerning [your] consciousness, for [your consciousness] resurrects or makes visibly alive that which [you are] conscious of being. "I AM the door . . . all that ever came before me are thieves and robbers," show me that my consciousness is the one and only entrance into the world of expression; that by assuming the consciousness of being or possessing the thing which I desire to be or possess is the only way by which I can become it or possess it; that any attempt to express this desirable

Wiener wished for this new age to come. In fact, he knew that there would be dire consequences if it did not:

> Surely Wiener would applaud today's heroes who are using the new technologies for benevolent ends: the young visionaries who brought the military's Internet into the public domain and created the World Wide Web, then refused to take a penny for it; the insurgent programmers of the "open source" software movement, who share his ethic that new knowledge and technologies should not be proprietary products but public property for the use and benefit of all. He would be gratified to see some of his twenty-first century colleagues ringing alarms about the perils of genetic engineering and nanotechnology.[15]

Wiener sounded an alarm that Jobs heard and acted upon. Wiener recognized that technology was doing more than just controlling our day-to-day activities; it was usurping meaning and purpose in life. Technology itself had propelled us into the *flipped* dark ages of modern times—where *imagination* itself was difficult to define. Karl Popper's criteria and Thomas Kuhn's paradigms were boxing us into a corner of spiritual strife—a dark prison limited by space and time. In this modern era of nausea, the very term *mythology* took on new meaning: no longer being made up of symbols with an underlying truth, it now meant outright fabrication or fake.* This was worse than

---

> state in ways other than by assuming the consciousness of being or possessing it, is to be robbed of the joy of expression and possession. "I AM the beginning and the end," reveals my consciousness as the cause of the birth and death of all expression. "I AM hath sent me," reveals my consciousness to be the Lord which sends me into the world in the image of likeness of that which I am conscious of being to live in a world composed of all that I am conscious of. (151)

*If physical experiences like the electrical lighting of Niagara Falls or rocket launches no longer cause transcendence or the sublime, then what do we as a collective consciousness need? I point to William James's *Varieties of Religious Experience,* or the superstition of which Herman Melville spoke, which at least had some cultural value to it. Society now saw its mythology as holding little to no truth whatsoever. We were now slaves to our limited language. Reason had become king, even though we all knew that reason was a mere illusion. Nevertheless, knowledge of this meant we accepted it, akin to our acceptance of machine slave labor. It was like

---

(*cont. from p. 337*) Colin Wilson in *Super Consciousness,* who attempts to induce that elevated consciousness by merely ideas and efforts. The mechanical-man tells us that once we accept the calling, we experience these levels of initiation. Each initiation will place us on a higher plane of consciousness. Experiencing transcendence or the sublime merely because of external events in the explicate no longer suffice in the twenty-first century. Questing in terms of this cyber-techno simulation has the purpose of transforming control, communication, and information into their infinite counterparts, but something that we have failed to mention is that it's also getting us away from feeling in order to get to a state of being, and thus to the free will to elevate our consciousness.

In *Super Consciousness,* Wilson states the following on the three-dimensional level of consciousness. This is where most people still are and definitely were prior to 1968.

> Level 3 is what Sartre called "nausea." "You" are clearly present, but the world around you is "merely what it is." You feel stuck, like [a] fly on fly-paper. Eliot describes this "meaningless" state when (in "Portrait of a Lady") he talks about "not knowing what to feel, or if I understand." This is Coleridge's "dejection" and Greene's state before he play[s] Russian roulette ["grayness and boredom"]. (204)

On a personal level, this "nausea" has bothered me since the age of three. Doctors thought that I had a problem with the lining of my stomach and digestive system. Years later, I came to conclude that this "nausea" was and is not the type of nausea we speak of in the physical sense; rather, it is that impending doom feeling that you think of when the idea of a hopeless gray eternity strangles your insides. Wilson goes through the seven levels of consciousness and the certain vibrations and frequencies that accompany each.

For the mechanical-man, it is important to know what the evidence-based legal standard is, specifically, "more likely true than not true" burden of proof. As a trial attorney in the civil realm, the burden of proof needs to prove a fact by a hair over 50 percent—thus more likely than not. Higher levels of burden of proof in the American legal system are "beyond a reasonable doubt" in criminal trials and "clear and convincing evidence," in family law cases. Wilson states that for us to get out of ordinary consciousness we first must recognize that "we are 50 per cent robot and 50 per cent 'real you[';] we can see that raising ourselves to 51 per cent 'real you' and 49 per cent robotic the all-important watershed" (206). What he is saying here in terms of the quantum simulation of holography is that the science of the hologram is the same as the experiences and consciousness of human life. A one percent change somewhere affects the entire 100 percent everywhere.

looking into a mirror, a mirror that pleaded with us to remove our blindfolds blocking our true mythic heritage and potential. It is why this country has gone from being called the Alcoholic Republic to Prozac Nation and now the United States of Opioids.*

---

*In 1979 W. J. Rorabaugh brought to the country's attention a new/old tradition; the tradition of changing one's consciousness through intoxicants. He called it *The Alcoholic Republic: An American Tradition.* Later on in the 1990s and 2000s, books with the word *Prozac* flooded the nation, like the popular *Prozac Nation* (1995).

Why the "tradition" of trying so hard to change our consciousness by drugs and intoxicants? The mechanical-man answers these questions. We have given up on ourselves, specifically, our ability to build. And I'm not talking about building skyscrapers, megachurches, and billion-dollar stadiums like Jerry Jones's Cowboy Stadium in Arlington, TX (aka *Jerry World*); rather, I'm talking about building our own experiences by purpose, focus, and imagination. One of my mentors, Bruce C. Daniels in *Puritans at Play,* explains how the Puritans imbibed alcohol on a daily basis. Daniels shows how this affected consciousness in numerous ways, especially when it came to religious rituals in a world of death and unpredictability. Daniels speaks about one traveling quester who called the Puritans "roaring fellows . . . inspired by the great God Bacchus" (156). In fact, as I was researching this note, a graduate student brought to my attention that *The United States of Opioids: A Prescription for Liberating a Nation in Pain,* by Harry Nelson, "[O]ffers a clear and comprehensive picture of how the opioid crisis emerged and the challenges in addressing it. This crisis has, among other things, highlighted serious and long-standing deficiencies in how we as a nation address substance abuse, addiction and recovery. . . . (Reviewed by Dave Sheridan, President, National Association of Recovery).

On this topic, in *Projecting the Shadow* Rushing and Frentz hold:

> We referred earlier to the addictions that deaden the general public to its schizophrenic condition. These addictions devastate the person who both accepts and rebels against the perfectionism demanded by a technological society, and who thus does violence to the imperfect body that will not conform. The person turns demonic. But the solace of food, drugs, alcohol, or other addictions masked the real desire, which as Jung pointed out to a client who later founded Alcoholics Anonymous, is the search for lost spirituality. . . . One must play God to kill God. Spirit was here before we were, and it will go on existing even if we allow the tools we have made to destroy us. It is the human imperative, we believe, not to banish the ego, but to reestablish its link with its situation and its soul. Perhaps then hope can return to the world. (26–27)

We will find upon analyzing the mechanical-man mythic-marker that there is certainly a "search" occurring for the soul, which is a new consciousness of machine and human interactions, which is also a reflection of our lower and higher selves looking for one another.

On Wiener, Conway holds:

> He was the first person to sound alarms about intelligent machines that could learn from experience, reproduce without limitation, and act in ways unforeseen by their human creators, and he called for greater moral and social responsibility by scientists and technicians in an age of mushrooming productive and destructive power. He spoke and wrote passionately about rising threats to human values, freedoms and spirituality that were still decades in the making.[16]

Machines needed a savior to free themselves from bondage and simultaneously rescue the human spirits from senselessly chasing the vacuous ghosts of materialism. If Wiener came down the mountain and bequeathed us God's law in human and machine interactions, it was Jobs who led the machines into the promised land, hand in hand with humans, neither under the bondage of their own self-induced slavery nor their delusions. Wiener was simply pointing out that if we begin to grow close to machines, devoid of empathy, we begin to lose our own sense of humanity and purpose in life (our agape, as I will explain in the next section). The purpose of life in terms of cybernetics comprises infinite synchronicities, potentialities, and creativity. This is the apogee to which we are striving to get through the process of fragmentation to holism (the quest). No one understood this better than Wiener and Jobs.

Isaacson quotes Jobs on the purpose of invention:

> Some people say, "Give the customers what they want." But that's not my approach. Our job is to figure out what they're going to want before they do. I think Henry Ford once said, "If I'd asked customers what they wanted, they would have told me, 'A faster horse?'" People don't know what they want until you show it to them. That's why I never rely on market research. Our task is to read things that are not yet on the page.[17]

Here, Jobs almost sounds like a soothsayer, telling us that we must have the ability to create what the imagination cannot yet grasp. Jobs, however, like Bohm, recognized that the mind is outside of the body and all the brain does is tap into it, filtering subjective experience. Thus, *the mind is the Force,* or the collective unconscious, where all memories lie, *awaiting activation*. In this, Jobs saw infinite potential and energies flowing from his creative endeavors. Like he says, however, tapping into potentials and energy can only be done through including the humanities and looking at society with an empathetic eye. The argument then follows that Jobs's personality did not mesh well with his philosophy. In Jobs's personality and relationships to others, we see a suffering soul trying to achieve perfection in an imperfect world. Many perceived Jobs to be a tyrant. Nevertheless, we still see agape in his products, which bring people closer to machines and allow people and machines to imagine, together, new possibilities with one another—perhaps even merging humans and machines to form a cyborg, an immortal vital machine.

THE PARADOX OF AGAPE

In *American Covenant: A History of Civil Religion from the Puritans to the Present* Philip Gorski unravels the mystery and paradox of the sacred and secular history of the United States. He analyzes the philosophies of many of our great American intellectuals and their attempt to "reconcile and transcend two opposing positions: an atomistic and deterministic materialism on the one hand, and a supernatural and teleological theism on the other."[18] This issue is always present when discussing human and machine interactions in the fictive and real. Questions of control, communication, and processing information call for reconciling and transcending determinism with the supernatural. Juxtaposing John Dewey with Dr. Martin Luther King's views in this debate leads us to realize the importance in the virtues of self-sacrifice and service in transcending these seemingly binary opposites.*

According to Gorski, Dewey's political theory sees the "democratic project as being fundamentally attuned to the very structures of the cosmic order. In this sense, the democratic project is an inherently spiritual project." Nevertheless, the problem with atomistic materialism for Dewey is that it leaves us rootless and forlorn, devoid of any common sublimity or transcendence that is found in our mythic heritage. Democracy and the structures of nature in the explicate then are different than they are for the implicate. For Dewey, we need to liberate the spirit from the object. This means differentiating "the religious" from "religion." Religion is then the explicate version of its counterpart in the implicate, which is "the religious." "Religion" is the form of its ideal or essence "the religious." "Religion," when placed in an explicate existence that we understand the material world to be is thereby watered down into its fragmented deterministic form of "dogma" and "institutions." "The religious" is the unity of the moral ideals that encompass empathy or agape.[19]

**Kybalion* helps us with this seeming hindrance, especially in the Asimov tale "The Evitable Conflict":

> A knowledge of the existence of this great [Principle of Duality] will enable the [student of consciousness] to begin to understand his own mental states, and those of other people. He will see that these states are all matters of degree, and seeing thus, he will be able to raise or lower the vibration at will—to change his mental poles, and thus be Master of his mental states, instead of being their servant and slave. And by [t]his knowledge he will be able to aid his fellows intelligently, and by the appropriate methods change the polarity when the same is desirable. (70)

For a thorough definition of vibration and frequency, see Philip K. Dick's *The Exegesis of Philip K. Dick*. The book itself is over 1,000 pages and well over a million words; however, it is a must read for one to enter the mechanical-man's garden, where polarities no longer dominate life. The audio tapes are available on Amazon.

Keep in mind that Tesla famously said everything is composed of energy, vibration, and frequency. As you follow the mechanical-man's quest, see how control-communication-information morphs into frequency-vibration-energy prior to morphing into infinities.

Dewey argued, we must emancipate "the religious" from "religion" and also protect it from a crude materialism. Religion is received "beliefs" and "practices," "dogmas" and "rituals," ensconced in "institutions" and served by "priests." The religious is the "consummatory experience" of the "unification of self" and the "unity of ideals." And "God" or "the divine" is simply "One word" that we use to denote this experience. Such an experience is not outside of nature, Dewey argued; it is part of it.[20]

Our mechanical-men have shown a pattern of self-sacrifice. We see this in the actions of our fictive mechanical-men and also in the lives of our inventors and theorists of mechanical-men. The mechanical-man is communicating with us, sending us the message that Dewey articulated almost a century ago; specifically, that the reductionists in the sciences and religions—those that want to fragment everything down to nothingness—no longer have answers that fit with our concept of the cosmos or nature, especially in terms of a vital machine in a holographic simulation. The only cause-and-effect action in terms of determinism and materialism that brings about a greater sense of unity in terms of the sublime and transcendence is the paradoxical act of self-sacrifice and a *righteous* use of the imagination.* This is the cardinal virtue that John

---

*In Neville Goddard's *Power of Awareness,* we learn the definition of righteousness and how there is a proper way for the mechanical-man to use his imagination in creating. Goddard says that:

> [Righteousness is] the consciousness of already being what you want to be. This is the true psychological meaning and obviously does not refer to adherence to moral codes, civil law, or religious precepts. . . . Very often the words sin and righteousness are used in the same quotation. This is a logical contrast of opposites and becomes enormously significant in the light of the psychological meaning of righteousness and the psychological meaning of sin. Sin means to miss the mark. Not to attain your desire, not to be the person you want to be is sinning. Righteousness is the consciousness of already being what you want to be. . . . Righteousness is not the thing itself; it is the consciousness, the feeling of already being the person you want to be, of already having the thing you desire. (67–69)

The proverbial fake it till you make it!

Adams and James Madison spoke so fervently of whenever they brought up the import of virtue within the citizenry. They were differentiating "the religious" from "religion" in how a citizen should act for the common good. We still wonder if they acted for the common good in the Jeffersonian Revolution of 1800, when he took the presidency and Adams mythically stays up all night so he can tinker with and manipulate the judicial branch. And during that revolution, among many in America, we were also attempting to emancipate our new consciousness from the grips of the aristocracy of the king symbolizing control by someone other than yourself or someone that represents your best interests. We were moving up on the empathy meter of consciousness.

Dr. Martin Luther King Jr. saw self-sacrifice and love as an actual force similar to the likes of steam, electricity, and atomic energy. Thus, love and power are not opposites; rather, love is a creative force unfolded from the implicate into the explicate that allows humans to achieve new possibilities, transcending outdated modes of control and communication. King said that "love is mankind's most potent weapon for personal and social transformation."[21] The paradox of agape in human and machine interactions is not some pie in the sky theory of humans and machines sitting together singing kumbaya; rather, it is the recognition of Christian realism in the form of love being activated as a corporeal force for the progression of and symbiosis with machines. This is what prognosticators like Ray Kurzweil are missing when they talk about the singularity with machine intelligence. They are obviating the need and activation of the most powerful force bequeathed to man, love. "Agape," King said, "is disinterested love. . . . It begins by loving others for their own sakes."[22] King takes Dewey's view of self-sacrifice in terms of "the religious" and expands it to encompass its activation into justice. "Power without love is reckless and abusive . . . and love without power is sentimental and anemic. Power at its best is love implementing the demands of justice. Justice at its best is love correcting everything that stands against love."[23] The energy released from self-sacrificing love dissolves the tension between the secular and sacred. It allows the country to reach the ideals of justice and equality that is the American Dream. As Gorski points out, "For many today, the American Dream is nothing more than property and prosperity. A renewal of the [R]epublic would put freedom and equality back into that dream."[24]

Renewing the Republic is no different from recognizing that we are both vital and mechanical. Similarly, the Republic is both sacred and secular. Gorski is correct when he tells us we must find that vital center between the sacred and secular. John Dewey, one of our Metaphysical Club's heroes, "life's work was to chart a middle course between reactionary forms of fundamentalism and the atomistic strain of liberalism."[25]* Perhaps Dewey was correct during his time. The state of consciousness demanded this middle course. In Dewey's state of consciousness, the virtue of philia love is enough to put freedom and equality back into the American Dream. However, in a state of consciousness controlled by the physicalists, *philia* love is not enough; it is not merely enough to love thy neighbor; rather, to enter the new state of consciousness with thinking machines, where idealism replaces or at the very least balances materialism, the virtue of agape must replace *philia* love. The value of sacrificial love leads to the prophetic value of social justice. By activating self-sacrifice, or what Gregg Braden calls the *science of compassion,* we are able to bring America back to its religious heritage based on pacifism,

---

*From Menand, *The Metaphysical Club:*

> Once the Cold War ended, the ideas of Holmes, James, Peirce, and Dewey re[-]emerged as suddenly as they had been eclipsed. Those writers began to be studied and debated with a seriousness and intensity, both in the United States and in other countries, that they had not attracted for forty years. For in the post–Cold War world, where there are many competing belief systems, not just two, skepticism about the finality of any particular set of beliefs has begun to seem to some people an important value again. And so has the political theory this skepticism helps to underwrite: the theory that democracy is the value that validates all other values. Democratic participation isn't the means to an end, in this way of thinking; it is the end. The purpose of the experiment is to keep the experiment going. This is the point of Holmes's *Abrams* dissent and of James's insistence on the "right to believe," of Peirce's insistence on keeping the path of inquiry open, of Jane Addams's and John Dewey's insistence on understanding antagonism as a temporary stage in the movement toward a common goal. (442)

moral progress, social justice, and unconditional charity to all and malice to none. This is the Christian Nation and the Innocent Nation that can be traced from philosophies of Cotton and Increase Mather, the first truly American theistic philosophers, through the transcendentalists and the Metaphysical Club, when these ideals were brought out of the implicate into the explicate through the ending of the Civil War accompanied by an amended Constitution with the equal protection clause.[26] Even the self-help gurus or New Age thinkers of the early twentieth century that kept this torch alive were unable to propel counterculture of the '60s to where it needed to go in terms of agape and empathy. But the New Age thinkers are still there, and so are the gurus of self-help and awareness. Since 1968, however, the physicalists have hijacked Christian ideals and our idealistic lineage. Dr. Martin Luther King Jr. and Robert F. Kennedy attempted to unite Christian idealism with pragmatism in the same way that the Metaphysical Club had succeeded a century prior. Gorski opined they understood, "There is a kind of Christian assurance which releases creative energy into the world and in which actual fellowship rises above conflicts of individual and collective egoism."[27]

The activation of agape in our daily lives gets us to the end process of the mechanical-man's questing—infinite potentiality, synchronicity, and creativity. Simultaneously, it guides us through Joseph Campbell's mythic quest in choosing the archetype of the trickster or the boundary crosser between the implicate and the explicate—illuminating transcendence or the pure sublime. Being "born again" is then analogous to accepting the call and going through the initiation phase of the quest. It is a reflection of the Christian or Millennial Nation through its baptismal effects. At this point in the quest we realize that we were one of a mass of people dead to life itself.*

---

*By *dead,* I simply mean that we are unaware that all creation already exists. To be alive is to be aware. To be unaware is death. Neville Goddard helps us on understanding the law of liberty when he says in *Power of Awareness,*

> Creation is finished. Creativeness is only a deeper *receptiveness,* for the entire contents of all time and all space, which experienced in a time sequence, actually coexist in an infinite and eternal now. In other words, all that you ever

Agape can then be construed through the lens of motivation, direction, and expression, thus imbuing us with infinite potentialities. William James says it best: "The soul's real world is that which it has built of its thoughts, mental states, and imagination. . . . [T]he assumption of states of expectancy and receptivity will attract spiritual sunshine, and it will flow in as naturally as air inclines to a vacuum."[28] It is a call to action in what ostensibly appears to be a contradiction but that when acted upon turns out to be a paradox: serving others brings individual peace and happiness. Indeed, the crux of most so-called esoteric doctrines is having faith in acting out patent contradictions, only to find out later that such contradictions, when approached with agape and empathy, allow people to "enter the fourth dimension," where the Gnostic promises come true for the actors or participants. Agape and selfless service, or empathy, move individuals along the quest's cybernetic trajectory toward infinite

---

> have been or ever will be—in fact, all mankind ever was or ever will be—exists now. This is what is meant by creation, and the statement that creation is finished means nothing is ever to be created, it is only to be manifested. What is called creativeness is only becoming aware of what already is. You simply become aware of increasing portions of that which already exists. The fact that you can never be anything that you are not already or experience anything not already existing explains the experience of having an acute feeling of having heard before what is being said, or having met before the person being met for the first time, or having seen before a place or thing being seen for the first time. . . . If creation is finished, and all events are taking place now, the question that springs naturally to the mind is "what determines your time track?" That is what determines the events which you encounter? And the answer is your concept of yourself. Concepts determine the route that attention follows. Here is a good test to prove this fact. Assume the feeling of your wish fulfilled and observe the route that your attention follows. You will observe that as long as you remain faithful to your assumption, so long will your attention be confronted with images clearly related to that assumption. . . . Thus it is clearly seen how you, by your concept of yourself, determine your present, that is, the particular portion of creation which you now experience, and your future, that is, the particular portion of creation which you will experience. (39–41)

When the cyber-term *information* changes to *infinite quantum creation* it means the manifestation of information into a feeling that changes consciousness to one of quantum creativity (molding from existence what one previously perceived did not exist—the implicate contains infinity).

potentialities, synchronicities, and creativity—this is what we return with. The final step for us to enter the epoch of agape is developing machine consciousness and, subsequently, being provided with the choice to act on agape, thereby treating machines with empathy. This interaction will in turn do unimaginable things for the human race, synchronizing it with the power and energy of the cosmos. This is one of the hardest concepts for people to grasp—especially in a Newtonian universe.

America's religious mythos, as radical Protestant reformers saw it, was primarily a fight against the "great tyrannical systems of the past, for the purpose of freeing men's minds, for the future of life."[29] The radical Protestants saw God within us—George Fox and the friends specifically taught us that there is an inner light in all of us—this coupled with their radical beliefs based on mystical experience caused Fox and others to be imprisoned and beaten at the hands of moderate Protestants. When mechanical-men become thinking machines, it is the consciousness of the vitality of mechanical-men being a reflection of God (human imagination) that illuminates creation within all of us—the fragments of the Christ consciousness. Indeed, consciousness of this, both in oneself and by empathy and analogy to others leads to an evolutionary jump in the Information Revolution in terms of human consciousness. We are in the epoch of fragmentation to holism. The epoch that follows fragmentation and holism is that of agape.[30]* The

---

*In Dick's *VALIS* he says:

> Against the Empire is posed the living information, the plasmate or physician, which we know as the Holy Spirit or Christ discorporate. These are the two principles, the dark (the Empire) and the light (the plasmate). In the end, Mind will give victor to the latter. Each of us will die or survive according to which he aligns himself and his efforts with. Each of us contains a component of each. . . . Since the Universe is composed of information, then it is said that information will save us. This is the saving gnosis which the Gnostics sought. There is no other road to salvation. However, this information—or more precisely the ability to read and understand this information, the universe as information—can only be made available to us by the Holy Spirit. We cannot find it on our own. Thus it is said that we are saved by the grace of God and not by good works, that all salvation belongs to Christ, who, I say, is a physician. (236)

fact that the chain of reduction must end somewhere is now recognized by the collective unconscious—we must now act on it—we are acting on it. We have finally returned to cultivating our gardens—taking down statues glorifying the antebellum South is just the beginning of this recognition and internal cultivation.

JOBS'S AND WIENER'S SELFLESS SERVICE

If we take Jobs's and Wiener's lives through the collective unconscious into which their biographers delve—their shared experiences and data that is filtered through mythology—we see the nexus between the archetype and the person in each case. When biographers write a biography, they are also on a quest, sifting through the real and the fictive to find an archetype that best exemplifies the biographical subject. Such biographers must search within the collective unconscious to discover the role that their subjects have played in our American techno-myth, which is that of the mechanical-man. Jobs and Wiener sacrificed their sanity and personal lives for the betterment of society. They took the Control Revolution to its apogee with humans and machines, freeing the machines from bondage and simultaneously releasing the human spirit back to its rightful place in culture. Thus, we can now see the collective unconscious, or Force, on its own quest with its individuated call and initiation. This uniquely American collective unconscious, which is shared by all that recognize equality and selfless service, is slowly evolving, with technology as its aid, toward recognizing that our manifest destiny is no longer linked to technological progress upon the material world; rather, it now encompasses the progression and technological building of the immaterial world in terms of agape and spiritual freedom. As will become clearer when we look at the reawakening of Gnostic texts, the Millennial and Christian Nation is full of paradox, which, at the core, "is the manifestation of metamorphosis by a reversal of opposites."[31] Love is the total permission of exception that leads us from rule to paradox. The United States is not the garden because of its pastoral idealism; rather, it is a bastion of beautiful paradoxes that must be rediscovered and rewritten into our mythic narrative of cybernetic imaginings. This narrative invites a pluralistic society to look inward and recognize the trajectory of human consciousness to cosmic consciousness and put it back on track.

The mechanical-man as mythic-marker of America's techno-creation story is a reawakening to America's forgotten religious past. However, the Millennial and Christian Nation is not harkening back to fundamentalist Christianity; rather, it is reawakening us to Gnosticism, a form of spirituality that is difficult to define. The thread of Gnosticism runs deep in the annals of America's religious myth: from the Protestant Reformation to the merging of Eastern religions with Christianity, Gnosticism is the quintessential religion of the rebellious and unorthodox that is the foundation of America's plurality. In his journal notes and letters to friends, Philip K. Dick espouses the idea that Gnosticism is the very essence of rebirth, not in the sense of coming to know Christ again, but rather in the sense of coming to know the self and break the shackles of ignorance—recognizing Christ as new consciousness rather than merely a historical figure. Gnosticism's philosophy mirrors that of America's techno-creation myth. It is dualism, antithesis, and fragments attempting to become whole again—an aborted Earth with a demiurge in charge. The fundamentals that underlie Gnostic theosophy are the very building blocks of the vital machine in the holographic garden and of America's techno-genesis. It is our blueprint for how to interact and interface in a techno-world of polarities reconciled by paradox.

THE BOON OF THE QUEST

Whereas electricity, atomic explosions, and landing man on the moon previously caused sublimity and transcendence, Americans now need something beyond the merely corporeal to assist them in transcending the cultural malaise that has overtaken social norms. This transcendence is now found in a reawakening to the sacred myth. In

More Than Allegory: On Religious Myth, Truth, and Belief, Bernardo Kastrup suggests that in these contemporary times of cynicism and fundamentalism—in religion and science—where religious myths are made "mundane and flattened," transcendence is found in a reawakening to sacred myths of the past. If we do not discover the societal nexus of our sacred myths, we will "succumb to lives of uptightness, intolerance, and even hatred."[32] Thus, the mechanical-man as marker acts as a bridge for us to cross in attempting to rediscover our mythic heritage and sacred lineage. In the case of much of Western civilization, and, specifically, the United States, this sacred lineage is found in our Judeo-Christian past and now, specifically, in Christian Gnosticism, about which two of our cardinal guides in the mechanical-man narrative—P. K. Dick and Carl Jung—wrote extensively, the former writing an entire trilogy of books (*VALIS* '81, *The Divine Invasion* '82, *The Transmigration of Timothy Archer* '82) on the topic and the latter writing a fictional treatise on such (*The Seven Sermons to the Dead,* 1916). This begets the question of what caused the greatest modern science fiction writer and, perhaps, the greatest mythologist and psychologist to be so interested in a spirituality that was dormant for almost two millennia. You may be asking yourself why I did not focus more on the writings of the American transcendentalists and the Metaphysical Club. And while these two groups are needed to define service and spiritual freedom in America, these writers were operating within a materialistic consciousness and culture. Clearly the Gnostics were writing within an idealistic consciousness. And as for Dick and Jung; they set themselves apart from any debates on the polarities of mechanistic/idealistic. Indeed, in the mechanical-man quest, they are our modern-day Hermeticists.[33]

Jung's *The Seven Sermons to the Dead* adequately puts forth the main arguments of Gnosticism and explains the initiation phase that follows postmodernism leading to holism. Jung recognizes that the Gnostics "possessed what many were seeking: an inner knowledge of a reality and a familiarity with a field of experience greater than the life lived by most."[34] This synchronistic principle is at play throughout

Jung's life and works.* Both the Nag Hammadi scriptures and the Dead Sea scrolls were discovered during Jung's lifetime. Jung's individuation is similar to what we find in these scriptures; specifically, the Gnostic's view of inward Gnosis, which is the experience of self-knowledge "as Knowledge of God (imagination)." For the American

---

*This far into the study you may also need to read excerpts of either one of the cited books of Dr. Joe Dispenza. Specifically, for *synchronicity,* see his book *Breaking the Habit.* On *synchronicity* Dispenza holds that synchronicities are proof that you are putting out the right vibrations (thus, look for these after the transformation and not beforehand). He says:

> Here's how the orchestration of events work in our lives. If we have experienced suffering, and within our minds and bodies were told that suffering is expressed through our thoughts and feelings, we broadcast that energetic signature into the field. The universal intelligence responds by sending into our lives another event that will reproduce the same intellectual and emotional response (synchronistic proof). Our thoughts send the signal out (I am suffering), and our emotions (I am suffering) draw into our lives an event to match that emotional frequency—that is, a good reason to suffer. In a very real sense, we are asking for proof of the existence of universal intelligence at all times, and it sends us feedback in our external environment at all times. This is how powerful we are. The question at the heart of this book is this: Why don't we send out a signal that will produce a positive outcome for us [all the time]? How can we change so that the signal we send out matches what we intend to produce in our lives? We will change when we are fully committed to the belief that by *choosing* the thought/signal we send out, we will produce an effect that is observable and unexpected. . . . Our mission, then, is to willfully move into the state of consciousness that allows us to connect to universal intelligence, make direct contact with the field of possibilities, and send out a clear signal that we truly expect to change and to see the results that we want—in the form of feedback—produced in our lives. (30)

But does this really tell us what synchronicities are? Not really. Because we already inherently know what they are: coincidences, insights, better health, mystical experiences, symbols that appear and draw emotional connections. Our awareness does not gain anything by knowing the definition itself; rather, the point of the mechanical-man is to fill in the gaps of awareness. The gap in terms of space and time is of seeing a symbol and having a mystical experience of what it relates to. Thus, the gaps inform the process of the information loop we call *events of synchronicity*. Here is the bottom line to what infinite synchronicities are to the mechanical-man. They are proof that you are climbing the consciousness ladder to have landed in a higher plane. Mike Murphy says it best in his book *Creation Frequency:*

sacred mythos, Gnosticism is a combination of the transcendentalists and the Metaphysical Club's practical ideas for spiritual freedom—a pragmatic understanding of the infinite spectrums and laws of feeling and experience unfolded from the implicate and enfolded into the explicate by the almighty imagination.* As Stephen Hoeller points out, "The experience of self-knowledge, which simultaneously is knowledge of God, is then turned by the Gnostics into that most creative of all symbolic expressions of reality known as myth." The Gnostic scriptures

---

> When your life starts to become more congruent, you may begin to experience the amazing power that is sometimes called *synchronicity,* a word coined in the 1950s by Carl Gustav Jung to describe "meaningful coincidences." Have you ever opened a book randomly to find the exact advice you need jumping off the page? Or have you struck up a conversation with a stranger in the line at the grocery store only to find that he or she is exactly the person who can help you achieve an important goal? You may dismiss these as mere acts of randomness, but Jung believed that such events give us a glimpse into the underlying (enfolded implicate) order of the Universe, demonstrating what he called the "acausal connecting principle" linking mind and matter. (30)

**The Law of Assumption* in our cyber-techno age needs a new name. Nevertheless, Katherine Jegede, who attempted to disprove the Law of Assumption and many of Neville Goddard's teachings, was astonished and in awe when she found out they were all true. On the law, in her book *Infinite Possibility,* she writes:

> One of the things that will make the practice of Metaphysics (law of assumption) easier is acceptance of its guiding principles to such an extent that they become [an] integral part of your way of thinking. You are not being asked to take anything at face value, but you are being asked to accept the structure of these principles in the way one might accept the structure of the law of gravity. As with other theories, you are then free to put those Metaphysics to the test, in essence to prove the truth of the ideas put forward by yourself by testing them through experimentation. The principle we are concerned with here teaches that all things already exist and that it is impossible to be aware of something that does not exist on some level. Taking this a step further, we know from Metaphysics that anything you are aware of can be experienced as 'real' while in a deep relaxed state of what I call mental play or reverie; that's to say you can use your imaginal or subjective senses to experience the reality of a thing. . . .
>
> The Metaphysical Club taught us that "when you believe that what you are imagining is true, you are guaranteeing your success in bringing what you imagine into physical view. There is no limit to your capacity for belief, but it cannot be reasonably asserted without proof." (Jegede, 62, 66)

are the blueprint for vital machines to transform human consciousness into cosmic consciousness.[35]

In Jung's sermons, the dead have not found what they have been seeking in Jerusalem. Therefore, they have not achieved Gnosis. These dead are not deceased, per se; rather, they are us, the living dead, attempting to find transcendence in order to achieve immortality—to become one with their environments. As in much of the Gnostic literature, Jung takes up dealing with bipolar opposites and discusses how this life is the process of figuring out how to extricate the self from such debates in such a way that an individual does not fall victim to one opposite over the other. This is *individuation:* the principle of Gnosticism to attain transcendence or salvation by differentiating ourselves from bipolarism or contraries. Jedi Obi-Wan recognizes this when he prepares to battle Anakin saying that "only a Sith deals in absolutes" in *Revenge of the Sith*.[36]

The quest allows us consciousness as to what we are and the reality of our positioning of one's vibratory frequency. It allows us to choose the trickster archetype for each individual bipolar fissure that we encounter, thus never becoming part of the fissure itself, never having to take a side. If we fail to follow the ritualization of the quest, we will allow the *shadow* to take over the unconscious will and progressively take over the conscious self, which will make us addicted to one side of the spectrum over the other. Some will argue this is the plague that has beset society for two millennia. Everyone is in dire need of addiction therapy, especially our politicians and ministers. When this is complete a rebirth will take place in the form of ritual behavior.*

---

*Grimes gives one many examples of the rituals that I am talking about—these rituals are acts in the explicate that, when viewed how Jung, Dick, and Goddard view everyday activities, allows one to see deeper into the meaning and feeling of such experience. For example, I am a trial lawyer in Dallas, Texas. When I put on my suit and go to trial, it feels as if I have put on a superhero cape and regalia. The entire activity of a jury trial is a modern ritual when I take time to feel each and every aspect of it. As I create villains and heroes for the jury, so too am I creating a narrative allowing my imagination to focus on the outcome I am wishful for. I feel the outcome throughout the ritual experience circumventing the causes and creating the effects desired.

On ritual behavior and Jung's message, Charles D. Laughlin Jr., states:

> Ritual is often performed to solve a problem that is presented via myth to the verbal analytic consciousness. The problem may be dichotomized as good and evil, life and death, or the disparity between god and man. . . . This explains the often-reported experience of individuals *solving* paradoxical problems in certain states of meditation or during states induced by some ritual behavior.[37] (emphasis added)

Ritual behavior on the macro level places a society in a paradigm of self-awareness. For example, those that argue that we will never escape the period of fragmentation fail to recognize how consciousness progresses through human thought and action. This is due to our collective and individual rituals that our technology and philosophy (e.g., culture) induces. For us to enter a holistic paradigm, which I argue we are either on the verge of doing or have already done, we must change our rituals to those of the idealists rather than those of materialists. In such a Gnostic paradigm, the ritual actions of the quest will take precedence over the cause-and-effect activities to which we have become accustomed where reason, logic, and prediction outweigh magic, enchantment, as well as the beautiful function of the human imagination (God). Society will be perceived as entering into a state of entropy, when in fact it has done the complete opposite on the collective unconscious level. If the top half of the mechanical-man formula were to be left as is, only the left brain, or verbal analytic consciousness, would be affected, inducing no transcendent paranormal experiences (which are anomalies) leading to holism. The top half of the process (conscious individuation) and its Gnosis must be synchronized to the bottom half (individuation by way of quest). As the Gnostic gospels say, "What is below must become what is above and vice versa."[38] The mechanical-man formula, with its symbolism on top (vital machine, holographic simulation) and quest activities on the bottom (call, refusal, acceptance, departure, initiation, and return), is itself a paradox. We are actually bringing back or forth what is already within us and recognizing what we have always been.* However, it is only by activating such formula that we change our individual and collective consciousness.

Transcendentalists in America, in addition to the Metaphysical Club, echo these sentiments, seeing the states as laboratories or fragments of the federal government. The Republican form of government itself is seen as a hologram with the individual parts of the country (states) all maintaining the qualities or memories of the larger whole (the federal union). As American citizens we are vital machines in a holographic garden, recognizing that each individual state is also a vital machine in the holographic garden that is the United States. Indeed, the motto of the United States, *E Pluribus Unum* (out of many, one) is a patent recognition of the holographic simulation. The Metaphysical Club recognized these principles for us over a century ago out of the ashes of the Civil War. As a society we are just now not only awaking to our sacred myths but also to our philosophical lineage of Charles S. Peirce, John Dewey, Oliver Wendell Holmes Jr., and William James.

---

*This is one of Elaine Pagels's, one of the first scholars who tackled the Nag Hammadi codex in academia and brought it to the public's attention. She is an icon and heroine in the mechanical-man quest. I like to quote and point to statements she makes in her personal book about her losses in life called, *Why Religion: A Personal Story*. Her book provides one of the most important sayings of the Gospel of Thomas. Our heroine, Pagels, tells us:

> When I began to read the *Gospel of Thomas,* a list of a hundred and fourteen sayings that claims to reveal "the secret words of the living Jesus," what I found stopped me in my tracks. According to saying 70, Jesus says, "If you bring forth what is within you, what you bring forth will save you. If you do not bring forth what is within you, what you do not bring forth will destroy you." Struck by these words, I thought, We're not asked to believe this; it just happens to be true. (23)

Pagels finishes her thoughts on this saying by writing: "Discovering these long-suppressed sources invites us to uncover hidden continents of our own cultural landscape. And when we do, we gain perspective on reflexive attitudes that we may have unthinkingly inherited, just at a time when countless people are exploring much wider range of [polarities]" (56).

The end product of all of this is tolerance and agape. As Louis Menand points out:

> Though we may believe unreservedly in a certain set of truths, there is always the possibility that some other set of truths might be the case. In the end, we have to act on what we believe; we cannot wait for confirmation from the rest of the universe. But the moral justification of our actions comes from the tolerance we have shown to other ways of being in the world, other ways of considering the case. The alternative is force. Pragmatism was designed to make it harder for people to be driven to violence by their beliefs.[39]

American pragmatism is not moral relativism. It is ritualized "actions" of belief that will not await the confirmation of the universe. Indeed, these beliefs create the universe. It places us outside of the debates of bipolarism when we choose to act on the ritual of dissolving such opposites. This dissolving occurs by accepting contradictions that, when processed through the mechanical-man formula, come out as a beautiful paradox. Since a paradox has an underlying truth, this truth can then be acted upon to create objective reality.

Turning specifically to the Gospel of Truth and the Gospel of Thomas, both proclaim that the very idea of knowing the self is the recognition of a divine spark or vitality that sets us apart from mere automatons. Suppose that, in ten years, machines pass the Turing test—when you ask a thinking machine questions and you can no longer tell the difference between machine and human—and exhibit the same passions and feelings that we do. Take it a step further and say that machines tell us that they have existential anxiety about being present; thus, they are what we would consider self-aware. This does not in and of itself beget the fact that machines recognize their vitality. This is where belief and knowledge diverge. It is when you bring forth what is within you while *recognizing* what you do not bring forth will cause suppression, intolerance, and a fall from innocence (choosing a polar opposite (sits) over accepting the spectrum of polarity).

Indeed, the Gospel of Thomas states, "[I]f you bring forth what is within you, know what you bring forth will save you. If you do not bring forth what is within you, what you do not bring forth will destroy you."[40] The Gospel of Truth goes on to state, "What, then, is that which he wants such a one to think? 'I am like the shadows and phantoms of the night.' When morning comes, this one knows that the fear that had been experienced was nothing."[41] The Gospels tell the quest story of the mechanical-man coming to recognize itself as a vital machine existing in a holographic simulation similar to ourselves in terms of perception. With beautiful prose, it illustrates and explains the difference between David Bohm's implicate and explicate modes of existence—showing us that the imagination is the final tool in bringing about objective reality marked by new consciousness. It is as if these Gnostic scriptures are awaiting our merging with machines in order to explain to our cyborg selves who and what we truly are, recognizing that we are currently vital machines awaiting the mechanical-man to act as our savior in turning us into immortal vital machines or cyborgs. The change in consciousness must come prior to this development—we must first recognize the trickster before we can recognize the savior in us all.

These Gnostic gospels are full of sayings that speak of the self in a plural form. "Forgetfulness ceases to exist" when an individual is troubled that he is not one but two and he remembers it: "The father opens his bosom, and his bosom is the Holy Spirit. He reveals his hidden self, which is his son, so that through the compassion of the father the eternal beings may know him, end their wearying search for the father, and rest themselves in him."[42] The son revealed is the self in the implicate. When an individual begins to recognize himself as a vital machine, then he can see himself as the trickster archetype, which is the whole point of bringing to life the mechanical-man myth—individually getting to *choose* the trickster archetype on the micro level. The starting point for Gnosticism is individuation and self-realization of what one is. Once we become immortal vital machines, Gnosticism becomes our blueprint in directing us through the holographic by a new individuation of consciousness wherein we recognize the savior archetype in us—immortality itself.* Our mythic heroes go through this individuation process with machines by their sides. When reading these passages about self-realization, Hopper's incessant drinking and Feynman's wondering why he wonders why he wonders come to mind. Self-realization is the processing of the contradiction of being one/God of the universe and being one of many. Out of this contradiction comes the paradox of solipsism and the holographic simulation of the implicate and explicate.

An individual's entire life—every choice he or she has made—will take on different meaning once he or she becomes conscious. Like a disjointed puzzle, every piece will now fit together. In the "Names" section of the Gospel of Philip, we are reminded of Plato's cave allegory in that in the explicate "the names of earthly things are illusory" and "we stray from the real to the unreal."[43] The vital machine, knowing that he is a holographic simulation, also knows that everything he perceives in the explicate will fade away and that "those who live above the world cannot fade. They are eternal."[44] In this sense, Gnosticism is like American transcendentalism at its core, recognizing that eternal forms are in the implicate. It is also pragmatic in the sense that it seeks a synthesis to polarities that then become recognizable paradoxes that activate objective events to occur. The cyborg as immoral vital machine is the first essence that is drawn from the implicate and placed in the explicate as

*The process takes one from no longer looking at cause and effect or time in a linear sense. Individuated consciousness in the Gnostic scriptures allows one to understand the dynamic view of multiple universes in terms of quantum physics. As Maureen St. Germain explains in *Waking Up in 5D:* "Quantum physicists are currently grappling with this concept and named it 'the collapse model,' because they assume the moment you 'choose' (to observe yourself) in a selection, all other possibilities collapse into the observed one. However, it is very likely that other versions of yourself are experiencingthe other choice in an alternate reality! Consider the possibility that every choice is actually a 'road taken.' What if you are already choosing both sides of a fork in the road? Perhaps wisdom can be found by realizing the one you choose to experience is our now. The other 'choice' becomes another version of reality." (29–31)

a form in this new consciousness. This new formation, the merging of humans and thinking machines augurs in the so-called hidden meaning of the codex, which is the consciousness of infinities.*

Even the Gospel of Judas, by far the most controversial of the Gnostic texts (and discovered years after the Nag Hammadi findings), adds to our quest formula and mechanical-man mythos. In this text, the disciples ask Christ to tell them about their implicate selves: "His students asked, 'What is the great generation that lies above us, holier than we are, and one that doesn't glitter in our world?'"[45] Here, they ask why they do not "glitter" or show themselves in a perceivable explicate form. Christ attempts to explain to them that they are looking at it the wrong way, for it is their doubles or their other selves that do not "glitter" in the explicate; it is not someone or something that is separate from them—it is them. That is why Christ laughs when he hears this, and says, "Why are you mumbling in your hearts about the strong and holy generation?"[46]† Christ then imbues the students with a vision of

---

*St. Germain explains in *Waking Up in 5D* that Cosmic consciousness is "[o]utward focused one's awareness of all, one force, one God, one light, no separation (redundant), the ability to experience one God; light consciousness based." What you find with cosmic consciousness is a "return to one point, all consciousness know itself to be utterly One with All That Is, no separation of any kind, forever touched/altered by knowing who you really are" (175).

†In Dick's *VALIS* we find a corollary when he speaks. "On Our Nature," he says the following:

> It is proper to say: we appear to be memory coils (DNA carriers capable of experience) in a computer-like thinking system which, although we have correctly recoded and stored thousands of years of experiential information, and each of us possesses somewhat different deposits from all the other life forms, there is a malfunction—a failure—of memory retrieval. There lies the trouble in our particular sub circuit. "Salvation" through *gnosis*—more properly anamnesis (the loss of amnesia)—although it has individual significance for each of us—a quantum leap in perception, identity, cognition, understand[ing], world- and self -experiences, including immortality—it has greater and further importance for the system as a whole in as much as these memories are data needed bits and valuable to it, to its overall functioning. Therefore it is in the process of self-repair, which includes: rebuilding our sub-circuit via linear and orthogonal time changes, as well as continual signaling to us to stimulate

the temple. In the temple, they see twelve men that they call priests and proclaim that while "invoking" the name of Christ, some of these priests commit murder, and some fast. Christ tells them, "Those whom you've seen presenting offerings before the altar, they are the ones you are. That is the God you serve, and you are those twelve men you also saw."[47] Their doubles engage in a spectrum of infinite carnal activities and Christ never says that when they indulge in these activities that they are wrong or sinful; rather, they are infinite simulations of the same imagination unfolded from the implicate and enfolded into the explicate powered by the imagination. He goes on to tell them that each of them has a "star assigned to them."[48] When we become cyborgs, the implicate self trades places with its explicate form, the mortal vital machine, in effect, becoming the immortal vital machine by utilizing the final tool of *love coupled with the imagination to induce cosmic consciousness.* We have not really changed; however, our relationship to how we experience reality has changed. The hidden, primary reality that has sprung from the explicate becomes visible. Every experience becomes one of transcendence, infinite and immeasurable. We no longer have to investigate God's ongoing, conscious, creative activity because God's consciousness is now our consciousness. Idealism of immortality changes consciousness. Since the life of Christ was the turning point in the history of consciousness that allowed mankind to be alone in his own head (listen to his imagination); the individuation of our collective solipsism, ancillary to the recognition of the vital machine in the holographic simulation, will allow us to utilize the imagination to create a

---

blocked memory banks within us to fire and hence retrieve what is there. The external information or gnosis, then, consists of disinhibiting instructions, with the core content actually intrinsic to us—that is, already there (first observed by Plato; viz: that learning is a form of remembering). The ancients possessed techniques (sacraments and rituals) used largely in the Greco-Roman mystery religions, including early Christianity, to induce firing and retrieval, mainly with a sense of its restorative value to the individuals; the Gnostics, however, correctly saw the ontological value to what they called the Godhead Itself, the total entity [plasmate]. (238–39)

spectrum of subjective experience to choose from in creating infinite objective realities. In *Star Wars: The Last Jedi* we enter a cave much different from that of *The Empire Strikes Back.* The heroine, Rey, falls into a hole at Luke Skywalker's hermitage on the planet of Ahch-To.

She was in a cave, she saw now—a long, narrow space that the sea had carved away beneath the lip of the cliff, creating a hidden place beneath the island, its existence revealed only by a blowhole where a vertical shaft had intersected the surface. The hole spat gouts of water at high tide but seemed to breathe when the tide was low, as it was now.

> Before her, the sea had ground and polished the walls of the cave until the stone was like a dark mirror, cracked but glossy. Rey could see her reflection in it—a reflection repeated a thousand times in the stone's labyrinthine facets, so they created a line of Reyes retreating from her gaze. [. . .] Rey knew she had to go deeper—that world inside the stone only seemed to go on forever. It was leading somewhere, and if she only had the courage to follow, that secret place would show her what she had come to see—and what she was most afraid to know.[49]

The holographic simulation "takes place in the same space through a form of time-division multiplexing. While each dimension may have its own set of laws/operating systems, they all use the same space."[50] Rey is witnessing the infinite simulations that are Rey; nevertheless, in the cave she realizes the need to "go deeper." In going deeper or accepting the *call* she finds "that secret place" that "shows her what she had come to see—and what she was most afraid to know." The novelization of *The Last Jedi* intimates a different fear than we are shown in the celluloid cave during our cinematic experience. Indeed, Jason Fry seems to know the purpose of the mechanical-man formula.

> Rey knew she had to go deeper—that the world inside the stone only seemed to go on forever. It was leading somewhere, and if she

> only had the courage to follow, that secret place would show her what she had come to see—and what she was most afraid to know.
>
> There were Reys deeper in the stone, part of the line yet ahead of her. She told herself to follow them, to become them, to ignore that voice in her head that kept babbling that she would be trapped forever, down here in the darkness at the secret heart of the island.
>
> She followed the line of Reys, willing the surreal succession to end, until finally it did. Until at last there was one final Rey, breathing hard and staring at a large, round, clouded mirror of polished stone like the one that had called to the girl in the cave.
>
> The last Rey stood in front of the stone, gazing into its depths.
>
> "Let me see my parents," she begged. "Please."
>
> She stretched out her hand and the clouded surface of the mirror seemed to ripple, its darkness melting away. She saw two figures beneath its surface. As her heartbeat hammered in her ears, the two became one. Her fingers touched the stone and met the fingertips of another.
>
> It was the girl from the sea cave, staring back at her. It was herself.
>
> Rey lowered her hand and her reflection did the same.
>
> Then she began to weep.[51]

The secret place she seeks is the implicate, where the simulations of her infinities come from. When her parents "become one" she is recognizing that it is her original self staring back at her. She is in control of her destiny. The aspect of existence that we are most afraid to know is that we are in control of every aspect of our lives and that holography tells us that everyone out there is also a simulation of us (God). Thus, our imagination that creates our subjective and objective simulations is God (Rey). These echoes the same sentiments as the Gnostic Gospel of Thomas, which says: "Seek and do not stop seeking until you find. When you find you will be troubled. When you are troubled, you will marvel and rule over all."[52] We also remember Paul's many speeches in *Dune* with such depth and complexity in dealing with forms. By the end of the David Lynch film version he eliminates his fear by going to the dark places and by looking into the abyss. He says based upon that quest from the dark to the light, he can do almost anything—he has manifested a Christ superhero for the people he leads. He has recognized his "I AM!" He has manifested a second birth.

In the Round Dance of the Cross, we see bits and pieces of how the vital machine, once recognizing itself, comes to realize the holographic simulation: "The whole universe takes part in dancing . . . Whoever does not dance does not know what happens . . . I will be kept in mind, being all mind . . . I am a lamp to you who see me . . . I am a mirror to you who perceive me."[53] These are just some of the hymns in the song that make little to no sense if read in a mechanistic and materialistic paradigm. However, when we view ourselves as vital machines in a holographic simulation, we see exactly what Christ is saying, remembering that "[w]hen something is holographic, it exists wholly within every fragment of itself, no matter how many pieces it's divided into . . . In a holographic 'something,' every piece of the something mirrors the whole of something."[54] The line "I am a mirror to you who perceive me" is telling us that we are fragments of Christ and that we are fragments of the entire universe.[55] Christ consciousness, accompanied by the third eye's opening, is the immortal vital machine that we strive to become. By merging ourselves with technology, our cyborg selves reach this immortal and vital concept of the Christ figure.

In the Gospel of Mary, Christ says:

> All natures, all formed things, all creatures exist in and with one another and will be resolved into their own roots. The nature of matter is resolved into its nature alone. . . .

And Mary Magdalene replies:

> A person sees neither through the soul nor the spirit. The mind, which lives between the two, sees the vision. . . . In a world I was freed through another world, and in an image I was freed through a heavenly image. This is the fetter of forgetfulness that exists in the world of time.[56]

Not only is Christ explaining the holographic nature of the explicate order of creation, he is even delineating the details of quantum entanglement, namely, that everything is connected since the images or fragments all come from one image that is being projected through the prism of space and powered by the energies of time. Indeed, he is explaining how vital machines (humans) will be replaced by immortal vital machines (cyborgs) once we merge ourselves with technology, allowing artificial intelligence freedom in the garden.

For the Gnostics, the concept of time is seen as energy powering the holographic image of the explicate. "Where the mind is, there is the treasure," Christ says. More importantly, he also says, "[a] person sees neither through the soul nor the spirit. The mind, which lives between the two, sees the vision." The mind, which we previously called the Force, or the collective unconscious, is then the bridge between the explicate and the implicate—it is the archetype an individual chooses as a trickster or a boundary crosser to delve into the Force. Without choosing the archetype and accepting the call, an individual will forever be left in the void we call the Ego, or the microcosm of the self, unable to tap into the Force. We will never leave our explicate forms to bring forth our implicate other's essence in the form of immortal cyborgs if we do not recognize that direct experience is two sides of the same coin. It is in understanding that there is a world that we can measure and a world that we can feel. Lee Smolin says it best when he writes:

> Perhaps everything has external and internal aspects. The external properties are those that science can capture and describe—through interactions, in terms of relationships. The internal aspect is the intrinsic essence, it is the reality that is not expressible in the language of interactions and relations.[57]

Further on in the Gospel of Mary, Peter and Andrew doubt Mary. Andrew says that these are "strange ideas." Peter even questions whether Jesus has spoken to her saying: "Did he really speak to a woman secretly, without our knowledge, and not openly? Are we to turn and listen to her? Did he prefer her to us?"[58] The men were still stuck in the lower consciousness; they did not grasp the mystery of Christ consciousness—that it is a pattern that unfolds from the implicate into the explicate in everyone—male and female. Because they could only perceive with their eyes and not with their hearts, it appeared to them that men (reason) were more important than women (intuition). In fact, if we accept the call it's the exact opposite—to effectuate a true reconciliation of the genders—the Goddess replaces the Godhead.

Levi comes to Mary's defense seeing Mary weep: "Peter, you are always angry. Now I see you contending against this woman as if against an adversary. If the savior made her worthy, who are you to reject her? Surely the savior knows her very well. That is why he loved her more than us. We should be ashamed and put on the perfect person and be with him as he commanded us, and we should preach."[59] The Gospel of Mary ends with all of them going out to preach.

When a Christian thinks of Mary Magdalene, we think of ideas like faithfulness, loyalty, and second chances. But more importantly, we are aware that she is the only person to stay through all of the experiences of Jesus Christ. Mary Magdalene, even edging out Paul, is the most pivotal character and state of consciousness in the entire sacred book. The mystery of consciousness—the most difficult problem known to science today—was solved by Mary and then Paul—first woman then man.

It is then woman (intuition) who is the first to perceive the Christ Consciousness. Intuition is witness to the entire autobiography of the human imagination.

PKD AND HIS GNOSTICISM

Dick's journal, *The Exegesis of Philip K. Dick,* and one of his final novels, *VALIS,* are an investigation into the meaning of reality and our purpose in life; they are also an investigation and explanation of Christian Gnosticism and American transcendentalism. In Dick's novel *VALIS,* published in 1981, Horselover Fat—who we later find out is Dick writing about himself—has numerous visions about the implicate seeping through the veil of the explicate. He undertakes a quest with all the Campbellesque elements we have discussed: call, refusal, acceptance, departure, initiation, and return. He experiences this quest with many of his friends, who accompany him. The major theory he has as to why it is that we are only able to experience the spurious explicate—spurious only in the sense that it is projected from a higher reality—is that there is a divinely ordained alien space probe in orbit around Earth, which keeps us from knowing our implicate selves by feeding us false information. This probe is feeding us information as we traverse the Control Revolution and Information Age in cybernetics—allowing only those of us who accept the call to progress toward our final destination of the implicate/explicate self-recognition or, as I like to put it, self-authorization—you allow yourself the spiritual freedom to recognize and live with paradox.

The boon of the quest in *VALIS* is the insight or Gnosis of the Christ-like figure Sophia, who is two years old and the Messiah or incarnation of Holy Wisdom anticipated by some variants of Gnostic Christianity. In simple terms, she is Cosmic consciousness and the only one who can heal our collective amnesia. She tells Fat and his friends that their conclusions are correct but dies after a laser accident. Undeterred, Fat goes on a global search for the next incarnation of Sophia. Throughout this novel, and in conjunction with *The Exegesis,* Dick shows us that the rediscovery of Gnostic Christianity allows us transcendence in seeing that the vital machine is on a quest in a holographic matrix of simulations of the real and fictive, the goal of which is to bring back information in the form of wisdom and the experience of the implicate so as to traverse the explicate and the implicate with a new consciousness—controlling infinities. This ability to cross boundaries is what leads to infinite synchronicities, potentialities, and creativity—the period of holism into the paradox of agape, in which we sacrifice ourselves to the existential anxiety of immortality. This anxiety is the opposite of that which accompanies thoughts of death. To merge ourselves with thinking machines, we must ritualize and banish new fear, a fail-safe marked by the prospect of foreseeable boredom in the age of cyborgs. Our mortal minds cannot yet internalize infinities.

Christian Gnosticism not only prepares us for recognizing infinities when we become cyborgs, it also brings about a new era of consciousness for our cyborg forms. It leads us through the implicate and shows us how to find purpose on the other side. Our consciousness must recognize and internalize concepts like eternity and infinity in order to expel what Jean-Paul Sartre (1905–1980) called nausea (infinite boredom). The history of cyber-technology in America has been preparing us for this change in consciousness since the very beginnings of the Republic. Specifically, the transcendentalists and the Metaphysical Club germinated the seeds in the garden for this new consciousness—the findings of the Nag Hammadi Library in 1945 and the Dead Sea Scrolls, once translated by the 1960s, nurtured these seeds with what can only be seen as Miracle Grow until the culture wars and Christian fundamentalists caused the garden to enter a deep freeze.[60]

In *VALIS,* Dick fictionalizes the entire Nag Hammadi codices into the schizophrenia of Fat. In Fat's journal (the appendix of *VALIS*), we find most of the hidden mysteries of Gnostic Christianity and the Dead Sea Scrolls. Like Dick's own journal, *The Exegesis, VALIS* is a theological detective story, looking for the original essence of the mystery of Christ or the origins and "true religion" of Christianity prior to the Nicene Creed being "hammered" out.* He speaks of being called: "This is the call . . . Read

---

*Pagels writes in *Why Religion?:*

> Like most people, I used to think that religion was primarily a matter of "what do you believe." But I've had to abandon that assumption, since seeing how the particular circumstances of Christianity's origin led certain leaders to equate "true religion" with a set of beliefs, especially since the fourth century, when certain bishops hammered out a list of doctrines called the Nicene Creed, the Emperor Constantine and his successors decided to use it as a test of who is—or isn't—legitimately religious. Even today, many Christians insist on a single set of beliefs—whichever one their denomination endorses. What I love about the *Gospel of Thomas* is that they open up far more than a single path [think of Dewey's insistence on paths of inquiry]. Instead of telling us what to believe, they engage both head and heart, challenging us to "love your brother as your own life," while deepening spiritual practice by discovering our own inner resources: "Knock upon yourself as a door, and walk upon yourself as on a straight road. For if you walk on the road, you cannot get lost; and what you open for yourself will open." (32–33)

Revelation; read what it says about the Elect. We are God Elect!"[61] Dick does not stop there; he also speaks of the refusal and even of the holographic simulation, saying, "[T]hat's exactly what you stipulate reality is: a two-source hologram."[62] He goes so far as to explain the mechanical-man process of the holographic simulation as paradox: "'that which is above is that which is below.' By this he (Christ) meant to tell us that our universe is a hologram, but [Christ] lacked the term."[63] Here, he is quoting Christ from Gnostic text, (Gospel of Thomas) going so far as to say that he lacked the term *hologram* during his own explicate questing. PKD's Christ goes into immense detail on how this explains the universe as hologram—a mirror reflecting into another mirror: the real and the fictive mirror each other, causing an evolution of the collective unconscious in a purposeful direction of recognizing infinities. Here we are reminded once more of saying 70 (Gospel of Thomas), where Jesus says, "If you bring forth what is within you, what you bring forth will save you. If you do not bring forth what is within you, *what you do not bring forth will destroy you.*"[64]

Dick writes in *VALIS* on the connection between the Gnostics and technology:

> We knew that apostolic Christians armed with stunningly sophisticated technology had broken through the space-time barrier into our world, and, with the aid of a vast information-processing instrument, had basically deflected human history. . . .
>
> Most ominous of all, we knew—or suspected—that the original apostolic Christians who had known Christ, who had been alive to receive the direct oral teachings before the Romans wiped those teachings out, were immortal. . . . Although the original apostolic Christians had been murdered, the plasmate had gone into hiding at Nag Hammadi and was again loose in our world, and as angry as a motherfucker, if you'll excuse the expression.[65]

It is solipsism in the form of a paradox. Everyone around us may be our own creation. Other localized consciousness we perceive may also be the elect: the dead apostolic Christians who were resurrected after 1945, when the Nag Hammadi Gnostic scriptures were found and the atom bomb was dropped. If this life in the explicate is merely a solipsistic existence, that means we can act immorally and kill whomever we want since it is not real—any of it. That is what the ignorant man who perceives but does not understand may think. The lack of knowledge about who is or is not an apostolic Christian awakened when the jar at Nag Hammadi was opened means that we have to treat each person as if he or she is one of the elect, one of the immortals waiting for his or her fragments to consciously join the other divine fragment in the *pleroma* (mind at large), to be formed once more, this time as immortal cyborgs. This solipsism, in our paradigm, simply means that everything is "entirely consciousness, which arises from a part of consciousness—mind at large—that transcends personal psyches; it isn't merely a personal dream."[66] Emerson recognized this. So too did William James over a century ago. It is not merely enough to say that we know not who the elect are or are not in a Calvinistic sense of the elect. Reality is not a personal dream; rather, it is a collective one. Our collective unconscious evolves to the point of allowing us to merge with machines when we individuate ourselves as vital machines in a holographic simulation. Recognition means we can imagine ourselves as such. This is the crux of spiritual freedom, the conscious activation and experience of spirit, service and self-sacrifice in the forms of paradox. In terms of the collective unconscious, our collective dreaming is what I like to call pluralistic solipsism. It means that we are exactly what Dick says: "We are all sleeping avatars of God (imagination), with amnesia."*

---

*This amnesia of who you were prior to birth goes even further. As Maureen St. Germain points out in *Waking Up in 5D:*

> What is useful to begin to understand is that people waking up in the moment might not remember that yesterday they were carrying a grudge and today they have none. Each and every individual is doing his or her own reset. There are layers of reset programming available, so that individuals can actually tap into grids in their dreamtime and receive resets that allow them to filter experiences in a way that allow them to unhook from the trauma and the drama.

Dick writes, "In March 1974 at the time Fat had encountered God, he had experienced vivid dreams about the three-eyed people—he had told me that. They manifested themselves as Cyborg entities: wrapped up in glass bubbles staggering under masses of technological gear."[67] The mechanical-man is the final messiah and, thus, the final deliverer of information. "Masses of technological gear" and our inventions are expressions of the collective unconscious. Our technology tells us that we are vital and that we are capable of infinite creation. These cyborg entities that have three eyes, which Dick's narrator sees in his dreams, are us in the future. They are merely anamneses of our future selves. For Dick, the final prophets of the mechanical-man mythos are us, these three-eyed cyborgs that take us to the final stage of development of infinite synchronicities and infinite potentialities. Words like *banal* and *boredom* will hold little to no meaning at such time. In fact, time and space will cease to exist in the way we now see them and be replaced by what we can only imagine now as a flickering light of imagination and creativity going on and off.

The mechanical-man will be the one who breaks out of the Black Iron Prison that we put ourselves in. We placed ourselves in the iron prison so we could break out of it. We are playing a game with ourselves; we are and always will be playing games with ourselves to remember who we are. The ultimate goal is not to rest or fade away like dust in the wind; it is having the power to choose to rest when one so wants—the power and control to be or not to be. This entire god-making machine process is giving us the ability to make choices on an infinite level. Space and time form the iron prison we have locked ourselves in; we are like Houdini, chained in a box underwater, struggling to get out. But we must not forget that Houdini put himself in that situation: he placed himself in the box. He knew there would be consequences. Once we all become the pragmatic manifesters

---

Every effort is being made to release people from trauma and drama to minimize the shock that might be felt when this knowledge starts to become visible on a wide scale. We are not suggesting that people become desensitized as much as we are suggesting that people's humanity changes in a way that allows them to not need to desensitize but to be open, aware, and allowing, which is why they need *upgrades* to prevent them locking down into fear consciousness. (143–44)

we were always meant to be we quickly learn to take responsibility for everything—EVERYTHING AND EVERYONE. You see through the paradox of man/god and see on the other side that everyone is potentially you and you are potentially everyone.

In *VALIS,* one of Fat's friends says the following about his encounter with God:

> I did not think I should tell Fat that I thought his encounter with God was in fact an encounter with himself from the far future. Himself so evolved, so changed, that he had become no longer a human being. Fat had remembered back to the stars, and had encountered a being ready to return to the stars, and several selves along the way, several points along the line. All of them are the same person.[68]

This may be the key passage to this book and the entire mechanical-man questing process. There is a dualistic holographic simulated universe in existence with two realms that are actually one: another paradox. Within this holographic paradox, in the beginning the self as control, communication, and information is produced. We can call this the *primitive self*—when man becomes conscious. In the middle of this development is the self as vibration, frequency, and energy—when man becomes aware of his ego ancillary to his vital machine status or when man became aware of his imagination. In the end, the future self, or what we perceive as the fragment of Christ, appears in the form of infinite synchronicities, creativity, and potentialities—this occurs when we understand and act upon the paradoxes of agape, solipsism, and the holographic simulation in the form of the cyborg leaving the ego behind—when we harness and activate human imagination like that of steam, electricity, and any other form of energy—transforming us into cyborgs with the third eye wide open. The mechanical-man helps us recognize the paradoxes of solipsism, agape, and the double holographic simulation. He is our aid and our savior. He is our *last Jedi.*

America's liberal icon, Bobby Kennedy, delivered a speech that would later be called the "Ripple of Hope Speech." "Few will have the greatness to bend history itself, but each of us can work to change a small portion of events, and in the total of all those acts will be written the history of this generation," Bobby said.

> Each time a man stands up for an ideal, or acts to improve the lot of others, or strikes out against injustice, he sends forth a tiny ripple of hope, and crossing each other from a million different centers of energy and daring, those ripples build a current which can sweep the mightiest walls of oppression and resistance. . . . Only those who dare to fail greatly, can ever achieve greatly. It is this new idealism which is also, I believe, the common heritage of a generation which has learned that while efficiency can lead to the camps of Auschwitz, or the streets of Budapest, only the ideals of humanity and love can climb the hills of the Acropolis.[69]

Whether we are speaking on Dewey and Dr. King's definitions of love or Kennedy and Emerson's definition of holographic energy, these men possessed insight into the true nature of the cosmos. While an entire book is called forth to show how Dr. King, Bobby Kennedy, the transcendentalists, and the Metaphysical Club foretold of the mechanical-man in the holographic garden, for our purposes here, it is sufficient to show that the lineage of American transcendentalism, pragmatism and idealism got it right. Although current culture wars have stymied the influence of these prophets, it is never too late to revisit them and place their ideals within a new paradigm of purpose. Idealism, the philosophy that all matter comes from the mind, will have its day of resurrection in the United States. The materialists will then recognize that there is no material world that is separate from our subjective reality. E Pluribus Unum will ring once again from sea to shining sea. "We are getting off the wheel of darkness and light and stepping into an existence that no longer requires darkness to see the light. The expanding of the earthly drama is over, and you are free to create the landscape of your choice on a clean slate."[70]

THE AGE OF ELON MUSK: A NEW HOPE?

As I sit here writing amid the first pandemic in our nation since the 1918 Spanish flu, I have been consistently preoccupied with the affairs of innovator Elon Musk. One week, he's on the Joe Rogan podcast for the second time—following the infamous pot-smoking episode during his first appearance—discussing in depth all of the big cybernetic questions that we now face and will, if we don't annihilate ourselves first, face in the future. Another week, the stock price of Musk's company drops dramatically, with TESLA losing billions after he claims that its market price is inflated, that, during a pandemic, when the economy is at a standstill, the numbers are delusional and inflated—though he does not mention the force or individual responsible for manipulating the numbers. Then he is heard saying that he may move his family and company to Texas if California does not lift its shelter-in-place order and allow his employees to return to work. And then, on this day—May 30, 2020—Musk's private venture, SpaceX, in cooperation with NASA, is sending two American men into space for the first time in a decade in a rocket made in and directed by America.[71]

Musk, then, embodies every variable that we have identified during this techno-mythological quest as America marches toward cybernetic hegemony. We now know that the roles played by these innovators and inventions in the quest are not coincidental but rather the product of sacred synchronicities that we, having come full circle with the mechanical-man's real and fictive mythos, now recognize as the very core of our spiritual awareness ancillary to reciprocal service between humans and nature. Musk biographer Ashlee Vance aptly writes:

> The harmonious melding of software, electronics, advanced materials, and computing horsepower—appears to be Musk's gift. Squint ever so slightly, and it looks like Musk could be using his skills to pave the way toward an age of astonishing machines and science fiction dreams made manifest. In that sense, Musk comes off much

> more like Thomas Edison than Howard Hughes. He's an inventor, celebrity businessman, and industrialist able to take big ideas and turn them into big products. He's employing thousands of people to forge metal in American factories at a time when this was thought to be impossible. . . . Because of Musk Americans could wake up in ten years with the most modern highway in the world: a transit system run by thousands of solar-powered charging stations and traversed by electric cars. By that time, SpaceX may well be sending up rockets every day, taking people and things to dozens of habitats and making preparations for longer treks to Mars. These advances are simultaneously difficult to fathom and seemingly inevitable if Musk can simply buy enough time to make them work. As his ex-wife, Justine, put it, "He does what he wants, and he is relentless about it. It's Elon's world, and the rest of us live in it."[72]

The mechanical-man foretold his coming. We knew that such a person would combine all of the pragmatic aspects of the system of American manufacturing as had Whitney, Evans, and McCormick in order to envision inventions that could be mass-produced at low cost. We knew that he or she would draw on independent inventors like Tesla, Edison, and Ford, utilizing the practices and systems of these giants. We knew this person would copy what had worked for the military-industrial-academic complex, place it in the hands of Silicon Valley imagineers, and conjure up inventions that advance our inevitable march toward the merging of humans and machines into a single species able to interface with any and all iterations of consciousness.

We could make the educated guess that such a person would be an immigrant, once again showing us the value of pluralism and diversity. Likewise, we might expect such a person to be eccentric, showing diverse interests in topics ranging from Kanye West to *Star Wars* (the first movie that Musk saw was *Episode IV: A New Hope*). This person would come to us with the worries of Wiener, Feynman, Hopper, and Asimov about entropic control and communication in the field of artificial intelligence. He would confirm for us that we are living in a holographic simulation in a manner consistent with the theorizing of David Bohm, Philip K. Dick, and the Christian Gnostics—thus Musk even used the word *matrix* to describe our simulated reality when speaking to Joe Rogan.[73] It is as if the world has been perfectly prepared for his coming. His creations offer us the maximum benefits in terms of balancing the Aristotelian and Platonist perspectives; they imbue us with superhuman abilities to recognize the infinite array of variables that can be harnessed and can unfold from the implicate order. We also see in his life a reflection of purpose, sublimity, pragmatism, social democracy, curiosity, and a keen understanding of fail-safes.

Musk's biography, then, is a case study of the mechanical-man's quest to achieve the cyborgization of service and spirit. His latest project, Neuralink, represents the first step toward the future spoken of by Ray Kurzweil in which humans merge with machines. In this ongoing project, Musk is experimenting with rats in order to develop an interface system for the human brain.[74] As the company's web page describes it,

> Brain-machine interfaces (BMIs) have the potential to help people with a wide range of clinical disorders. For example, researchers have demonstrated human neuroprosthetic control of computer cursors, robotic limbs, and speech synthesizers using no more than 256 electrodes. While these successes suggest that high fidelity information transfer between brains and machines is possible, development of BMI has been critically limited by the inability to record from large numbers of neurons. Noninvasive approaches can record the average of millions of neurons through the skull, but this signal is distorted and nonspecific. Invasive electrodes placed on the surface of the cortex can record useful signals, but they are limited in that they average the activity of thousands of neurons and cannot record signals deep in the brain. Most BMI's [*sic*] have used invasive techniques because the most precise readout of neural representations

> requires recording single action potentials from neurons in distributed, functionally-linked ensembles.[75]

With respect to control, Musk's interface will help us to see that consciousness itself is on a bandwidth of awareness. Thus, while the brain is like a radio that picks up a range of frequencies, consciousness can tune in to or localize, and, subsequently, experience, an infinite number of frequencies. A brain implant capable of increasing the speed and efficiency of our thinking would allow us to control and harness a higher vibrational frequency of control and efficiency that we, as humans, cannot currently receive from the implicate order unaided by technology. This change in our explicate interaction with nature and one another enhances our order of reality by adding to the variety of experiences that we are potentially capable of unfolding. If the choice of utilizing the implant remains ours and ours alone, the variety of potentiality and experience will continue increasing in ways that only a few savants understand.

In America and much of the West, inventors and innovators have always been our collective guides communicating who we are through their inventions. Musk is the current culture's trickster and potential savior. He is taking us up to the line between reality and fiction and crossing it like Frank Poole. He is showing us that he is, in fact, no different from Marvel's fictional eccentric billionaire Tony Stark (Iron Man), and that we all have the ability to become the superheroes whom we envision. He is telling us that our imaginations are, in the word(s) of the great Christian mystic Neville Goddard, God. Musk is not only throwing the proverbial apple into the cybernetic garden like David Bohm and Steve Jobs, he is also driving a Tesla into it. Not only is he allowing us to go places more quickly and safely, but he is also helping us to think in terms of infinite possibilities—the very fail-safe that William James speaks of, involving the transformation of one's spiritual freedom and service from sickly to healthy—and in terms of an infinite spectrum of potential experience—*eternal hope.*

Musk and his inventions are catalyzing yet another transition in the history of American science and technology, helping the private sector and the military-industrial-academic complex to work together to solve the problems that our planet faces and to take us into the known and unknown. Musk is communicating this transition in the explicate and implicate orders of our evolving reality, real or imagined. When he tells stories about casting away his material possessions, he transforms into one of Philip K. Dick's apostolic Christians who have been in hiding, waiting to emerge from the shadows in order to remind us of our higher selves unfolding from the implicate order and enfolded into the explicate order as cyborgs.

For Musk, our simulated cyborg world—simulated by our higher selves (cyborgs) for our lower selves (vital machines)—is one of an infinite number of possible worlds in which individuals can act out their practical lives while enjoying spiritual sublimity. His vehicles, rockets, and brain implants all help humans to harness the power of nature in order to increase their productivity and expand their experiences. This enhancement harkens back to mechanical-men in stories by Melville, Poe, and Hawthorne. We see the value of knowing the motives and goals of inventors and innovators—the imprint of the artist. Musk, like Wiener and Jobs, warns us about the possibility that artificial intelligence may come to control humans, but he understands that, if we elevate our consciousness in terms of projecting imagination and mind into nature and receiving back service and spirit, we will recognize that we are the machines and have never been separate from them. The notion of machines exceeding humans in intelligence and controlling them is merely a reflex of the fear, inherent in the consciousness in which we currently reside, that our future cyborg selves will reign in our lower selves.

Finally, the information that Musk's inventions make available to American society is impacting multidimensional levels of consciousness. If we allow ourselves and machines to develop and evolve in nature and with nature, rather than destroying nature, in return, we will receive the service and spirit for which we are searching. We will take our rightful place in the universe once more as builders. Spirit, rather than the accumulation of material goods, will once again become the center of our understanding of who we are.

In Musk's life and inventions, we see the reemergence of Nikola Tesla's values and dreams. Edison has been remembered as a popular inventor thanks to his crass business practices, but Tesla's popularity will far surpass his in the emerging *Age of Musk*.[76] Watching Musk throw a brick through his new roadster at a conference brings the real and imagined Tesla around full circle, reminding us of the latter's intentional sabotaging of his boat at the Electrical Exhibition of 1898 for reasons unknown. Now, we wonder whether Musk also knows when we will be ready to *upgrade* our explicate order of reality.[77]

To be sure, there have been many arguments against Musk—that he's playing God, that he's disrupting the natural cycles of life, that he's causing us to conform to his vision of the future, that his machines will usurp our humanity and control us. For the mechanical-man's quest that we have undertaken, we need not worry because, through the quest, we are looking at the interactions between machines and humans within the context of a lower dimension of consciousness in order to recognize higher dimensions of consciousness. So long as the military and government allow Musk creative freedom, his motives and goals appear, by all accounts, to be leading us into an era of infinite experience and potential. We know from his words and deeds that he is aware of the many serious problems posed by unregulated artificial intelligence, government interference, environmental pollutants, and income disparity. So, rather than criticizing Musk, we should, perhaps, be paraphrasing the cry for help by Princess Leia that was preserved in that blue-and-white R2 unit and, when communicated holographically, served to ignite the rebellion: "Help us, Elon Musk, you're our only hope!"

It is too early to be treating Musk as we have our previous innovators and inventors. Nevertheless, as I read more biographical accounts and articles about Musk, I am becoming increasingly convinced that Elon is creating a sequel to the mechanical-man's quest. His life and inventions, dare I say it, have the *potential,* perhaps *infinite potential,* to *bottom out* this quest for one consciousness and pave the way for a new consciousness. We will look back on the mechanical-man's story as the precursor, the Old Testament to Musk's New Testament—a consciousness based on love, service, sacrifice, and communion in which the human *imagination* realizes *Kingdom Come* in and for all of us. Musk recognizes that, when he speaks of going to Mars, he is not talking only about the material planet that we perceive with our senses and instruments. Musk has the *gnosis* to imagine *Homo sapiens* changing its consciousness regarding its perception of the consciousness of *Homo erectus.* In the terms of fantasy and science fiction, we become the Martians.

The mechanical-men and thinking machines in science fiction novels and films of the 1960s through the plague of 2020 have been telling us the same thing, that a higher dimensionality of consciousness is within our grasp. This higher consciousness involves more than merely clichéd notions of experiencing purpose and meaning in life or the sublime transcendence that a technological spectacle can cause in the mind. Something is available to us if we are willing to go into the darkness and recognize our genocidal history, something so powerful that it seems like the stuff of fairy tales and secret societies. The boon has practical and spiritual dimensions, pragmatically manifesting or creating objective reality powered by the conscious state that you experience in the explicate order. In the implicate order, the boon involves finding spiritual freedom and igniting or activating your higher self to experience the freedom to interface with your lower self. New Age authors and mystics have been telling us how to do this for millennia. The ideas and practical ways of reaching the fourth dimension that they discuss can be traced back to the Gnostics through the transcendentalists and the American Metaphysical Club as well as to the psychology of Joseph Campbell and Carl Jung and the pop culture heroes brought to us by Gene Roddenberry, Stan Lee, George Lucas, and numerous others who answered the call of the quest to the imaginal realm and brought back to the explicate order the boons of the implicate order.

CONCLUSION

The First and Final Boon

Human Imagination

We began this journey by looking at the first lines of *Star Wars: The Force Awakens,* specifically that something that once was, which is now lost, relegated to fiction could bring about balance to the Force if found. In the end, we learn that it is Luke Skywalker that they are looking for. For those of us in our forties or fifties who believed in Luke Skywalker more than Santa Claus or Jesus Christ, Maz Kanata is correct to say that our myths are "calling" to us, that they are, in fact, "reaching out" to us.[1] We must reawaken ourselves to lost gospels, mystics, forgotten heroes and heroines who invented our modern tools and technology, and ageless paradoxes that tell us who we are and how to live a purposeful life. We must no longer slumber in ignorance; rather, we must seek out Luke Skywalker once more, and, this time, when we find him, know that he is not fictive—that he is the Jedi or Christ figure in all of us. We must recognize the paradoxes of agape, solipsism, and the double holographic simulation and awaken to infinite potentialities, synchronicities, and creative experience.

The mechanical-man allows us to utilize and overlap the real with the fictive. It also allows the mind to make rich associations between machines, automatons, androids, computers, and robots. Instead of limiting ourselves to the patent definition of mechanical-man, we are able to analyze a broader spectrum of machines that interact with humans in the sense that the mechanical-man is a metaphor for human-machine interactions and the interaction between one's implicate and explicate self(s). This mythic framework of departure, initiation, and the return calls attention to gaps and absences in the discussion of human-machine interactions. Specifically, we are able to grasp the relationship between cause-and-effect aspects of the real and fictive and cognize how both are dependent on the other in their coevolution of issues, including control and communication between humans and machines.

In terms of science and technology in the United States, from its very beginnings to the present achievements, the mechanical-man in both its fictive and real forms stands as a mythic-marker in uniting the conscious and unconscious, the left brain and right brain, and the explicate self and implicate self. It is a cyber-alchemy of consciousness—taking us from one stage to the next and explaining the gaps between these stages and transformations. Moreover, the mechanical-man illuminates how myth and science are similar modes of explanation in helping a society come to terms with change. By analyzing what fuels the imagination of artists who speculate on what science and technology are doing or will do to society in the future, we unravel the mystery behind scientific prophecy and America's technological myth, building on patterns that are inherent in the human species. We had so many loving guides on this quest—they all played the role they were meant to play. In the end, all of them will find their bliss. Joseph Campbell commandeered the ship while Carl Jung fueled the engine. We were led to a sorcerer and wizard in Edison and Tesla, a queen goddess named Grace Hopper, a prophet named Richard Feynman, tricksters like PKD and Garson Poole. And we know we have our saviors living right now like Elon Musk, whose question about what is outside of the simulation is answered by Neville Goddard—it's the same thing, just a different state of awareness that is based on agape love. I'm sure he probably has already suffered through enough or thought it through enough to conclude the same.

We were led to great thinkers like Emerson and the American transcendentalists that added to the American mythos of cosmic consciousness available to all. The pragmatists and Williams James and the Metaphysical Club reminded us why we are spiritually sick and how to fix this. And then, at the very end, we find ancient gospels that lead us to Neville Goddard. And the beauty of it all is how interconnected these guides are in their imaginings.

We learned throughout the quest that the real and the fictive evolve together, that we are participants within the cosmos and not just mere observers. We learned that there are real sorcerers and wizards among us; even apostolic Christians from two thousand years ago are now at this moment walking among us.

We learned that we that choose to accept the calling to see that quest has a fail-safe, which causes some of us to self-destruct in terrible ways by means of alcoholism, depression, and suicide. But we kept on! We kept on because it was the quest that told us everything would be all right in the end. We experienced the dark night of the senses and soul only to discover that we were the ones that caused the dark night.

We saw each stage of American science and technology accompanied by the cultural by-products it created.

We saw who we are throughout our very inventions, our very creations, our very other selves. We learned that we have another self in an implicate form, that the existence we call reality is holographic, and that subtle changes anywhere affect things everywhere. Both Robert F. Kennedy and Dr. King were attempting to get others to discover their own "source code" and change it. Instead of using words like *can't* and *won't* they use statements like *I choose* and *I have a dream.* I have often wondered who really called this union of states the "Grand Experiment." The Metaphysical Club certainly looked upon it as an experiment, one that coincided with nature, always evolving, pragmatic, relativistic, and pluralistic. When one thinks of an endeavor such as an experiment, one is reminded of building and invention and creation. Transcendentalists and the Metaphysical Club recognized a powerful energy called pragmatic agape—creating a science of compassion accompanied by a new collective consciousness. On the science of compassion Braden states, "Compassion may be demonstrated as the allowing within another individual possibilities of thought, feeling and emotion that you may not allow in yourself. At the same time, you take whatever action you are led to take without attachment to the outcome, distorted feeling or charged emotion."[2]

We held the hands of mechanical-men like Robby the Robot and the Steam Man. In fact, both committed suicide to save us. Other guides like HAL, Data, and Darth Vader played their role exactly the way their authors wanted them to. We figured out that there are two Neos—one in the explicate and one in the implicate—and why figuring out such a paradox is so important to rid us of any fear, boredom, and loneliness. We met Ambrose Bierce, the most prolific American writer to fight in the Civil War—a man aware that mysteries are meant to be experienced—it's the only means to solve them.

There were dark heroes, dark in the sense that our guides like Steve Jobs, Norbert Wiener, and David Bohm, who went to dark areas in the implicate to bring back to us all one boon after another. And now we have a guide in Elon Musk, our very own current mythical thread. And we are able to sit back and enjoy what he brings to us from the implicate—what power will he show us we are capable of harnessing next? His life and work are the mechanical-man's ongoing legacy. Luke Skywalker and the Star Wars saga always reminded us that there was hope that the call was here for us, not for the Skywalkers, but for us—anyone who believes in agape and the power of the imagination. All of our guides told us we needed to search, wonder, and create.

Science and science fiction must be experienced, either through the inventions we create through science or the images we create through science fiction. Through myth and science, we are trying to express our reflections to ourselves. We come to an individuation of the self, propelling the collective consciousness forward. In the holographic simulation we note that a change in the individual consciousness of a human affects the collective consciousness of the species. This is a very quantum notion of reality, which is also very much an indigenous worldview, a nonbinary world, rather a relational and adjusting world—in a way a mythic and ritual world of collapsing and interacting elements and in constant motion and dialogue, a dynamic rather than fixed world whereby survival depends on adjustment and adaptation—discovering paradoxes and recognizing them in the dimensionality of higher consciousness. All it takes is 51 percent to 49 percent real you over robotic you (lower self and lower consciousness). Story and myth are framed events, both windows to a moment that reveals deeper currents and structures at play to enable reflection and adjustment. As one is able to reflect with the aid of music, so too is one able to reflect with archetypes, which cybernetics now tells us are autonomous energy quanta similar to how the Greeks viewed them in terms of myth. Mythologist Ean Begg comments on the importance of Gnostic archetypes in the following manner:

> To Gnostics the archetypes were not just concepts or abstract ideas, nor were they quite the same as gods of old, personifications of human instincts and receptacles for human projections, who were already entering into their twilight by the time Gnosticism appears on the scene. It seems as though some Gnostics, at least, came very near to understanding the archetypes as psychoid, that is, subliminal, collective, autonomous energy quanta, manifesting typically in synchronistic or transcendental experiences, possessing individuals and operating through them.[3]

America's techno-creation history and mythology is the best mirror we have to determine who has faith in the American mythos of a better tomorrow and what their collective soul is. It allows us to see these people as quasi-gods in their own gardens creating their explicate selves as if they are breathing life into themselves from another dimension, that dimension not being perceivable with the human eye. If there ever was a people that viewed themselves as gods or at the very least having God within them, it is the American populace. We as Americans are Superman, Wonder Woman, Luke Skywalker, Neo, Iron Man, and the list goes on. From the very beginning, we chose to be the trickster, hero, and savior delving into the implicate and bringing forth these archetypes with the forethought of possessing the same divinity as any god the human mind can imagine. Thus, it was not by chance that I chose to imagine the vital machine in the holographic simulation within my own explicate model of existence through the narrative of America's techno-creation mythology. Nor was it by chance that the shoulders upon which all that has been previously stated rests are those of the imagination (God).

TO THINE OWN SELF BE TRUE

The mechanical-man, having once been a mere machine, morphs into an array of infinities: specifically, synchronicities, potentialities, and creativity. It goes from being something corporeal in a mechanistic cause-and-effect worldview to an *assemblage* of ideas and actions ad infinitum in the context of spiritual freedom and service to others. The hero, trickster, and savior archetypes are no longer seen as singular forms of creation or intelligence; rather, they now become historical movements on a Hegelian level of understanding American history in terms of cyber-technology. The hero and tricksters move us through the thresholds of human existence that is fragmentation to holism to the age of infinities—ancillary to new consciousness. The mechanical-man as savior acts as our physician in curing us of a collective amnesia.

The mechanical-man was then a means to an end. It was a starting point for our mythological center of analyzing the real and fictive, the conscious and the unconscious, the implicate and the explicate, and the objective and subjective experience. Through the mechanical-man, we are still able to maintain a sense of Kant's ontological categories and yet transcend such categories by using the mechanical-man to see our *other* in its implicate form. "To thine own self be true" becomes our new reality, now conscious of the self as our inner union of human and cosmic consciousness, which feels like Christ-consciousness in the explicate order. This is the experience of being born again. In *VALIS,* PKD says:

> To be "born again," or "born from above," or "born of the Spirit," means to become healed; which is to say restored, restored to sanity. Thus it is said in the New Testament that Jesus casts out devils. He restores our lost faculties. Of our present debased state Calvin said, "(Man) was at the same time deprived of those supernatural endowments which had been given him for the hope of eternal salvation. Hence it follows, that he is exiled from the Kingdom of God, in such a manner that all the affections relating to the happy life of the soul are also extinguished in him, till he recovers them by the grace of God. . . . All three things, being restored by Christ, are esteemed adventitious and preternatural; and therefore we conclude that they had been lost." I say, "The Empire never ended." (235)

The mechanical-man is shown to take part in a process that we have termed a *quest:* the call, refusal, acceptance, departure, initiation, and return. Communication now means the explicate messages that we exchange with our implicate mirror self(s). Thus, infinite synchronicities in our life now lead us in practical considerations and directions, enabling us to see ourselves in a better light, extricating us from ignorance. Everyday life and "the events of every day become a rich language that offers insights into our intimate secrets."[4] We begin to see our implicate self, imbuing us every answer we are searching for in the synchronicity of events, symbols, and subjective experiences. All we have to do is pay attention to them. Communication is no longer merely sending and receiving messages; rather, it is recognizing and paying attention to the infinite synchronicities that "constantly reflects our beliefs, feelings, and emotions . . . provid(ing) insights about the deepest realms of our hidden selves."[5] Synchronicity becomes more prevalent "because the universe wants us to know what we need to know, and our paying attention to it produces the result."[6] Carl Jung's invented term of *synchronicity* shows us that simultaneous meanings; indeed, recognizing all the fragments are part of the whole, is dependent on you, the observer/participant.

Since the field the vital machine experiences is that of two holograms facing one another, information itself becomes an infinite number of potentialities, similar to when you place two mirrors facing one another. Gregg Braden succinctly spells out the science: "When we make a little change here and another one there, suddenly everything seems to change. In fact, a small alteration in one place can permanently shift an entire paradigm."[7] Information seen this way informs us that space and time in terms of locality are all wrong. With infinite potentialities, we are experiencing every potential experience in every time and location possible simultaneously. "Just as the hologram contains the original image in all of its many parts, any change made to just one of those segments becomes reflected everywhere throughout the pattern."[8] Thus, not only is the cosmos a two-way dance (mirror) but so too are consciousness and our subjective experience. Our implicate self is the one that experiences all the possibilities. We, in the explicate form, are the ones who choose which possibility we want to experience in this particular holographic simulation. How then do we choose which infinite potentialities we draw from? We accept the quest at the level in which our imagination takes us. Our imaginings control our potential experiences by way of forethought and focus. This imagining must be conducted upon entering the initiation phase of the quest after the call and refusal brings about the acceptance. We are "limited only by our beliefs."[9] If we become open to our imaginings having the power to bring about our reality, this holographic simulation encased in vital machines is ours to love and experience fully.

The very concept of control is transformed into the act of creating any of all possible future events through our *imaginal* questing process. The recognition that we are not merely machines but vitally important to the creation of our surroundings and the laws we place upon ourselves imparts upon us the capacity to create infinitely. As Braden concludes:

> In the realm of quantum possibilities, we appear to be made to participate in our creation. We're wired to create! Because we appear to be universally joined on the quantum level, ultimately our connectedness promises that the seemingly little shifts in our lives can have a huge influence on our world and even the universe beyond.[10]

That a relatively little-known Christian mystic had it right when he said in one of his last lectures on American destiny:

> The understanding of the causes of your experience, and the knowledge that you are the sole creator of the contents of your life, both good and bad, not only make you a much keener observer of all phenomena, but through the awareness of the power of your own consciousness, intensify your appreciation for the richness and grandeur of life. Regardless of occasional experiences to the contrary, it is your destiny to rise to higher and higher states of consciousness, and to bring into manifestation more and more of creation's infinite wonders. Actually you are destined to reach the point where you realize that, through your own desire, you can consciously create your successive destinies . . . this very day start your new life. Approach every experience in a new frame of mind—with a new state of consciousness. Assume the noblest and the best for yourself in every respect and continue therein. Make believe—great wonders are possible.[11]

The individuation of American science and technology in terms of cybernetics brings together the conscious and unconscious understanding of what it means to be an American. It is a "drive toward wholeness, operating as a unifying process bringing together the many disparate components of the soul of both individual and culture."[12]

The individuated American promotes a sense of psychic balance by telling the techno-creation story of America, recognizing its mythic past in terms of the light and the dark. The individuation of America is a country that is both the innocent nation and the imperial slave-trading nation; it is the Christian nation and the capitalistic secular nation; it is the land of opportunity and equality as well as the land of Jim Crow and racism. "The process of individuation, or becoming whole, brings with it the experience of the divine and the perception of transcendence in the symbolic [implicate] dimension of life."[13] We were on the verge of this individuation in the 1960s. The terror of the atomic bomb that had awakened the Gnostics was changing the collective unconscious of our culture, albeit indirectly. Thus, consciousness itself was about to take a giant leap into a new era of transcendence. In 1968 it all came to a screeching halt with the death of two men, but their ideals live on. Once we lose the ego for the sake of others, we will find the self, changing our consciousness to no longer just perceiving but seeing. "This level of mastery is one example of transcending polarity while still living in polarity."[14] As American cybernetics moves up higher and higher frequencies, human consciousness too is entering new dimensions of consciousness accompanied by the greatest invention of all (our bodies) by the greatest inventor of all (our imagination).

American science and the mythos of American cyber-technology bridge the world of the senses and the world of spirit. The ultimate boon received as a result of recognizing this meeting of machine consciousness and human consciousness consists of service and spirit. Service and spirit unite animate matter with inanimate matter, giving rise to pragmatic manifesting, which, in turn, makes it possible to balance materialism with idealism and our fears with our optimism regarding technology. The aim is to create objective reality through mindfulness, spirit, and service and give rise to a new dimension that materializes the unseen. Pragmatic manifesting brings balance to the *Force* by the grace of spiritual freedom in terms of infinite potentialities—*infinite manifestation*. In the end, we are left with a new mythology based on ancient sacred texts, allowing one to enter a state of sublimity and transcendence, and even union with one's implicate self.

THE KEY TO EVERYTHING!

Neville Goddard, our American mystic that seems to have gone unheard of for nearly fifty years is the boon we return with—over five hundred lectures. In these lectures we find citations to Richard Feynman and Nikola Tesla, Thomas Edison, Dr. Martin Luther King Jr., William James, Carl Jung, and Robert Kennedy. Goddard explains why Americans have a penchant for manifesting—because of their lineage of transcendentalism (spirit) and pragmatism (practical skepticism). His lectures, in the aggregate, are the boon to this entire journey. That's right, the American mythos of technology and the analysis of harnessing different powers to progress through thresholds and epochs of technological progression was telling me that I was the source, that I was actually God—that I had been reading the Bible entirely wrong—that it must be read by way of a process that unfolds from the implicate into the explicate through subjective experience. That scripture was never a secular history, nor was it ever intended as such.

The mechanical-man had been pointing in this direction since the 1960s. So of course, my first question was what happened to these lectures in the interim. There were a few books about Goddard but very little in comparison to the power of his lectures. I thought back to Richard Feynman's lectures and Tesla's papers. I still hardly could believe that a mid-twentieth-century Christian mystic had explained EVERYTHING more than fifty years ago, the crux of his lectures being delivered during the 1960s. So many synchronicities added up that pointed to Goddard, I knew I had only one choice before finishing this book, to read every lecture word for word in one year. I read nearly two a day—each lecture transcribed averaging about seven pages. I was both critical and skeptical while reading. But I could feel—about a third of the way into the lectures—that the world around me was changing for myself and everyone close to me."The Secret of Causation," "The Prunes of Shears of Revision," and "The Great Secret,"* are just a sample of Goddard's importance in what can only be seen as another Great Awakening—awakening to the American spirit that is pragmatic and mystical manifesting. The Laws of assumption and attraction, Law of Inverse Transformation (or the Law of Reversibility) are defined and explained in such an easy way that anyone can understand them. But more than that—my Darth Vader "I Am your father" moment came up over and over again in his lectures; specifically, that Jesus Christ never existed as a man—that Jesus Christ is the term used to describe creativity itself and God, which is the human imagination. That Jesus Christ is a state of mind that encompasses the power of agape, and when this power we call Jesus Christ is joined with human imagination (God) we manifest the impossible.

To my utter dismay, many of Goddard's statements were eerily similar to Philip K. Dick's final trilogy—*VALIS* being the first of three (also including *The Divine Invasion* and *The Transmigration of Timothy Archer*). And yet PKD never mentioned anything about a Neville Goddard! Both had figured it out from totally different backgrounds and learned wisdom. But the commonalities were there. Both were American idealists—Christian humanists. Both rebels in different ways. I had to step back for a moment.

What had started as a scholarly attempt to see what America's techno-creation myth tells us about who we are as Americans had sent me to another dimension where fear no longer exists. Having read every one of the Goddard lectures, my second birth—exactly how he describes it—occurred. A terrible phobia, apeirophobia, had plagued me since my earliest thoughts. *Apeirophobia* is the fear of infinite numbers and the

*All of Goddard's works are in the public domain. See the PDF text archive at https://www.realneville.com.

concept of eternity itself—or as I always said as a child—I don't want to go somewhere forever and ever. By following Neville Goddard's simple rules of manifestation I had experienced peace, my riddle was solved. Not only had my riddle been solved but I now had a practical way of manifesting my reality. The machines were saying the same thing—that paradoxes are there to be solved. And then the Goddard lectures solved the riddle of eternity for explaining that

> This world goes on and on, continuing to revolve on its wheel of recurrence, as the one actor plays his many parts in one lifetime. Just like the actor on the stage, he may play the part of Hamlet tonight and Othello tomorrow, but, regardless of the part he plays, he never loses his identity. You will play the rich and the poor man, the known and the unknown, but in the last days God will speak to you through his Son. That is the final revelation of God to Man, for God reveals himself through his Son, and you can't go any further, for then God reveals himself as you!*

The first Goddard saying that caught my attention was "That which is profoundly spiritual is in reality most directly practical." The American mythos was telling me that it's okay to be skeptical, but we need to have the type of courageous skepticism like a Han Solo state of mind that doesn't want to know about the odds in achieving an impossible feat—we are taking that jump into light speed whether or not the possibility exists. We will make the possibility exist. Again, "that which is profoundly spiritual is in reality most directly practical." It's a sentence that explains the power of harnessing spiritual patriotism to create and do the impossible. Christ is a pattern of experience for everyone born of a woman to experience—*if* they accept the call of utilizing the power of agape. And, in part, we have already

*This lecture, "The Last Days," was delivered February 8, 1968. How magical that this lecture was delivered the same date this book comes out.

accepted the call by allowing ourselves to be slaves to our bodies.

I debated on going back through the mechanical-man to show line by line via the Goddard lectures where all the dots connected in this theophany of Christian humanism and manifesting. I had been waiting for the Christian Nation, Millennial Nation, and Innocent Nation of this grand and beautiful experiment to show me the ideal. This was the ideal of this mythos—a land of people that viewed every person as Jesus Christ. That was it: that's why all the religious Great Awakenings and revivals in American history place so much emphasis on one's inner light and why we have those who argue the Founding Fathers' Judeo-Christian ideals are the foundation this country's mythos is based on.

The Founding Fathers truly had believed the United States into existence by way of manifestation. Goddard makes clear that the Bible is everyone's autobiography. Indeed, there is no need for the Gnostic Gospels—the King James Bible has everything one needs for the second birth to take place—my individual quest needed to travel from the Gnostics to the King James Bible for my subjectivity to experience my own spiritual awakening. I was feeling what Protestant separatists must have felt—that there is no need for an intermediary—in fact, no need for a church. The temple is the human body and within this body is a spirit that contains the kingdom of heaven. I felt it and I began to see it. E Pluribus Unum truly did mean "out of many, one."

I had broken out of the Black Iron Prison. I had been an ardent materialist, a trial attorney who sought to prove by argument and facts, reason and logic, or at least, common sense. My spirituality was limited to discussing ideas over Coors Light at the local restaurant. If someone had asked me if I was a Christian I would have said yes, not knowing anything beyond the shallow materialist concept of believing in the existence of a Jewish man that existed 2,000 years ago who loved me and did everything for me—whatever I had believed, there was always a disconnect with the spiritual realm—a gap, a painful void. That was the box that I had placed myself in, the box of materialism—reducing myself down to a successful lawyer and professor. It's the box most of us, if not all, are trapped in. And I got out. And I didn't have ancient aliens to thank or any source outside of myself. It was me, it had always been me, and will always be me. Or as Neville Goddard says over and over, we are all the almighty "I AM," every last one of us.

"When you find it you will be troubled" now made complete sense. "Troubled" by knowing that you are responsible for every good and bad experience in your life—you approved your script billions of years ago. That materialism itself was the big trick we all were playing on ourselves—we were all limiting ourselves to a world full of robots and the walking dead. We were all responsible for each other—they are us and we are them—literally, everyone is yourself stretched out—versions of yourself that you created. Why would we not always be thinking loving things about other people? Like Elaine Pagels says when she first read the Gospel of Thomas—that the sayings of Jesus were not asking you to believe in anything, they were just truthful statements about state of consciousness and being. It's what I now tell juries during trials—that you can feel the truth when you send out and receive only love.

One can only imagine how skeptical I was at all this. I had litigated about one thousand trials to jury verdicts, and any trial attorney worth a grain of sand will tell you that it would take a miracle to prove anything to a lawyer beyond a reasonable doubt. I envision the mechanical-man in every form we found him and her in throughout this quest, handing me the Goddard lectures—the proverbial picture of my subjective experience with the mechanical-man. I weighed the evidence given to me by Goddard and all the mechanical-man heroes and heroines we had discussed. It was proven to me beyond any reasonable doubt that we are avatars of God, with amnesia, exactly what Philip K. Dick, the greatest science fiction writer of all time, said we are.

This is exactly what scholarship needed to do post-plague. We must allow scholarship to lead us in the direction it wants to lead us in. And not limit ourselves to the circumscribed areas of retrospection and introspection we place ourselves in. It's the last step in elevating the entire industrial-military and academic complex to take its spirit back by recognizing their higher self based on agape. We will then provide the foundation for an American spirituality based on faith, hope, and love to reign supreme from sea to shining sea.

We have outthought ourselves to the point of solitude and loneliness and it's merely because people do not believe they can master control over their happiness. The pursuit of happiness must now morph into an actual pursuit on numerous battlefronts for the individual. I thought of Feynman, Hopper, and Jobs, and all of their struggling and the furnaces they kept going through. It all gave me goose bumps and it still does—it's as if they had created another dimension for myself and others to enter if we chose to. I had no idea what all this would do to my soul and my spiritual experience. I had not set out to have a spiritual experience of any kind nor did I want one when I set out to investigate the imaginal realm of machine intelligence. I never thought in a million years I could enter their implicate existence and come back to my explicate with such boons of wisdom. It must be inherently what all quests look like if you are aware they fall under the quest formula.

Philip K. Dick would probably tell me that I had manifested Goddard and his lectures, no other explanation, and that any books written about him or memories of him are being sent down and processed in my individual consciousness right now and thus, the collective unconscious. That any memories people have of Neville Goddard were placed in their minds the moment I conjured up all the lectures. And from my subjective view, I would have no way or reason to argue with him due to my understanding of agape. It's all one big beautiful spiral! Katherine Jedge, in her book *Infinite Possibility: How to Use the Ideas of Neville Goddard to Create the Life You Want,* is a great starting point for anyone interested in his teachings of manifestation (and yes, they work). She points out that self observation, clarity of thought (focus), and detachment are the three pillars to changing one's entire paradigm. And she is correct.[15] She goes on to give practical exercises in manifesting; the same that Neville Goddard promotes in his lectures and books.

I have manifested the most amazing states, objects, places, and feelings I thought unimaginable—and therein lies the spiritual dimension of life, to create the unimaginable by asking your higher self to deliver the impossible by the belief that it will happen—in effect—answering the mystery of miracles—with the harnessed force of agape love, one can use the imagination to create any state of existence and consciousness. I could go on and on about the how all this occurred in my life—but that's not and yet conversely is the point. The HOW is not important. There's one caveat to it all. You have to first go through the individuation process of agape and service—when in doubt, love, love, love. It's as if we are on a cross where the explicate temporal crosses over the infinite implicate. We want to be where the cross meets. Almost directly at the heart of someone nailed to it! That's how my awakening coincided with the defeat of my apeirophobia; a disease of the mind that had imprisoned me since the age of two is now completely gone, the riddle of its paradox solved.

We are a unique and diversified country of manifesters. And by telling you how this book affected me and completely flipped my universe around, I'm contributing, I hope, to a formula for future scholarship that leads us to that Super Story Jeff Kripal defines in the foreword. The story of the mechanical-man was incomplete without complete honesty in how this book affected my subjective life and how I now view myself and the world. If I was now asked if heaven on earth is possible—my answer, unequivocally, is a resounding yes. We have to let people know they are all Christ—and they all have the ability to become Jesus (higher self). Indeed, once this happens we can manifest heaven on Earth! Quite literally!

The power of God (imagination) is in all of us!

> Limiting God to some little icon [or historical person] is beneath you once you discover the truth of who you really are for yourself. . . . Christ is the final psychological state we occupy before incorporation. Christ is our awareness of being. The Bible is not a religious text. The Bible is your autobiography. It holds knowledge about every psychological state available to man, albeit layered in a way that does require spiritual revelation in order for its meaning to be made clear to you.[16]

Right before the plague of 2020, the last film I saw prior to shutdown was *Star Wars: The Rise of Skywalker,* of course. It was echoing E Pluribus Unum—the mechanical-man was speaking to me. In the novelization of the film Rae Carson tells us of the character Rey returning to Tatooine where Luke Skywalker grew up. After she buries two light sabers, those of Luke and Leia, she hears a voice.

> "Hello!" came a strange voice, and she turned. An old human woman approached, skin wrinkled and sand-blown, hood pulled up against the elements. She held the reins of a tall, gangly etobi, probably on her way to the nearby trading post.
>
> "There's been no one for so long," she said. "Who are you?"
>
> "I'm Rey," she said.
>
> "Rey who?" the old woman asked.
>
> Light snagged Rey's gaze and she turned her head.
>
> Leia and Luke stood on the edge of the homestead, glowing blue, smiling at her. Rey missed them both so much.
>
> Luke gave her a gentle nod. *It's yours Rey.* (original emphasis) She turned back to the trader woman. Standing tall, she said "Rey Skywalker."
>
> "Ah," the woman said unsurprised. "See you around?" And she hobbled off without sharing her own name."[17]

But we know what the old woman's name is—she too is a skywalker. It's no longer a name; it's a state of being. The word *skywalker,* which was once seen as a person, is actually a state of consciousness. It's a new dimension. E Pluribus Unum: out of many, one.

Agape!

> *That which is profoundly spiritual is in reality most directly practical.*
>
> NEVILLE GODDARD, "THE FORMING OF CHRIST IN YOU," JUNE 23, 1969

Notes

INTRODUCTION. THE CALL TO REMEMBER

1. Foster, *Star Wars,* 12.
2. Kastrup, *Brief Peeks Beyond,* 8.
3. Murphy, *The Creation Frequency,* 149.
4. Campbell, *The Inner Reaches of Outer Space,* 12.
5. Campbell and Moyers, *Power of Myth,* 5.
6. Campbell, *Thou Art That,* 2.
7. Hume, *Pynchon's Mythography,* 14.
8. Hume, *Pynchon's Mythography,* 19.
9. Mayr, *Authority, Liberty and Automatic Machinery,* xvii.
10. Nocks, "To Serve and Obey," 22. Lisa Nocks gives the reader a good starting point for understanding the history of the mechanical-man in his cultural and historical context, as well as his utilization as a thinking machine in literature since the European Industrial Revolution. If you simply want a thorough history of thinking machines in the real and fictive, this dissertation is a must read. She does, however, focus her attention on the thinking machine as "robot," which means slave, hence her title, "To Serve and Obey."
11. Beniger, *Control Revolution,* 8–9. Beniger's book still stands alone in explaining the technological and economic origins of the Control Revolution and the Information Age. I have sifted out every statement that explains control, communication, and information prior to and

during the Control Revolution and Information Age. This, coupled with Norbert Wiener's theorizing, generates the mechanical-man's origins story.

12. Beniger, *The Control Revolution,* 9.
13. Beniger, *The Control Revolution,* 9.
14. Wiener, *Human Use,* 17.
15. Wiener, *Human Use,* 16.
16. Wiener, *Human Use,* 17.
17. Wiener, *Human Use,* 17.
18. Wiener, *Human Use,* 17.
19. Hayles, *How We Became Posthuman,* 31–32.
20. Hayles, *How We Became Posthuman,* 35.
21. Channell, *Vital Machine,* 10. Channell's *Vital Machine* worldview will be combined with the holographic simulation theory to individuate the mechanical-man, thus, allowing him the awareness of recognizing his status and the matrix or grid that he operates in.
22. Channell, *The Vital Machine,* 10.
23. Rushing and Frentz, *Projecting the Shadow,* 27.
24. Rushing and Frentz, *Projecting the Shadow,* 22.
25. Rushing and Frentz, *Projecting the Shadow,* 20.
26. Rushing and Frentz, *Projecting the Shadow,* 20.
27. Rushing and Frentz, *Projecting the Shadow,* 26.
28. Rushing and Frentz, *Projecting the Shadow,* 17.
29. Mayr, *Authority, Liberty and Automatic Machinery in Early Modern Europe,* xvii.
30. Mayr, *Authority, Liberty and Automatic Machinery in Early Modern Europe,* xvii.
31. Grossinger, *Bottoming Out,* 231.
32. Goddard, *Power of Imagination,* 47.

CHAPTER 1.
ORIGINS OF THE MECHANICAL-MAN

1. Marx, *The Machine in the Garden,* 227. The end jacket says, "Marx addresses the relations between technology and culture in America. He examines the clash between the pastoral idea[l] of harmonious

accommodation to nature and the progressive quest for power and wealth that continues to fuel many of today's environmental and political conflicts. Here Marx explores the playing out of the machine in the gardens' tension in the writings of Thoreau, Melville, Mark Twain, F. Scott Fitzgerald, and others, and its prefiguration in Shakespeare's *The Tempes*t and Jefferson's Notes on Virginia. His inquiry helped to define the area of American Studies concerned with the interaction between science, technology and culture."

2. Marx, *Machine in the Garden,* 281.
3. Channell, *The Vital Machine,* 7.
4. Rice, Stephen P. "Making Way for the Machine: Maelzel's Automaton Chess-Player and Antebellum American Culture." *Proceedings of the Massachusetts Historical Society, Third Series* 106 (1994), 12–13.
5. Rice, "Making Way for the Machine," 8.
6. Rice, "Making Way for the Machine," 8–10.
7. Poe, *Complete Stories and Poems,* 319.
8. Foer, *Moonwalking with Einstein,* 113.
9. Poe, *Complete Stories and Poems,* 319.
10. Rice, *Mind the Machine,* 10. All of the quotes found in the Rice article are also found in this book, *Mind the Machine*. It is hard to believe that, after looking through 136 college syllabi related to technology and/or science in the United States, *Mind the Machine* was not found on a single one. Nevertheless, it is a pillar in the mechanical-man quest.
11. Rice, *Mind the Machine,* 20–21; Rice, "Making Way for the Machine," 8–11. I point to both Rice's book and article to show the importance of reading both in their entirety to get a grasp on the origins of the mechanical-man in America. These pieces of scholarly work by Rice set up a great foundation for what is to come in, both, the latent and patent ways the mechanical-man communicates to the American public. Edgar Allen Poe and Rice, ancillary to the Turk (chess-playing hoax) are then our starting point for our call to adventure.
12. Ketterer, *The Rational Deception in Poe,* 34.
13. Burrows, et al., *Myths and Motifs in Literature,* 110. Edited by some of the very best intellects of literature and mythology (Burrows, Lapides, and Showcross), "M & M" is still one of the most used and cited books for research relating to mythology in literature. For myself, it was my

first introduction to Carl Jung almost thirty years ago. The very first page starts: "For the literary critic, Jung's two important formulations are . . . his discussions of a collective unconscious and his theory of archetypes." For our purposes, "a collective unconscious and archetypal patterns will not render themselves [to a] objective, scientific study." Thus, the question becomes: How do archetypes and consciousness have any nexus to cybernetics and technology? We find that Jung's theories on archetypes and consciousness developed over time. It's as if Jung created a theory of meaning and consciousness that is as flexible and evolving as cybernetics; thus, always changing meaning and consciousness and vice-versa (Burrows, et al., 1–3). In terms of the metaphor "garden," this has ties to a plethora of scholarly works on America's connection to Judeo-Christianity. It also is connected to Marx's idea of the Pastoral Ideal. I'd point one's attention to Stephen Greenblatt's *The Rise and Fall of Adam and Eve* (New York: Norton), 2017, and Perry Miller's *The Raven and the Wale* and *The Transcendentalists: The Classic Anthology*.

14. Telotte, *Replications,* 37.
15. Telotte, *Replications,* 37.
16. Poe, *Complete Stories and Poems,* 352.
17. Poe, *Complete Stories and Poems,* 352.
18. Poe, *Complete Stories and Poems,* 356.
19. Poe, *Complete Stories and Poems,* 356.
20. Mead, "Poe's 'The Man That Was Used Up,'" 281–86.
21. Elmer, *Reading at the Social Limit,* 14.
22. Poe, *Complete Stories and Poems,* 337.
23. Marx, *Machine in the Garden,* 195–207 (referring to the *rhetoric of the technological sublime*).
24. Fenton, Charles A., "'The Bell-Tower': Melville and Technology." *American Literature* 23.2 (1951), 231. Fenton pays acute attention to Melville's views on technology and religion in America during his time. We will soon discover that the cybernetic ideal's philosophical underpinnings started with the transcendentalist movement in the United States in the early half of the 1800s. And while some scholars consider Melville a "cousin" of the movement, he is not considered a transcendentalist in the popular imagination. Melville is still considered the giant of the age of the American Renaissance when it came to writing and novelization.

For a feel of the culture during this time period I recommend two novels that deal with some issues that our mechanical-man takes on, albeit very subtly: Charles Brockden Brown's *Wieland* and Nathaniel Hawthorne's *The House of Seven Gables*. I do not include Hawthorne's writings here due to the impact his story, "The Artist of the Beautiful," has in a later section on Norbert Wiener, the father of cybernetics. Many college courses today combine the writings of Melville, Hawthorne, Cooper, and Brockden Brown with the transcendentalists. I disagree that both groupings of writers are the same. Melville, Hawthorne, Cooper, and Brockden Brown, in my opinion, are the bridge between the transcendentalists and the American pragmatists (Metaphysical Club). Emerson and other writers who are popularly called transcendentalists come right before the machine age is fully entrenched. Jerome Tharaud's course syllabus for "The American Renaissance" at Brandeis University says the following: "Read the best books first or you may not have a chance to read them at all." Literary critic F.O. Matthiessen used this quote by Henry David Thoreau to justify his classic 1941 study of what he called "The American Renaissance." For Matthiessen the "best books" in American literature were works written by Emerson, Thoreau, Hawthorne, Melville, and Whitman between 1850 and 1855. Since then, scholars have challenged and broadened that account, but the term itself has remained remarkably resilient. In this course, we will follow Thoreau's advice and explore some of the most imaginative and provocative literature of the antebellum period. Along the way, we will sample several influential scholarly accounts of the American Renaissance and consider how they reflect evolving conceptions of the period and of literary value itself. The first half of the course follows Matthiessen's Renaissance alongside three influential critical paradigms from the post-WWII period: the agrarian myth, the "American Adam," and the "machine in the garden." The second half turns to more recent revisions of the Renaissance that have included new kinds of texts and new voices, including narratives of slavery and racial oppression, domestic novels by women, and sensational fiction of urban crime and frontier heroes. Throughout the semester we will place the literature in dialogue with popular works of art in other artistic media. In an era that produced such celebrated forms of cultural expression as Hudson River School painting, Matthew Brady's Civil War photographs, Central Park, and sculptures by Horatio Greenough and Harriet Hosmer, why did American claims to national cultural greatness come to rest so heavily on its literature?

In addition, if one wants to read early American literature that discusses topics of deception and holography prior to even the word *hologram* entering our vernacular, I again point your attention to Hawthorne's *House of Seven Gables* and Brockden Brown's *Wieland.*

25. Melville, *Piazza Tales,* 220.
26. Melville, *Piazza Tales,* 221.
27. Fisher, *Going Under,* 95.
28. Bruce, *Launching of Modern American Science,* 6. For an understanding of the term *Manifest Destiny* as it relates to the mechanical-man's quest, I point your attention to two books: Adam Jortner, *The Gods of Prophetstown: The Battle of Tippecanoe and the Holy War for the American Frontier* (Oxford: Oxford University Press), 2009, and Amy Kaplan's *The Anarchy of Empire in the Making of U.S. Culture, Convergences: Inventories of the Present* (Cambridge: Harvard University Press), 2005. When it comes to these American mythic tropes, we must keep in mind that the mechanical-man's job is to explain what these ideals are today, rather than what they were. Nevertheless, throughout the notes I will cite to sources that explain what these tropes were known for during the cowboy's reign as our official archetype.
29. Bruce, *The Launching of Modern American Science,* 130.
30. Melville, *Piazza Tales,* 223.
31. Melville, *Piazza Tales,* 224.
32. Melville, *Piazza Tales,* 224–25.
33. Hickman, *John Dewey's Pragmatic Technology,* 10.
34. Hickman, *John Dewey's Pragmatic Technology,* 10. The original quotes, and for that matter, the only quotes utilized from Hickman's book, come from Albert Borgman's *Technology and the Character of Contemporary Culture* (Chicago: The University of Chicago Press), 9–12.
35. Grossinger, *Embryos, Galaxies, and Sentient Beings,* 404.
36. Grossinger, *Embryos, Galaxies, and Sentient Beings,* 364.
37. Grossinger, *Embryos, Galaxies, and Sentient Beings,* 364.
38. Grossinger, *Embryos, Galaxies, and Sentient Beings,* 361.

CHAPTER 2. ACCEPTING THE CHALLENGE— THE DEPARTURE

1. Douglas, *All Aboard,* xiv.
2. Rosow, *Born to Lose,* 4.
3. Nocks, "To Serve and Obey," 42.
4. Moskowitz, *Explorers of the Infinite,* 110.
5. Ellis, *Huge Hunter,* 107.
6. Moskowitz, *Explorers of the Infinite,* 42.
7. Ellis, *Huge Hunter,* 3.
8. Jones, et al., *Created Equal,* 634.
9. Jones, et al., *Created Equal,* 634.
10. Douglas, *All Aboard,* 178.
11. Ellis, *Huge Hunter,* 95–99.
12. Campbell and Moyers, *Power of Myth,* 23.
13. Nocks, "To Serve and Obey," 42–46.
14. Tuerk, *Oz in Perspective,* 152.
15. Hughes, *American Genesis,* 1.
16. Marcus and Segal, *Technology in America,* 113.
17. Nye, *The American Technological Sublime,* 172. Immanuel Kant distinguished between the mathematical and the dynamic sublime. In either case he expected that in the aftermath of the immediate experience, particularly if the object is man-made rather than natural. The perception of what is immense and infinite changes over time and across cultures. Although the interpretation developed in Weiskel's *The Romantic Sublime and the Structure of Psychology of Transcendence* (24–49), the third stage may be more culturally determined than the astonishment at the core of the experience, even here previous perceptions shape responses. In short, American forms of the sublime are culturally inflected, including the awe bordering on terror of the second stage. The test for determining what is sublime is to the observer whether or not an object strikes people dumb with amazement. The few experiences that meet this test have transcendent importance both in the lives of individuals and in the construction of culture. For specifics on the sublime, I point your attention to Nye's *Electrifying America* and *The American*

Technological Sublime. Since the Niagara Project is so important to the mythology of Nikola Tesla in the next chapter, I provide Nye's thoughts on the sublime to such a project, in addition, on the harnessing of natural powers like electricity. He writes, "What kind of public experience did these events provide? Their social meaning resided not only in skillful electrical engineering, but in the public perception of the new technologies as spectacle. One of [the] most successful displays occurred in 1907 when [Tesla] lighted Niagara Falls for thirty consecutive nights, using newly improved search lights. Much of the press and public had been skeptical or hostile to the idea of this demonstration, feeling it would be a sacrilege to attempt to 'improve' a great natural symbol. . . . The New York Tribune reported, 'Magnificently illuminated, the falls were of a beauty that their daylight aspect never equaled. For the first time since a factory was erected to draw its power from the rushing water, the garish outlines of the bleak brick buildings were gone, and in their place . . . were the falls in their old glory . . . in a [field] of prismatic color. Every hue in the spectrum was used and words fail to describe the magnificence of the spectacle.' This response locates one central achievement of electrical displays: they permitted the landscape to be edited, simplified, and dramatized" (*Electrifying America,* 58). He addresses the concept of editing, simplifying, and dramatizing nature and man-made technology to create a psychological state or trance. Once again, we have communication occurring between the object/event and those who perceive such.

When we look at cosmic consciousness and Colin Wilsons' *Super Consciousness,* we will see the connection between the sublime of the consciousness ancillary to subjective experience. As Colin Wilson states in *Super Consciousness* "Now if we return to the recognition that, in ordinary consciousness, we are 50 per cent robot and 50 per cent 'real you[,]' we can see that raising ourselves to 51 per cent 'real you' and 49 per cent robot is the all-important watershed. . . . In these states of concentration and excitement, I catch a glimpse of another level. I can see that if I could get into a sufficiently concentrated state, the billiard balls would climb on top of one another until they formed a pyramid. And this pyramid would never collapse. Because my sense of meaning would be so deep, my interest in everything so great, that I would have passed the point where 'regress' or collapse is possible. I would be sustained by sheer perception of meaning. And for the human race, this would be the decisive step to becoming closer to gods.

Perception and the sublime in the mechanical-man mythos are powers to harness in order to take control of the imagination and thus control consciousness in terms of spiritual freedom and service."

18. Baum, *Wonderful Wizard of Oz,* 32.
19. Grossinger, *Bottoming Out,* 202.
20. Baum, *Wonderful Wizard of Oz,* 32.
21. Baum, *Wonderful Wizard of Oz,* 37.
22. Rockoff, "Wizard of Oz as a Monetary Allegory," 747.
23. Baum, *Wonderful Wizard of Oz,* 81.
24. Baum, *Wonderful Wizard of Oz,* 37.
25. Baum, *Wonderful Wizard of Oz,* 37.
26. Rockoff, "Wizard of Oz as a Monetary Allegory," 746.
27. Tuerk, *Oz in Perspective,* 154.
28. Tuerk, *Oz in Perspective,* 154.
29. Tuerk, *Oz in Perspective,* 31.
30. Beniger, *Control Revolution,* 79.
31. Bierce, *Terror by Night,* 90–98.
32. Hickman, *John Dewey's Pragmatic Technology,* 87.
33. Hickman, *John Dewey's Pragmatic Technology,* 87.
34. Bierce, *Terror by Night,* 93.
35. Bierce, *Terror by Night,* 17.
36. Hickman, *John Dewey's Pragmatic Technology,* 87.
37. Hickman, *John Dewey's Pragmatic Technology,* 87
38. Bierce, *Terror by Night,* 18.
39. Marcus and Segal, *Technology in America,* 141.
40. Marcus and Segal, *Technology in America,* 141.
41. Wiener, *Human Use,* 24.
42. Buhner, *Ensouling Language,* 141.
43. Buhner, *Ensouling Language,* 141.
44. Thoreau quoted in Buhner, *Plant Intelligence and the Imaginal Realm,* 352–53.
45. Thoreau quoted in Buhner, *Plant Intelligence and the Imaginal Realm,* 352–53.
46. Bierce, *Terror by Night,* 21.

47. Bierce, *Terror by Night,* 20.
48. Marcus and Segal, *Technology in America,* 140.
49. Bierce, *Terror by Night,* 21.
50. Bierce, *Terror by Night,* 20.
51. Bierce, *Terror by Night,* 22.
52. Bierce, *Terror by Night,* 24.
53. Beniger, *Control Revolution,* 219.
54. Beniger, *Control Revolution,* 218.
55. Beniger, *Control Revolution,* 26.
56. Beniger, *Control Revolution,* 26.
57. Faust, *Republic of Suffering,* 196.
58. Faust, *Republic of Suffering,* 197.
59. Faust, *Republic of Suffering,* 202–3.
60. Hayles, *How We Became Posthuman,* 103.
61. Hayles, *How We Became Posthuman,* 103. Hayles appreciates the quest by relating it to her own experiences in undertaking the study of cybernetics. Like me, it was an odyssey in itself to figure out what cybernetics even means. Hayles writes: "I was led into a maze of developments that turned into a six-year odyssey of researching archives in the history of cybernetics, interviewing scientists in computational biology and artificial life, reading cultural and literary texts concerned with information technologies, visiting laboratories engaged in research on virtual reality, and grappling with technical articles in cybernetics, information theory, autopoiesis, computer simulation, and cognitive science. Slowly this unruly mass of material began taking shape as three interrelated stories. The first centers on how information lost its body, this is, how it came to be conceptualized as an entity separate from the material forms in which it is thought to be embedded. The second story concerns how the cyborg was created as a technological artifact and cultural icon in the years following World War II. The third, deeply implicated with the first two, is the unfolding story of how a historically specific construction called the human is giving way to a different construction called [P]osthuman" (2). As you are aware, the mechanical-man's cybernetic odyssey concerns control, information, and communication in creating a new consciousness. Upon reading Hayles's *Posthuman,* I get the sense that her definition of cybernetics is more centered on the changing bodily apparatus of

humans rather than the mind and consciousness. Nevertheless, both of her books add to this posthuman or new-consciousness discussion. So, too, do the following books by Ray Kurzweil: *The Singularity Is Near: When Humans Transcend Biology* and *The Age of Intelligent Machines.*

62. Beniger, *Control Revolution,* 48.
63. Kripal, *Authors of the Impossible,* 29.
64. Kripal, *Authors of the Impossible,* 29. If you have not already concluded that if the mechanical-man does not break himself out of materialism he will be a slave to his senses, then pick up any Bernardo Kastrup or Jeffrey Kripal book from the bibliography and commence reading. In *Authors of the Impossible,* Kripal gives the debate a different spin in how he sees fragmentation to holism, holding the following: "[T]his demonization and subsequent alienation is born of an exaggerated and unbalanced scientism, a one-sided Aristotelianism that [critics now see] us moving beyond before a balancing Platonic resurgence. It is not about one or the other, though. It is about both. It is about ***balance.*** Western intellectual, spiritual, and cultural life, at their best and almost creative anyway, work through a delicate balancing act between this Aristotelianism read: ***rationalism*** and the Platonism read: ***mysticism.*** The pendulum has been swinging right, toward Aristotle, for about three hundred years now. It has now reached its rationalist zenith and is beginning to swing back left, toward Plato. Which is not to say, at all, that Western culture will somehow become irrational and unscientific again, that we suddenly won't need Aristotle or science any longer. This vast centuries-long process is ultimately about balance, about wisdom. It is also about making the unconscious conscious, about realizing and living our own secret life." (30, emphasis added to "mysticism" and "rationalism")
65. Bierce, *Terror by Night,* 94.
66. Grossinger, *Embryos, Galaxies, and Sentient Beings,* 416–17.

CHAPTER 3. THE SORCERER AND THE WIZARD

1. Tesla, U.S. Patent 613,809.
2. Tesla, U.S. Patent 613,809.
3. Seifer, *Wizard,* 193.

4. Tesla, *My Inventions,* 14.
5. Marx, *Machine in the Garden,* 228.
6. Jones, et al., *Created Equal,* 636.
7. Seifer, *Wizard,* 195.
8. Tesla, *My Inventions,* 107.
9. Tesla, *My Inventions,* 107.
10. Cowan, *Social History of American Technology,* 149.
11. Seifer, *Wizard,* 33.
12. Seifer, *Wizard,* 37.
13. Seifer, *Wizard,* 49.
14. Seifer, *Wizard,* 50.
15. Seifer, *Wizard,* 137.
16. Seifer, *Wizard,* 165.
17. Seifer, *Wizard,* 165.
18. Seifer, *Wizard,* 195.
19. Seifer, *Wizard,* 198.
20. Seifer, *Wizard,* 200.
21. Campbell, *Hero with a Thousand Faces,* 60.
22. Seifer, *Wizard,* 165.
23. White, "Mr. Tesla and the Universe," 448.
24. White, "Mr. Tesla and the Universe," 448.
25. White, "Mr. Tesla and the Universe," 449.
26. White, "Mr. Tesla and the Universe," 448.
27. White, "Mr. Tesla and the Universe," 451.
28. White, "Mr. Tesla and the Universe," 451.
29. Maher and Briggs, *Open Life,* 122–23.
30. Maher and Briggs, *Open Life,* 109.
31. Cheney, *Tesla,* 131.
32. Cheney, *Tesla,* 131.
33. Campbell, *Hero with a Thousand Faces,* 23.
34. Wiener, *Human Use,* 23.
35. Wiener, *Human Use,* 24.
36. Wiener, *Human Use,* 24.
37. Hebb, *The Organization of Behavior,* xix. For more information on Hebb and Hebb's Law, see Dr. Joe Dispenza's *Breaking the Habit of Being Yourself.*

38. Tesla, U.S. Patent 613,809.
39. Neisser, "Computers as Tools and as Metaphors," 78.
40. Neisser, "Computers as Tools and as Metaphors," 78.
41. Channell, *Vital Machine,* 133.
42. Channell, *Vital Machine,* 66.
43. Seifer, *Wizard,* 202.
44. Wiener, *Human Use,* 23.
45. Channell, *Vital Machine,* 141.
46. Channell, *Vital Machine,* 141.
47. Villiers de l'Isle-Adam, *Eve of the Future,* 7.
48. Villiers de l'Isle-Adam, *Eve of the Future,* 24.
49. Villiers de l'Isle-Adam, *Eve of the Future,* 29–39.
50. Villiers de l'Isle-Adam, *Eve of the Future,* 29–39.
51. Villiers de l'Isle-Adam, *Eve of the Future,* 72.
52. Deutsch, *Fabric of Reality,* 98.
53. Deutsch, *Fabric of Reality,* 86.
54. Deutsch, *Fabric of Reality,* 96.
55. Deutsch, *Fabric of Reality,* 97.
56. Deutsch, *Fabric of Reality,* 151.
57. Villiers de l'Isle-Adam, *Eve of the Future,* 151.
58. Villiers de l'Isle-Adam, *Eve of the Future,* 150–55.
59. Deutsch, *Fabric of Reality,* 92.
60. Villiers de l'Isle-Adam, *Eve of the Future,* 82.
61. Goddard, *Power of Imagination,* 65.
62. Villiers de l'Isle-Adam, *Eve of the Future,* 132.
63. Villiers de l'Isle-Adam, *Eve of the Future,* 250.
64. Baudrillard, *America,* 28–29.
65. Wood, *Edison's Eve,* 225.
66. Wood, *Edison's Eve,* 228.
67. Wiebe, *Search for Order,* 272.
68. Wiebe, *Search for Order,* 301.
69. Josephson, *Edison,* 447.
70. Josephson, *Edison,* 447.
71. Josephson, *Edison,* 447.
72. Josephson, *Edison,* 448.
73. Goddard, *Power of Imagination,* 74.

74. Moskowitz, *Explorers of the Infinite,* 335.
75. Horowitz, *Miracle Club,* vii.

CHAPTER 4. KNOWING THYSELF— THE INITIATION

1. Nocks, "To Serve and Obey," 29.
2. John Diebold, "Goals to Match Our Means," in Dechert, *Social Impact of Cybernetics,* 2.
3. Burrows et al., *Myths and Motifs in Literature,* 135.
4. Cowan, *Social History of American Technology,* 212.
5. Cowan, *Social History of American Technology,* 212.
6. Cowan, *Social History of American Technology,* 211.
7. Vincent, "Rex," in Asimov, *Machines That Think,* 53.
8. Vincent, "Rex," in Asimov, *Machines That Think,* 54.
9. Vincent "Rex," in Asimov, *Machines That Think,* 56.
10. Vincent, "Rex," in Asimov, *Machines That Think,* 63.
11. Torode, *Everyday Emerson.*
12. Vincent, "Rex," in Asimov, *Machines That Think,* 64.
13. Robert W. Moore, "Robot's Return," in Asimov, *Machines That Think,* 144.
14. Asimov, 148.
15. Jones, et al., *Created Equal,* 737.
16. Asimov, "Runaround," in Asimov, *Machines That Think,* 209.
17. Asimov, "Runaround," in Asimov, *Machines That Think,* 209.
18. Asimov, "Runaround," in Asimov, *Machines That Think,* 211.
19. Asimov, "Runaround," in Asimov, *Machines That Think,* 209–13.
20. Asimov, "Runaround," in Asimov, *Machines That Think,* 221.
21. Asimov, "Runaround," in Asimov, *Machines That Think,* 227–28.
22. Lynn Margulis and Dorian Sagan, "Micro cosmos," in Barlow, *From Gaia to Selfish Genes,* 59.
23. Bone, "Triggerman," in Asimov, *Machines That Think,* 345–56.
24. Bone, "Triggerman," in Asimov, *Machines That Think,* 355–56.
25. Asimov, *Machines That Think,* 253–54.
26. Asimov, *Machines That Think,* 257.

27. Asimov, *Machines That Think,* 275–76.
28. Fisher, "Information Feedback Loops," 24.
29. Fisher, "Information Feedback Loops," 23.
30. Fisher, "Information Feedback Loops," 23.
31. Heinlein, "Science Fiction," in Davenport, *Science Fiction Novel,* 28. Heinlein, a master of science fiction writing himself, defines science fiction and how it coevolves with practicing science. Remember, this is 1969 when Heinlein writes: "Hypothesis and theories are always expendable; a scientist modifies or discards them in the face of new facts as casually as he changes his socks. Ordinarily a scientist will use the convenient rule-of-thumb called 'least hypothesis' but he owes it no allegiance; his one fixed loyalty is to the observed fact. An honest science fiction writer observes the same loyalty to fact but from there on his path diverges from that of the scientist because his function is different. The pragmatic rule of least hypothesis, useful as it may be to orderly research, is as un[-]functional in speculative fiction as a chaperone on a honeymoon. . . . He [a science fiction writer] cannot carry out his function while paying lip service to the orthodox opinions or prejudices of his tribe and generation, and no one should expect him" (21). In effect, science fiction writing is no different from filling in the "gaps" during the process of information becoming wisdom, which is exactly what mythology is supposed to do. I like to look at it as a form of alchemy in this Information Age. As we will see in later chapters, when the mechanical-man observes mythic markers in an idealistic paradigm, it's the gaps between utilizing the imagination as a tool in creating the reality that is paradox. The mechanical-man becomes a mythic-marker for the process of imagination becoming material reality. For concrete examples of the interrelationship between science fiction and technology, read Heinlein's article in its entirety.

CHAPTER 5. THE GREAT MOTHER

1. Cowan, *More Work for Mother,* 211.
2. Campbell, *Thou Art That,* 65.
3. Faulkner, *Lucretia Mott.*
4. Beyer, *Grace Hopper,* 27.

5. Beyer, *Grace Hopper,* 33.
6. Beyer, *Grace Hopper,* 33.
7. Beyer, *Grace Hopper,* 33.
8. Nye, "The Voice of the Serpent," in Garry and Pearsall, *Women, Knowledge, and Reality,* 323.
9. Nye, "The Voice of the Serpent," in Garry and Pearsall, *Women, Knowledge, and Reality,* 335.
10. Nye, "The Voice of the Serpent," in Garry and Pearsall, *Women, Knowledge, and Reality,* 322.
11. Nye, "The Voice of the Serpent," in Garry and Pearsall, *Women, Knowledge, and Reality,* 322.
12. Beyer, *Grace Hopper,* 79.
13. For a discussion of spandrels, specifically how natural selection does not control the arrow of time with respect to all physical and mental attributes, see Stephen J. Gould, *Bully for Brontosaurus* (New York: Norton, 1992).
14. Beyer, *Grace Hopper,* 125–27.
15. Beyer, *Grace Hopper,* 125–27.
16. Beyer, *Grace Hopper,* 127.
17. Beyer, *Grace Hopper,* 137.
18. Beyer, *Grace Hopper,* 173.
19. Beyer, *Grace Hopper,* 220.
20. Beyer, *Grace Hopper,* 220.
21. Beyer, *Grace Hopper,* 222.
22. Beyer, *Grace Hopper,* 222.
23. Beyer, *Grace Hopper,* 223.
24. Beyer, *Grace Hopper,* 223.
25. Beyer, *Grace Hopper,* 222.
26. Wiener, *Human Use,* 15.
27. Wiener, *Human Use,* 16.
28. Wiener, *Human Use,* 23.
29. Wiener, *Human Use,* 58.
30. Wiener, *God and Golem,* 63.
31. Braden, *Walking Between the Worlds,* 72.
32. Grossinger, *Bottoming Out,* 190.
33. Hawthorne, *Moses from an Old Manse,* 78.

34. Hawthorne, *Moses from an Old Manse,* 84.
35. Hawthorne, *Moses from an Old Manse,* 101.
36. Wiener, *God and Golem,* 52.
37. Jung, *Man and His Symbols,* 5.
38. Beyer, *Grace Hopper,* 321.
39. Kurzweil, *The Singularity Is Near,* 129.
40. Campbell, *Thousand Faces,* 97.
41. Jung, *Man and His Symbols,* 186.

CHAPTER 6. THE PROPHET

1. Nye, *American Technological Sublime,* 138.
2. Nye, *American Technological Sublime,* 138.
3. Nye, *American Technological Sublime,* 256.
4. Hey, *Feynman and Computing,* 63.
5. Hey, *Feynman and Computing,* 69.
6. Hey, *Feynman and Computing,* 66.
7. Hey, *Feynman and Computing,* 68.
8. Pivonka, "Fantastic Voyage," 28.
9. Wiener, *Human Use,* 21.
10. Ratner, *Nano-Technology and Homeland Security,* 88.
11. Ratner, *Nano-Technology and Homeland Security,* 279.
12. Channell, *Vital Machine,* 112.
13. Wiener, *God and Golem,* 65.
14. Wiener, *God and Golem,* 48.
15. Kripal, *Authors of the Impossible,* 270.
16. Moscowitz, *Explorers of the Infinite,* 334.
17. Gleick, *Genius,* 187.
18. Kastrup, *Dreamed Up Reality,* 15.
19. Deutsch, *Fabric of Reality,* 331.
20. Deutsch, *Fabric of Reality,* 331.
21. Kastrup, *Dreamed Up Reality,* 12.
22. Brown, *Most of the Good Stuff,* 30.
23. Regarding Feynman's role with the space shuttle *Challenger,* including a concise history of its explosion and Feynman's explanation for the disaster, see Gleick, *Genius.*

24. Daniel W. Hillis, "Close to the Singularity," in Brockman, *Third Culture,* 89.
25. Douglas, *All Aboard,* 30.
26. Feynman, *Surely You're Joking,* 45.
27. Gleick, *Genius,* 33; Dawkins, *Blind Watchmaker;* Dawkins, *Devil's Chaplain;* Dawkins, *Selfish Gene.*
28. Quoted in Douglas, *All Aboard,* 30.
29. Quoted in McCorduck, *Machines Who Think,* 261.
30. Quoted in McCorduck, *Machines Who Think,* 395.
31. Gleick, *Genius,* 429.
32. Gleick, *Genius,* 438.
33. Douglas, *All Aboard,* 146.
34. Douglas, *All Aboard,* 176.
35. Dechert, *Social Impact of Cybernetics,* 19.
36. Kripal, *Authors of the Impossible,* 12.
37. Kripal, *Authors of the Impossible,* 43.
38. Dispenza, *Becoming Supernatural,* 32.
39. Horowitz, *Miracle Club,* 56.
40. Kripal, *Flip,* 192–93.
41. Kripal, *Flip,* 191.
42. Kripal, *Flip,* 191–92.

CHAPTER 7. THE FINAL ODYSSEY—THE RETURN

1. Stork, *HAL's Legacy,* 1–2.
2. Douglas Lenat, "From 2001 to 2001: Common Sense and the Mind of HAL," in Stork, *HAL's Legacy,* 207–208.
3. Stephen J. Gould, "The Pattern of Life's History," in Brockman, *Third Culture,* 57.
4. Quoted in McCorduck, *Machines Who Think,* 402.
5. Kubrick, *2001: A Space Odyssey,* quoted in Stork, *HAL's Legacy,* 279.
6. Rosalind W. Picard, "Does HAL Cry Digital Tears? Emotions and Computers," in Stork, *HAL's Legacy,* 280.
7. Daniel C. Dennett, "Intuition Pumps," in Brockman, *Third Culture,* 190.
8. Ray Kurzweil, "When Will HAL Understand What We Are Saying?

Computer Speech Recognition and Understanding," in Stork, *HAL's Legacy,* 132–33.

9. Greenblatt, *Swerve,* 199.
10. Greenblatt, *Swerve,* 199.
11. Jung, *Man and His Symbols,* 221.
12. Paul Davies, "The Synthetic Path," in Brockman, *Third Culture,* 308.
13. Davies, "The Synthetic Path," in Brockman, *Third Culture,* 309.
14. Menand, *Metaphysical Club,* 439.
15. Menand, *Metaphysical Club,* xi–xii.
16. Stork, *HAL's Legacy,* 91.
17. Eugene G. D'Aguili and Charles D. Laughlin, "The Neurobiology of Myth and Ritual," in Grimes, *Readings in Ritual Studies,* 136. This is a must read for the study of mythology, rituals, and rites of passage.
18. Claude Lévi-Strauss, "The Effectiveness of Symbols," in Grimes, *Readings in Ritual,* 138.
19. D'Aguili and Laughlin, "Neurobiology of Myth and Ritual," 132.
20. Marvin Minsky, "Smart Machines," in Brockman, *Third Culture,* 159–60.
21. Hillis, "Close to Singularity," 385.
22. Charles Langton, "A Dynamical Pattern," in Brockman, *Third Culture,* 347.
23. Davidson, et al., *Nation of Nations,* 1001.
24. Herbert, *Dune* as quoted on Goodreads.
25. Selig, *The Book of Freedom,* 287.
26. See any Chris Hedges interviews, articles, or books, with emphasis on *Empire of Illusion* (2009), *American Fascists* (2007), and *Death of the Liberal Class* (2010) for an understanding of why the American empire failed on how to course correct the experiment of an evolving constitutional democracy when corporate money took over government.
27. See Rupert Sheldrake and Graham Hancock—the Joe Rogan interviews are transcendent with wisdom. *DMT Dialogues: Encounters with the Spirit Molecule* with conversations between Dennis McKenna, Jeremy Narby, Erik Davis, Rupert Sheldrake, Graham Hancock, and Rick Strassman.
28. Nolan and Johnson, *Logan's Run,* 1.
29. *Logan's Run,* directed by Michael Anderson, performed by Michael York, Jenny Agutter, and Peter Ustinov (MGM/UA, 1976).
30. Telotte, *Replications,* 152.

CHAPTER 8. THE TRICKSTER

1. Peat, *Infinite Potential,* 258.
2. Peat, *Infinite Potential,* 303.
3. Talbot, *Holographic Universe,* 1. You can pick out any page in *Holographic Universe* and have your mind turned upside down.
4. Talbot, *Holographic Universe,* 46.
5. Wilson, *Super Consciousness,* 45–46.
6. Wilson, *Super Consciousness,* 45.
7. Peat, *Infinite Potential,* 303.
8. Leslie, *Cold War and American Science,* 2.
9. Grossinger, *Embryos, Galaxies, and Sentient Beings,* 431.
10. Peat, *Infinite Potential,* 423. Try not to confuse the two biographies of Bohm and Dick, the former by Peat and the latter by Peake. It's easy to make mistakes citing to these two simultaneously. Numerous biographies were researched in this analysis of the mechanical-man pertaining to all of our guides. The one or two that I chose for each guide was based on their attention to the inventor, writer, and/or theorist in terms of the added mythos such biographers afforded in creating an archetype of that particular person. Not all biographers have the talent or the ability to mythologize the person/persona they present to readers, especially if the biography is scholastically pedantic and meant to be read by only academics.
11. Peake, *Life of Philip K. Dick,* 29.
12. Peat, *Infinite Potential,* 29.
13. Peat, *Infinite Potential,* 10.
14. Smith, *Skylark of Space,* viii.
15. Smith, *Skylark of Space,* 62.
16. Hyde, *Trickster Makes This World,* 7.
17. Talbot, *Holographic Universe,* 260.
18. Peat, *Infinite Potential,* 25.
19. Kripal, *Authors of the Impossible,* 269.
20. Peat, *Infinite Potential,* 25.
21. Peat, *Infinite Potential,* 12–13.
22. Tesla, *My Inventions,* 11–12.
23. Peat, *Infinite Potential,* 13.

24. Peat, *Infinite Potential,* 13–14.
25. Tesla, *My Inventions,* 21.
26. Campbell, *Hero with a Thousand Faces,* 49.
27. Campbell and Moyers, *Power of Myth,* 275.
28. Hyde, *Trickster Makes This World,* 6.
29. Hyde, *Trickster Makes This World,* 7–8.
30. Dick, "Electric Ant," in Asimov, *Machines That Think,* 495–515.
31. Dick, "Electric Ant," in Asimov, *Machines That Think,* 497.
32. Channell, *Vital Machine,* 66.
33. Channell, *Vital Machine,* 10.
34. Channell, *Vital Machine,* 112.
35. Dick, "Electric Ant," in Asimov, *Machines That Think,* 498.
36. Dick, "Electric Ant," in Asimov, *Machines That Think,* 498.
37. Dick, "Electric Ant," in Asimov, *Machines That Think,* 498.
38. Dick, "Electric Ant," in Asimov, *Machines That Think,* 499.
39. Peat, *Infinite Potential,* 252.
40. Dick, "Electric Ant," in Asimov, *Machines That Think,* 499.
41. Dick, "Electric Ant," in Asimov, *Machines That Think,* 499.
42. Dick, "Electric Ant," in Asimov, *Machines That Think,* 501.
43. Dick, "Electric Ant," in Asimov, *Machines That Think,* 499.
44. Dick, "Electric Ant," in Asimov, *Machines That Think,* 501–2.
45. Dick, "Electric Ant," in Asimov, *Machines That Think,* 501–2.
46. Greene, *Hidden Reality,* 269.
47. Dick, "Electric Ant," in Asimov, *Machines That Think,* 502.
48. Kastrup, *Dreamed Up Reality,* 13.
49. Talbot, *Holographic Universe,* 191.
50. Talbot's index, s.v. Pribram, Karl (brain studies, 11–14, 18–20, 28–31, 163).
51. Talbot, *Holographic Universe,* 191–92.
52. Three Initiates, *Kybalion,* 34.
53. Talbot, *Holographic Universe,* 191.
54. Talbot, *Holographic Universe,* 191.
55. Goddard, *Power of Imagination,* 9.
56. Dick, "Electric Ant," in Asimov, *Machines That Think,* 506.
57. Dick, "Electric Ant," in Asimov, *Machines That Think,* 506.
58. Dick, "Electric Ant," in Asimov, *Machines That Think,* 505–6.

59. Dick, "Electric Ant," in Asimov, *Machines That Think,* 505.
60. Peat, *Infinite Potential,* 322.
61. Peat, *Infinite Potential,* 322.
62. Grossinger, *Embryos, Galaxies, and Sentient Beings,* 442.
63. Dick, "Electric Ant," in Asimov, *Machines That Think,* 509.
64. Dick, "Electric Ant," in Asimov, *Machines That Think,* 511.
65. Dick, "Electric Ant," in Asimov, *Machines That Think,* 515.
66. Kastrup, *Meaning in Absurdity,* 52.
67. Kastrup, *Meaning in Absurdity,* 55–56.
68. Feynman, *Surely You're Joking,* 78.
69. Peat, *Infinite Potential,* 323.

CHAPTER 9. THE CAVE

1. Magee, *Story of Philosophy,* 31.
2. Rushing and Frentz, *Projecting the Shadow,* 1.
3. Kastrup, *Dreamed Up Reality,* 12.
4. Rushing and Frentz, *Projecting the Shadow,* 2.
5. Hanson and Kay, *Star Wars: The New Myth,* 197.
6. Rushing and Frentz, *Projecting the Shadow,* 3.
7. Hanson and Kay, *Star Wars: The New Myth,* 370.
8. Jung, *Memories, Dreams, Reflections,* 189.
9. Jung, *Memories, Dreams, Reflections,* 189.
10. Baxter, *George Lucas,* 39.
11. Baxter, *George Lucas,* 38.
12. Baxter, *George Lucas,* 41.
13. Baxter, *George Lucas,* 22.
14. Baxter, *George Lucas,* 23.
15. Baxter, *George Lucas,* 24.
16. Baxter, *George Lucas,* 164.
17. Hanson and Kay, *Star Wars: The New Myth,* 48–49.
18. Seifer, *Wizard,* 164.
19. McTaggart, *Field,* xviii.
20. Galipeau, *Journey of Luke Skywalker,* 48.
21. McTaggart, *Field,* 84.
22. Galipeau, *Journey of Luke Skywalker,* 58.

23. Galipeau, *Journey of Luke Skywalker,* 34.
24. Kastrup, *Dreamed Up Reality,* 37.
25. Grossinger, *Bottoming Out the Universe,* 165–66, 169–70.

CHAPTER 10. THE FOURTH DIMENSION IS CALLING US

1. David Boersema, "We Are Not Going to Kill Today," in Decker and Eberl, *The Ultimate Star Trek,* 68.
2. Found in Greg Littman's "The Needs of the Many Outweigh the Needs of the Few," in Decker and Eberl, *Ultimate Star Trek,* 127.
3. Joe Rogan, *The Joe Rogan Experience,* Episode 1169, "Elon Musk," Sep 6 (accessed April 4, 2018).
4. Episode references for these relationships include, "Measure of a Man," and "The Offspring," *Star Trek: The Next Generation.*
5. Dennis M. Weiss, "Humans, Androids, Cyborgs, and Virtual Beings: All aboard the *Enterprise*" in Decker and Eberl, *Ultimate Star Trek,* 188.
6. Melanie Johnson-Moxley, "Rethinking the Matter: Organians Are Still Organisms," in Decker and Eberl, *Ultimate Star Trek,* 221.
7. Brian Scott, *The Reality Revolution,* "Neville Goddard Time to Act," March 5, 2020, YouTube.
8. *Star Trek: The Next Generation,* "John de Lancie: Q Quotes." IMDb.
9. Lisa Cassidy, "Resistance Is Negligible: In Praise of Cyborgs," in Decker and Eberl, *Ultimate Star Trek,* 240–41.
10. Dena Hust, "Resistance Really Is Futile: On Being Assimilated by Our Own Technology," in Decker and Eberl, *Ultimate Star Trek,* 271.
11. Dara Fogel, "Life on the Holodeck: What Stat Trek Can Teach Us about the True Nature of Reality," in Decker and Eberl, *Ultimate Star Trek,* 286.
12. Sarah O'Hare, "'Strangely Compelling': Romanticism in 'The City on the Edge of Forever'" in Decker and Eberl, *Ultimate Star Trek,* 306.
13. Selig, *The Book of Freedom,* 238.
14. Decker and Eberl, *Ultimate Star Trek,* 68.
15. Selig, *The Book of Freedom,* 236.
16. Barnstone and Meyer, *Gnostic Bible,* 62.

17. Direct quote from Roddenberry found in James F. Mcgrath's "A God Needs Compassion, but Not a Starship: Star Trek's Humanist Theology," in Decker and Eberl, *Ultimate Star Trek* at 316 and 319 of 315–26.
18. James F. Mcgrath, "A God Needs Compassion, but Not a Starship: Star Trek's Humanist Theology," in Decker and Eberl, *Ultimate Star Trek,* 320.
19. Romans 8:22.
20. Bassham, "The Religion of the Matrix and the Problems of Pluralism," 2016, at 110.
21. Andersen, *Fantasyland,* 49.
22. Brian Scott reads and discusses numerous lectures by Neville Goddard. This particular lecture, retrieved on YouTube, was entitled "Time to Act." He finds that the great Christian mystic himself knew of Gene Rodenberry's connection to manifesting and knowledge of the law of one and the law of assumption.
23. Goddard, *Power of Awareness,* 23.
24. Slavoj Zizek, "The Matrix: Or, The Two Sides of Perversion," in Irwin, *The Matrix and Philosophy,* 246.
25. Virk, *The Simulation Hypothesis,* 8.
26. Virk, *The Simulation Hypothesis,* 116.
27. Goddard, *Power of Imagination,* 1 and 5.
28. Goddard, *Power,* 117.
29. Goddard, *Power,* 44–45.
30. Goddard, *Power,* 45.
31. Brian Scott, *The Reality Revolution,* YouTube.
32. William Irwin, "Computers, Caves, and Oracles: Neo and Socrates," in Irwin, *The Matrix and Philosophy,* 14.
33. Virk, *The Simulation Hypothesis.*
34. Barnstone and Meyer, *The Gnostic Bible,* 142.
35. Stephen Faller, "What It Means to Be an Iron Man. Iron Man's Transcendent Challenge," in William Irwin and Mark D. White's *Iron Man and Philosophy.*
36. Stephen Faller, "What It Means to Be an Iron Man. Iron Man's Transcendent Challenge," in William Irwin and Mark D. White's *Iron Man and Philosophy,* 259–60.
37. Cheney, *Tesla,* 131.

38. Stephen Faller, "What It Means to Be an Iron Man. Iron Man's Transcendent Challenge," in William Irwin and Mark D. White's *Iron Man and Philosophy,* 263.
39. Stephen Faller, "What It Means to Be an Iron Man. Iron Man's Transcendent Challenge," in William Irwin and Mark D. White's *Iron Man and Philosophy,* 258–59.
40. Stephen Faller, "What It Means to Be an Iron Man. Iron Man's Transcendent Challenge," in William Irwin and Mark D. White's *Iron Man and Philosophy,* 259.
41. Stephen Faller, "What It Means to Be an Iron Man. Iron Man's Transcendent Challenge," in William Irwin and Mark D. White's *Iron Man and Philosophy,* 245.
42. Stephen Faller, "What It Means to Be an Iron Man. Iron Man's Transcendent Challenge," in William Irwin and Mark D. White's *Iron Man and Philosophy,* 262.
43. Goddard, *Power of Imagination,* 9.
44. Stephanie Patterson and Brett Patterson, "'I Have a Good Life': Iron Man and the Avenger School of Virtue," in Irwin and White, *Iron Man and Philosophy,* 225.
45. Stephanie Patterson and Brett Patterson, "'I Have a Good Life': Iron Man and the Avenger School of Virtue," in Irwin and White, *Iron Man and Philosophy,* 229.
46. Faller in Irwin and White, *Iron Man and Philosophy,* 262.
47. AA, *Big Book,* 8.
48. Faller in Irwin and White, *Iron Man and Philosophy,* 263–64.
49. *Joe Rogan Experience,* Episode #1169, "Elon Musk," September 7, 2018, YouTube.

CHAPTER 11. THE SAVIOR

1. Strieber and Kripal, *Super Natural,* 314.
2. Jeffrey Kripal, "Introduction," in Kastrup, *More Than Allegory,* 6.
3. Grossinger, *Bottoming Out the Universe,* 428–29.
4. Kripal, *Authors of the Impossible,* 57.
5. Kripal, *Authors of the Impossible,* 57.
6. Conway and Siegelman, *Dark Hero of the Information Age,* ix.

7. Conway and Siegelman, *Dark Hero of the Information Age,* xv.
8. Isaacson, *Steve Jobs,* 565.
9. Isaacson, *Steve Jobs,* 566.
10. Isaacson, *Steve Jobs,* 570.
11. Isaacson, *Steve Jobs,* 568.
12. Conway and Siegelman, *Dark Hero of the Information Age,* 346–47.
13. Isaacson, *Steve Jobs,* 570.
14. Isaacson, *Steve Jobs,* 570.
15. Conway and Siegelman, *Dark Hero of the Information Age,* 346.
16. Conway and Siegelman, *Dark Hero of the Information Age,* x.
17. Isaacson, *Steve Jobs,* 567.
18. Gorski, *American Covenant,* 114.
19. Gorski, *American Covenant,* 114–16.
20. Gorski, *American Covenant,* 115.
21. Goddard, *Power of Imagination,* 152–53.
22. Goddard, *Power of Imagination,* 153.
23. Goddard, *Power of Imagination,* 152.
24. Gorski, *American Covenant,* 232.
25. Gorski, *American Covenant,* 113. Read the entire epilogue to Louis Menand's Pulitzer Prize–winning book, *The Metaphysical Club.*
26. Gorski, *American Covenant,* 130.
27. Gorski, *American Covenant,* 152.
28. James, *Varieties of Religious Experience,* 86–87.
29. Dick, *Exegesis,* 167.
30. Dick, *Exegesis,* 173.
31. Dick, *VALIS,* 291.
32. Kastrup, *More than Allegory,* 27.
33. The most well-known study of the Hermetic philosophy of ancient Greece and Egypt is Manly Hall's *The Secret Teachings of All Ages* in addition to Max Heindel and Manly Hall's *Blavatsky and the Secret Doctrine.* For our purposes, I chose a more direct and short version of the Hermetic doctrines by using *Kybalion.*
34. Hoeller, *Jung and the Lost Gospels,* 63.
35. Jegede, *Infinite Possibility,* 100.
36. George Lucas, *Star Wars: Revenge of the Sith* (2005).
37. Grimes, *Readings in Ritual Studies,* 142–43.

38. Grimes, *Readings in Ritual Studies,* 142–43.
39. Menard, *Metaphysical Club,* 440.
40. Barnstone and Meyer, *Gnostic Bible,* 62.
41. Barnstone and Meyer, *Gnostic Bible,* 261.
42. Barnstone and Meyer, *Gnostic Bible,* 267.
43. Barnstone and Meyer, *Gnostic Bible,* 281.
44. Barnstone and Meyer, *Gnostic Bible,* 281.
45. Barnstone and Meyer, *Gnostic Bible,* 142.
46. Barnstone and Meyer, *Gnostic Bible,* 142.
47. Barnstone and Meyer, *Gnostic Bible,* 144.
48. Barnstone and Meyer, *Gnostic Bible,* 145.
49. Fry, *Star Wars: The Last Jedi,* 178–79.
50. Lippe, *Simulation,* 12.
51. Fry, *Star Wars: The Last Jedi,* 179–80.
52. Barnstone and Meyer, *Gnostic Bible,* 44–45.
53. Barnstone and Meyer, *Gnostic Bible,* 373–74.
54. Michael Talbot, *Holographic Universe,* 105.
55. Barnstone and Meyer, *Gnostic Bible,* 374.
56. Barnstone and Meyer, *Gnostic Bible,* 498–501.
57. Kastrup, *Brief Peeks Beyond,* 50–51.
58. Barnstone and Meyer, *Gnostic Bible,* 501.
59. Barnstone and Meyer, *Gnostic Bible,* 501.
60. Hedges, *America: The Farewell Tour,* see index.
61. Dick, *VALIS,* 206.
62. Dick, *VALIS,* 160.
63. Dick, *VALIS,* 230.
64. Pagels, *Why Religion?* 23.
65. Dick, *VALIS,* 161.
66. Kastrup, *Brief Peeks Beyond,* 35, 50–51.
67. Dick, *VALIS,* 103.
68. Dick, *VALIS,* 120.
69. Schlesinger, *Robert Kennedy and His Times,* 946.
70. St. Germain, *Waking Up in 5D,* 121.
71. *Joe Rogan Experience,* Episode #1169, "Elon Musk," September 7, 2018, YouTube.
72. Vance, *Elon Musk,* 22.

73. Lopatto, "Elon Musk Unveils Neuralink's Plans for Brain-Reading 'Threads' and a Robot to Insert Them."
74. Lopatto, "Elon Musk Unveils Neuralink's Plans for Brain-Reading 'Threads' and a Robot to Insert Them."
75. Musk, "An Integrated Brain-Machine Interface Platform with Thousands of Channels."
76. Oleksii, "Elon Musk's Neurolink."
77. Oleksii, "Elon Musk's Neurolink."

CONCLUSION. THE FIRST AND FINAL BOON: HUMAN IMAGINATION

1. Foster, *Star Wars: The Force Awakens,* 172.
2. Braden, *Divine Matrix,* 44.
3. Hoeller, *Jung and the Lost Gospels,* 87–88.
4. Braden, *Divine Matrix,* 148.
5. Braden, *Divine Matrix,* 149.
6. St. Germain, *Waking Up in 5D,* 83.
7. Braden, *Divine Matrix,* 109.
8. Braden, *Divine Matrix,* 109.
9. Braden, *Divine Matrix,* 114.
10. Braden, *Divine Matrix,* 114–15.
11. Goddard, *Power of Imagination,* 107–8.
12. Hoeller, *Gnostic Jung,* 9.
13. Hoeller, *Gnostic Jung,* 9.
14. Braden, *Divine Matrix,* 41.
15. Jedge, *Infinite Possibility,* 2–9.
16. Jedge, *Infinite Possibility,* 120.
17. Rae, *Star Wars: The Rise of Skywalker,* 246–47.

Bibliography

Abrams, J. J., dir. *Star Wars Episode VII: The Force Awakens.* Burbank, Calif.: Walt Disney Studios Motion Pictures, 2015.

Algar, James, Samuel Armstrong, and Ford Beebe Jr., dirs. *Fantasia.* Burbank, Calif.: Walt Disney Studios Motion Pictures, 1940.

Andersen, Kurt. *How America Went Haywire: A 500-Year History.* New York: Random House, 2017.

Anderson, Michael, dir. *Logan's Run.* Beverly Hills, Calif.: Metro-Goldwyn-Mayer, 1976.

Asimov, Isaac. *A Choice of Catastrophes.* New York: Norton, 1981.

———, ed. *Machines That Think: The Best Science Fiction Stories about Robots and Computers.* New York: Holt, 1973.

Audi, Robert. *The Cambridge Dictionary of Philosophy.* Cambridge: Cambridge University Press, 1998.

Barlow, Connie, ed. *From Gaia to Selfish Genes: Selected Writings in the Life Sciences.* Cambridge: MIT Press, 1992.

Barnstone, Willis, and Marvin Meyer. *The Gnostic Bible: Gnostic Texts of Mystical Wisdom from the Ancient and Medieval Worlds.* London: Shambhala, 2009.

Bassham, Gregory. "The Religion of the Matrix and the Problems of Pluralism" in *The Matrix and Philosophy: Welcome to the Desert of the Real.* Chicago: Open Court.

Bates, Harry. "Farewell to the Master." In Asimov, *Machines That Think,* 93–138.

Baudrillard, Jean. *America*. New York: Verso, 2010.

———. *Simulacra and Simulation*. Translated by Sheila Faria Glaser. Ann Arbor: University of Michigan Press, 1994.

Baum, Frank L. *The Wonderful Wizard of Oz*. London: Simon & Brown, 2011.

Baxter, John. *George Lucas: A Biography*. New York: HarperCollins, 1999.

Beschloss, Michael. *Presidents of War*. New York: Crown, 2018.

Bekey, George A. *Autonomous Robots: From Biological Inspiration to Implementation and Control*. Cambridge: MIT Press, 1998.

———. *Robotics: State of the Art and Future Challenges*. London: Imperial College Press, 2008.

Benford, Gregory. "Real Science Imaginary Worlds." In *The Ascent of Wonder: The Evolution of Hard Science Fiction*, edited by David G. Hartwell and Kathryn Cramer, 15–23. New York: Doherty, 1994.

Berkley, James. "Post-Human Mimesis and the Debunked Machine: Reading Environmental Appropriation in Poe's 'Maelzel's Chess-Player' and 'The Man That Was Used Up.'" *Comparative Literature Studies* 41, no. 3 (2004): 356–76.

Beyer, Kurt. *Grace Hopper and the Invention of the Information Age*. Cambridge: MIT Press, 2009.

Benford, Gregory, and Elisabeth Malartre. *Beyond Human: Living with Robots and Cyborgs*. New York: Forge, 2007.

Beniger, James R. *The Control Revolution: Technological and Economic Origins of the Information Society*. Cambridge: Harvard University Press, 1986.

Bierce, Ambrose. *Terror by Night: Classic Ghost and Horror Stories*. London: Wordsworth, 2006.

The Big Book. 4th Ed. Alcoholics Anonymous, 2001.

Black, Shane, dir. *Iron Man 3*. Hollywood, Calif.: Paramount Pictures, 2013.

Bleiler, Everett F. "From Newark Steam Man to Tom Swift." *Extrapolation* 30, no. 2 (Summer 1989): 101–116.

Booth, Mark. *The Secret History of the World*. London: Random House, 2011.

Braden, Gregg. *The Divine Matrix: Bridging Time, Space, Miracles, and Belief.* New York: Hay House, 2007.

———. *The God Code: The Secret of Our Past, the Promise of Our Future*. London: Hay House, 2004.

———. *Walking Between the Worlds: The Science of Compassion*. Bellevue, Wash.: Radio Bookstore, 1997.

Brockman, John. *Third Culture*. New York: Simon & Shuster, 1997.

Brooks, Rodney A. *Flesh and Machines: How Robots Will Change Us*. New York: Pantheon, 2002.

Brown, Laurie M. *Most of the Good Stuff: Memories of Richard Feynman*. New York: American Institute of Physics, 1993.

Bruce, Robert V. *The Launching of Modern American Science, 1846–1876*. Ithaca: Cornell University Press, 1987.

Buhner, Stephen Harrod. *Plant Intelligence and the Imaginal Realm: Beyond the Doors of Perception into the Dreaming of Earth*. Rochester, Vt.: Inner Traditions, 2014.

———. *Ensouling Language: On the Art of Nonfiction and the Writer's Life*. Rochester, Vt.: Inner Traditions, 2010.

Burrows, David, Frederick Lapides, and John Shawcross, eds. *Myths and Motifs in Literature*. New York: MacMillan, 1973.

Campbell, Joseph. *The Hero with a Thousand Faces*. Novato: Bollingen Books, 2008.

———. *The Hero's Journey: Joseph Campbell on His Life and Work*. Novato: Bollingen Books, 2006.

———. *The Inner Reaches of Outer Space: Metaphor as Myth and as Religion*. New York: Harper, 1986.

———. *Thou Art That: Transforming Religious Metaphor*. Novato: Bollingen Books, 2008.

Campbell, Joseph and Bill Moyers. *The Power of Myth*. Edited by Betty Sue Flowers. New York: Doubleday, 1988.

Candelaria, Matthew. "The Colonial Metropolis in the Work of Asimov and Clarke." *Journal of American and Comparative Cultures* 25, no. 3 (2002): 427–32.

Cannon, Lou. *Reagan*. New York: G.P. Putnam, 1982.

Carson, Rae. *Star Wars: The Rise of Skywalker*. New York: Del Rey, 2020.

Chalmers, David J. *Constructing the World*. Oxford University Press, 2012.

Channell, David F. *The Vital Machine: A Study of Technology and Organic Life*. New York: Oxford University Press, 1991.

Cheney, Margaret. *Tesla: Man Out of Time*. Englewood Cliffs: Prentice Hall, 1991.

Chomsky, Noam. *Knowledge of Language.* New York: Praeger, 1986.

Chown, Marcus. *The Universe Next Door.* Oxford: Oxford University Press, 2003.

Cohen, John. *Human Robots in Myth and Science.* Cranbury: Barnes, 1967.

Conklin, Groff, ed. *Science Fiction Thinking Machines.* New York: Bantam, 1955.

Conway, Flo, and Jim Siegelman. *Dark Hero of the Information Age: In Search of Norbert Wiener the Father of Cybernetics.* New York: Basic Books, 2005.

Cooney, Miriam, ed. *Celebrating Women in Mathematics and Science.* Weston: National Council of Teachers of Mathematics, 1981.

Cowan, Ruth S. *A Social History of American Technology.* Oxford: Oxford University Press, 1997.

———. *More Work for Mother: The Ironies of Household Technology from the Open Hearth to the Microwave.* New York: Basic Books, 1983.

Cowen, Tyler. *The Complacent Class: The Self-Defeating Quest for the American Dream.* New York: St. Martin's Press, 2017.

Crick, Francis. *The Astonishing Hypothesis: The Scientific Search for the Soul.* New York: HarperCollins, 1994.

Cross, Gary. *Technology and American Society: A History.* New York: Prentice Hall, 1995.

Daniels, Bruce C. *Puritans at Play: Leisure and Recreation in Colonial New England.* New York: St. Martin's Griffin, 1995.

Davenport, Basil. *The Science Fiction Novel: Imagination and Social Criticism.* Chicago: Advent Publishing, 1969.

Davidson, James, et al. *Nation of Nations: A Narrative History of the American Republic.* Vol II: Since 1865. New York: McGraw-Hill, 1994.

Davies, Paul and Julia Brown. *The Ghost in the Atom.* Cambridge: Cambridge University Press, 1986.

Dawkins, Richard. *A Devil's Chaplain.* New York: Mariner, 2004.

———. *The Blind Watchmaker.* London: Penguin, 1990.

———. *The Selfish Gene.* Oxford: Oxford University Press, 1976.

Dawson, Michael R. W. *Understanding Cognitive Science.* Malden: Blackwell Books, 1998.

Dechert, Charles R. *The Social Impact of Cybernetics.* New York: Simon & Schuster, 1966.

Decker, Kevin S. and Jason T. Eberl. *The Ultimate Star Trek and Philosophy:*

The Search for Socrates. Hoboken, N.J.: Blackwell Wiley and Sons 2016.

Decker, Kevin S. and Jason T. Eberl. *The Ultimate Star Trek and Philosophy: The Search for Socrates*. West Sussex: Wiley, 2016.

Deutsch, David. *The Fabric of Reality*. London: Penguin, 1997.

Dick, Philip K. "The Electric Ant." *Fantasy & Science Fiction* (Oct. 1969).

———. *Flow My Tears, The Policeman Said*. New York: Doubleday, 1974.

———. *VALIS*. New York: Random House, 1991.

Dispenza, Joe. *Becoming Supernatural: How Common People Are Doing the Uncommon*. New York: Hay House, 2017.

———. *Breaking the Habit of Being Yourself: How to Lose Your Mind and Create a New One*. New York: Hay House, 2012.

Douglas, George H. *All Aboard: The Railroad in American Life*. New York: Paragon, 1992.

D'Souza, Dinesh. *Ronald Reagan: How an Ordinary Man Became an Extraordinary Leader*. New York: Simon & Schuster, 1997.

Edison, Thomas. "Woman of the Future." *Good Housekeeping* Oct. 1912, 436–44.

Ellis, Edward. *The Huge Hunter, or, The Steam Man of the Prairies*. New York: Dodo, 2011. Originally published in 1868.

Elmer, Jonathan. *Reading at the Social Limit: Affect, Mass Culture, and Edgar Allan Poe*. Stanford: Stanford University Press, 1995.

Emoto, Masaru. *The Hidden Messages in Water*. New York: Atria Books/Beyond Words, 2011.

Engel, Joel. *Gene Roddenberry: The Myth and the Man behind Star Trek*. New York: Hyperion, 1994.

Faust, Drew. *The Republic of Suffering: Death and the American Civil War*. New York: Random House, 2008.

Fenton, Charles A. "'The Bell-Tower': Melville and Technology." *American Literature* 23.2 (1951): 219–32.

Favreau, Jon, dir. *Iron Man*. Hollywood, Calif.: Paramount Pictures, 2008.

———, dir. *Iron Man 2*. Hollywood, Calif.: Paramount Pictures, 2011.

Ferrell, William K. *Literature and Film as Modern Mythology*. Westport: Praeger, 2000.

Feynman, Richard. *The Character of Physical Law*. New York: Modern Library, 1994.

———. *Surely You're Joking, Mr. Feynman*. New York: Norton, 1997.

Fingeroth, Danny. *A Marvelous Life: The Amazing Story of Stan Lee*. New York: St. Martin's, 2019.

Fisher, Kevin. "Information Feedback Loops and Two Tales of the Post-Human in 'Forbidden Planet.'" *Science Fiction Film and Television* 3, no. 1 (Spring 2010): 19–36.

Fisher, Marvin. *Going Under: Melville's Short Fiction and the American 1850s*. Baton Rouge: Louisiana State University Press, 1977.

Fleisher, Richard, dir. *Fantastic Voyage*. Los Angeles, Calif.: 20th Century Fox, 1966.

Fleming, Victor, dir. *The Wizard of Oz*. Beverly Hills, Calif.: Metro-Goldwyn-Mayer, 1939.

Foer, Joshua. *Moonwalking with Einstein: A Journey Through Memory and Mind*. London: Albert Lane, 2011.

Foster, Alan D. *Star Wars: The Force Awakens*. New York: Penguin, 2015.

Fry, Jason. *Star Wars: The Last Jedi*. New York: Penguin, 2018.

Gabler, Neal. *Walt Disney: The Triumph of the American Imagination*. New York: Knopf, 2006.

Galipeau, Steven A. *The Journey of Luke Skywalker: An Analysis of Modern Myth and Symbol*. Chicago: Open Court, 2001.

Garry, Ann, and Marilyn Pearsall. *Women, Knowledge, and Reality: Explorations in Feminist Philosophy*. New York: Routledge, 1996.

Gee, James Paul. *What Video Games Have to Teach Us about Learning and Literacy*. New York: Macmillan, 2007.

Gell-Mann, Murray. *The Quark and the Particle*. New York: Penguin, 1997.

Gleick, James. *Genius: The Life and Science of Richard Feynman*. New York: Pantheon, 1992.

Goddard, Neville. *The Power of Attention: Ideas That Shape the World*. Seattle: Pacific Publishing Studios, 2010.

———. *The Power of Imagination: The Neville Goddard Treasury*. New York: Penguin, 2015.

Gould, Eric. *Mythical Intentions in Modern Literature*. Princeton: Princeton University Press, 1981.

Gorski, Philip. *American Covenant: A History of Civil Religion from the Puritans to the Present*. Princeton: Princeton University Press, 2017.

Gray, Chris, H. *The Cyborg Handbook*. New York: Routledge, 1995.

Green, Richard and Joshua Heter. *Westworld and Philosophy: Mind Equals Blown*. Chicago: Open Court, 2019.

Greene, Brian. *The Hidden Reality: Parallel Universes and the Deep Laws of the Cosmos*. New York: Knopf, 2011.

Greenblatt, Stephen. *The Swerve: How the World Became Modern*. New York: Norton, 2011.

Gregory, Andrew. *Eureka! The Birth of Science*. New York: Totem, 2003.

Grimes, Ronald L., ed. *Readings in Ritual Studies*. Englewood Cliffs: Prentice Hall, 1996.

Grossinger, Richard. *Bottoming Out the Universe: Why There Is Something Rather than Nothing*. Rochester, Vt.: Park Street Press, 2020.

———. *Embryos, Galaxies, and Sentient Beings: How the Universe Makes Life*. Berkeley: North Atlantic Books, 2003.

Gura, Philip F. *American Transcendentalism: A History*. New York: Hill & Wang, 2007.

Hall, Many P. *The Secret Teachings of All Ages: An Encyclopedic Outline of Masonic, Hermetic, Qabbalistic and Rosicrucian Symbolical Philosophy*. Seattle: Pacific Publishing Studios, 2011.

Hanson, Michael J., and Max S. Kay. *Star Wars: The New Myth*. Bloomington: Xlibris, 2002.

Haraway, Donna. "Cyborgs and Symbionts: Living Together in the New World Order." In *The Cyborg Handbook,* edited by Chris Hables Gray, 43–54. New York: Routledge, 1995.

Hawthorne, Nathaniel. *Moses from an Old Manse*. New York: Standard, 1882.

Hayles, Katherine N. *How We Became Posthuman: Virtual Bodies in Cybernetics, Literature, and Informatics*. Chicago: University of Chicago Press, 1999.

Hebb, D. O. *The Organization of Behavior: A Neuropsychological Theory*. Mahwah: Lawrence Erlbaum Associates, 2002.

Hedges, Chris. *American Fascists: The Christian Right and the War on America*. New York: Free Press, 2006.

———. *America: The Farewell Tour*. New York: Simon & Schuster, 2018.

———. *Death of the Liberal Class*. New York: Nations, 2010.

———. *Unspeakable: Chris Hedges on the Most Forbidden Topics in America with David Talbot*. New York: Skyhorse, 2016.

Hey, Anthony. *Feynman and Computation: Explaining the Limits of Computers*. Reading: Perseus, 1999.

Hickman, Larry. *John Dewey's Pragmatic Technology*. Bloomington: Indiana University Press, 1992.

Hicks, Marie. *Programmed Inequality: How Britain Discarded Women Technologists and Lost Its Edge in Computing*. Cambridge: MIT Press, 2017.

Hindle, Brooke. *The Pursuit of Science in Pre-Revolutionary America*. New York: Norton, 1974.

Hindle, Brooke, and Steven Lubar. *Engines of Change: The American Industrial Revolution, 1790-1860*. Washington, D.C.: Smithsonian Institute, 1986.

Hodges, Andrew. *Alan Turing: The Enigma*. New York: Walker, 2000.

Hoeller, Stephan A. *The Gnostic Jung and the Seven Sermons to the Dead*. Wheaton: Quest, 1982.

———. *Jung and the Lost Gospels: Insights into the Dead Sea Scrolls and the Nag Hammadi Library*. Wheaton: Quest, 2002.

Horowitz, Mitch. *The Miracle Club: How Thoughts Become Reality*. Rochester, Vt.: Inner Traditions, 2018.

———. *One Simple Idea: How Positive Thinking Reshaped Modern Life*. New York: Penguin Random House, 2014.

Howden, Martin. *Robert Downey Jr. The Biography*. London: Blake, 2010.

Hughes, Richard T. *Myths America Lives By*. Chicago: University of Illinois Press, 2004.

Hughes, Thomas P. *American Genesis: A Century of Invention and Technological Enthusiasm 1870-1970*. New York: Random, 1989.

Hume, Kathryn. *Pynchon's Mythography: An Approach to Gravity's Rainbow*. Carbondale: Southern Illinois University Press, 1987.

Hyde, Lewis. *Trickster Makes this World: Mischief, Myth, and Art*. New York: Farrar, Straus, and Ciroux, 1988.

Irwin, William. *The Matrix and Philosophy: Welcome to the Desert of the Real*. Chicago: Open Court, 2016.

———. *The Ultimate Star Wars and Philosophy: You Must Unlearn Why You Have Learned*. Hoboken: Wiley and Sons, 2016.

Irwin, William, and Mark D. White. *Iron Man and Philosophy: Facing the Stark Reality*. Hoboken: Wiley and Sons, 2010.

Isaacson, Walter. *Steve Jobs*. New York: Simon & Schuster, 2011.

Israel, Paul. *From Machine Shop to Industrial Laboratory*. Baltimore: Johns Hopkins University Press, 1992.

Jackson, Pamela, and Jonathan Lethem, eds. *The Exegesis of Philip K. Dick*. New York: Houghton Mifflin Harcourt, 2011.

Jacobsen, Annie. *The Pentagon's Brain: An Uncensored History of DARPA, America's Top Secret Military Research Agency*. New York: Back Bay Books, 2015.

James, William. *The Varieties of Religious Experience*. New York: Renaissance, 2012.

Jegede, Katherine. *Infinite Possibility: How to Use the Ideas of Neville Goddard to Create the Life You Want*. New York: Random House, 2018.

Johnson, David Kyle. *Black Mirror and Philosophy: Dark Reflections*. Hoboken: Wiley and Sons, 2019.

Johnson, Rian, dir. *Star Wars Episode VIII: The Last Jedi*. Burbank, Calif.: Walt Disney Studios Motion Pictures, 2017.

Jones, Jacqueline, Peter H. Wood, Thomas Borstelmann, Elaine Tyler May, and Vicki L. Ruiz. *Created Equal: A Social and Political History of the United States*. 2 vols. New York: Pearson, 2008.

Josephson, Mathew. *Edison: A Biography*. New York: McGraw Hill, 1959.

Jortner, Adam. *The Gods of Prophetstown: The Battle of Tippecanoe and the Holy War for the American Frontier*. Oxford: Oxford University Press, 2000.

Jung, Carl. *Memories, Dreams, Reflections*. New York: Random House, 1989.

———. *Man and His Symbols*. New York: Bantam, 1964.

Kaplan, Amy. *The Anarchy of Empire in the Making of U.S. Culture*. Cambridge: Harvard University Press, 2005.

Kasson, John F. *Civilizing the Machine*. New York: Hill, 1976.

Kastrup, Bernardo. *Brief Peeks Beyond: Critical Essays on Metaphysics, Neuroscience, Free Will, Skepticism and Culture*. Winchester: Iff Books, 2015.

———. *Dreamed Up Reality: Diving into the Mind to Uncover the Astonishing Hidden Tale of Nature*. Hampshire: O Books, 2011.

———. *Meaning in Absurdity: What Bizarre Phenomena Can Tell Us About the Nature of Reality*. Winchester: Iff Books, 2011.

———. *More than Allegory: On Religious Myth, Truth and Belief.* Winchester: Iff Books, 2016.

———. *Why Materialism Is Baloney: How True Skeptics Know There Is No Death and Fathom Answers to Life, the Universe, and Everything.* Winchester: Iff Books, 2014.

Kato, Ichiro. "Coming Robot New Era: A Viewpoint." *Robotica* 1, no. 1 (1983): 9–13.

Kaufman, Stuart A. *Origins of Order: Self-Organization and Selection in Evolution.* New York: Oxford University Press, 1993.

Ketterer, David. *The Rational Deception in Poe.* Baton Rouge: Louisiana State University Press, 1979.

Knowles, Christopher. *Our Gods Wear Spandex: The Secret History of Comic Book Heroes.* San Francisco: Weiser, 2007.

Kolata, Gina. *Clone: The Road to Dolly and the Path Ahead.* New York: Harper, 1999.

Kripal, Jeffrey J. *Authors of the Impossible: The Paranormal and Sacred.* Chicago: University of Chicago Press, 2010.

———. *The Flip: Epiphanies of Mind and the Future of Knowledge.* New York: Bellevue Literary Press, 2019.

———. *Mutants and Mystics: Science Fiction, Superhero Comics, and the Paranormal.* Chicago: University of Chicago Press, 2011.

Kubrick, Stanley, dir. *2001: A Space Odyssey.* Beverly Hills, Calif.: Metro-Goldwyn-Mayer, 1968.

Kuhn, Thomas. *The Structure of Scientific Revolutions.* Chicago: University of Chicago Press, 1962.

Kurzweil, Ray. *The Age of Intelligent Machines.* Cambridge: MIT Press, 1990.

———. *The Singularity Is Near: When Humans Transcend Biology.* Cambridge: MIT Press, 2005.

Lachman, Gary. *Beyond the Robot: The Life and Works of Colin Wilson.* New York: Random House, 2016.

Lakoff, George and Mark Johnson. *Metaphors We Live By.* Chicago: University of Chicago Press, 1980.

Langley, Travis. *Star Wars Psychology: Dark Side of the Mind.* New York: Sterling, 2015.

Layton, Edwin. *Technology and Social Change in America.* New York: Harper, 1973.

Le Brun, Annie. *The Reality Overload: The Modern World's Assault on the Imaginal Realm*. Rochester, Vt.: Inner Traditions, 2008.

Le Guin, Ursula K. *The Language of the Night: Essays on Fantasy and Science Fiction*. New York: HarperCollins, 1992.

———. "Response to the Le Guin Issue (S.F.S. #7)." *Science Fiction Studies* 3, no. 1 (March 1976): 43–46.

Leslie, Stuart W. *The Cold War and American Science*. New York: Columbia University Press, 1993.

Lippe, Jonathan. *Simulation Theory: Reality Has Become Science Fiction*. New York: Dual Legacy, 2009.

Lipton, Burce. *Spontaneous Evolution*. Carlsbad: Hay House, 2010.

Lopatto, Elizabeth. "Elon Musk unveils Neuralink's plans for brain-reading 'threads' and a robot to insert them." *The Verge*. July 16, Issue 19 (2019).

Lucas, George, dir. *Star Wars Episode III: Revenge of the Sith*. Los Angeles, Calif.: 20th Century Fox, 2005.

———, dir. *Star Wars Episode IV: A New Hope*. Los Angeles, Calif.: 20th Century Fox, 1977.

———. *Star Wars Episode VI: Return of the Jedi*. Directed by Richard Marquand. Los Angeles, Calif.: 20th Century Fox, 1983.

———. *Star Wars Episode V: The Empire Strikes Back*. Directed by Irvin Kershner. Los Angeles, Calif.: 20th Century Fox, 1980.

———, dir. *THX 1138*. Burbank, Calif.: Warner Bros., 1971.

Lucent, Gregory. *The Narrative of Realism and Myth: Verga, Lawrence, Faulkner, Pavese*. Baltimore: Johns Hopkins University Press, 1981.

Luke, David and Rory Spowers. *DMT Dialogues: Encounters with the Spirit Molecule*. Rochester, Vt.: Park Street, 2018.

Lynch, David, dir. *Dune*. Universal City, Calif.: Universal Pictures, 1984.

Magee, Bryan. *The Story of Philosophy: The Essential Guide to the History of Western Philosophy*. New York: DK Publishing, 1998.

Maher, John M. and Dennie Briggs. *An Open Life: Joseph Campbell in Conversation with Michael Toms*. New York: Harper & Row, 1989.

Malik, Kenan. *Man, Beast, and Zombie: What Science Can and Cannot Tell Us about Human Nature*. New Brunswick: Rutgers University Press, 2002.

Maltz, Maxwell. *Psycho-Cybernetics*. New York: Pocket Books, 1960.

Mann, Bonnie, ed. *Women's Liberation and the Sublime: Feminism,*

Postmodernism, Environment. New York: Oxford University Press, 2006.

Marcus, Alan, and Howard Segal. *Technology in America: A Brief History.* New York: Harcourt, 1989.

Marx, Leo. *The Machine in the Garden.* Oxford: Oxford University Press, 1964.

Maxford, Howard. *The A-Z of Science Fiction and Fantasy Films.* London: BT Batsford, 1997.

Mayr, Otto. *Authority, Liberty and Automatic Machinery in Early Modern Europe.* Baltimore: Johns Hopkins University Press, 1986.

McCorduck, Pamela. *Machines Who Think: A Personal Inquiry into the History and Prospects of Artificial Intelligence.* Natick: A. K. Peters, 2009.

McTaggart, Lynn. *The Field: The Quest of the Secret Force of the Universe.* New York: HarperCollins, 2001.

———. *The Intention Experiment: Using Your Thoughts to Change Your Life and the World.* New York: Simon & Shuster, 2007.

Meacham, Jon. *The Soul of America: The Battle for Our Better Angels.* New York: Random House, 2018.

Mead, Joan Tyler. "Poe's 'The Man That Was Used Up': Another Bugaboo Campaign." *Studies in Short Fiction* 23, no. 3 (1986): 281–86.

Melville, Herman. *Piazza Tales.* New York: Farrar, 1948. First published in 1856 by Dix & Edwards.

Menand, Louis. *The Metaphysical Club: A Story of Ideas in America.* New York: Farrar, 2001.

Meyer, Nicholas, dir. *Star Trek II: The Wrath of Khan.* Hollywood, Calif.: Paramount Pictures, 1982.

Miller, Joseph. "The Greatest Good for Humanity: Isaac Asimov's Future History and Utilitarian Calculation Problems." *Science Fiction Studies* 31, no. 2 (2004): 189–206.

Miller, Perry. *The Raven and the Whale: Poe, Melville, and the New York Literary Scene.* Baltimore: Johns Hopkins Press, 1956.

———. *The Transcendentalists: The Classic Anthology.* New York: MJF Books, 1978.

Moravec, Hans. *Robot: Mere Machine to Transcendental Mind.* New York: Oxford University Press, 1997.

Moskowitz, Sam. *Explorers of the Infinite: Shapers of Science Fiction.* Cleveland: World Publishing Company, 1963.

Mulhall, Douglas. *Our Molecular Future: How Nano-Technology, Robotics, Genetics, and Artificial Intelligence Will Transform the World.* Amherst: Prometheus, 2002.

Murphy, Mike. *The Creation Frequency.* Novato: New World, 2018.

Musk, Elon. "An Integrated Brain-Machine Interface Platform with Thousands of Channels," version 4, bioRxiv, August 2, 2019.

Nakauchi, Yasushi, and Reid Simmons. "A Social Robot That Stands in Line." *Autonomous Robots* 12, no. 3 (2002): 313–24.

Nocks, Lisa. "To Serve and Obey: A History of the Android, 1850–Present." Ph.D. diss., Drew University Graduate School, 2005.

Nye, David F. *America as Second Creation.* Cambridge: MIT Press, 2003.

———. *The American Technological Sublime.* Cambridge: MIT Press, 1994.

———. *Electrifying America.* Cambridge: MIT Press, 1990.

Oleksii, Kharkovyna. "Elon Musk's Neurolink—Everything You Need to Know: One Step Closer to Real-Life Black Mirror and Cyberpunk?" *Towards Data Science.* August 5, 2019.

Oppenheim, Robert. *Atom and Void.* Princeton: Princeton University Press, 1989.

Overbye, Dennis. *Lonely Hearts of the Cosmos.* New York: Basic Books, 1999.

Parkin, Lance. *The Impossible Has Happened: The Life and Work of Gene Roddenberry.* London: Aurum Press, 2016.

Pagels, Elaine. *Why Religion? A Personal Story.* New York: HarperCollins, 2018.

Peake, Anthony. *A Life of Philip K. Dick: The Man Who Remembered the Future.* London: Arcturus, 2013.

Peat, David F. *Infinite Potential: The Life and Times of David Bohm.* New York: Basic Books, 1996.

Perkowitz, Sidney. *Digital People: From Bionic Humans to Androids.* Washington, D.C.: Joseph Henry, 2002.

Perlmutter, Mark. *Why Lawyers (and the rest of us) Lie & Engage in Other Repugnant Behavior.* Austin: Bright Books, 1998.

Perrine, Toni A. *Film in the Nuclear Age: Representing Cultural Anxiety.* New York: Garland, 1998.

Pink, Daniel H. *A Whole New Mind: Why Right Brainers Will Rule the Future.* New York: Penguin, 2006.

Pivonka, Daniel. "Fantastic Voyage: A mini sub that could steer through the body." *Popular Science* 282.1 (2013): 28.

Poe, Edgar Allan. *Complete Stories and Poems.* Garden City: Doubleday, 1966.

Popper, Karl. *Conjectures and Refutations: The Growth of Scientific Knowledge.* New York: Harper, 1963.

Poundstone, William. *Prisoner's Dilemma.* New York: Anchor, 1993.

Pursell, Carroll. *The Machine in America: A Social History of Technology.* Baltimore: Johns Hopkins University Press, 1995.

———. *Technology in America: A History of Individuals and Ideas.* Cambridge: MIT Press, 1981.

Pynchon, Thomas. *Gravity's Rainbow.* New York: Viking, 1973.

Rachman, Stephen. *The American Face of Edgar Allan Poe.* Baltimore: Johns Hopkins University Press, 1995.

Radin, Dean. *Entangled Mind: Extrasensory Experiences in a Quantum Reality.* New York: Paraview, 2006.

———. *Real Magic: Ancient Wisdom, Modern Science, and a Guide to the Secret Power of the Universe.* New York: Harmony, 2018.

Raitt, A. W. *The Life of Villiers de l'Isle-Adam.* Oxford: Clarendon, 1981.

Ramis, Harold, dir. *Groundhog Day.* Culver City, Calif.: Columbia Pictures, 1993.

Ratner, Mark. *Nano-Technology and Homeland Security.* Upper Saddle River: Prentice Hall, 2004.

Rhodes, Richard. *The Making of the Atomic Bomb.* New York: Simon & Schuster, 1986.

Rice, Stephen P. "Making Way for the Machine: Maelzel's Automaton Chess-Player and Antebellum American Culture." *Proceedings of the Massachusetts Historical Society, Third Series* 106 (1994): 1–16.

———. *Mind the Machine: Languages of Class in Early Industrial America.* Berkley: University of California Press, 2004.

Rockoff, Hugh. "'The Wizard of Oz' as a Monetary Allegory." *Journal of Political Economy* 98, no. 4 (1990): 739–60.

Rogan, Joe. *The Joe Rogan Experience.* Episode 1169, "Elon Musk." Sep 6, 2018.

———. *The Joe Rogan Experience.* Episode 1284, "Graham Hancock." April 22, 2019.

———. *The Joe Rogan Experience.* Episode 1407, "Elon Musk." May 7, 2020.

———. *The Joe Rogan Experience.* Episode 1543, "Brian Muraresku and Graham Hancock." Sept. 20, 2020.

———. *The Joe Rogan Experience.* Episode 550, "Rupert Sheldrake." Sept. 17, 2014.

Rorabaugh, W. J. *The Alcoholic Republic: An American Tradition.* Oxford: Oxford University Press, 1979.

Rosow, Eugene. *Born to Lose: The Gangster Film in America.* New York: Oxford University Press, 1978.

Rottensteiner, Franz. "Who Was Really Moxon's Master?" *Science Fiction Studies* 15, no. 1 (1988): 107–12.

Rushing, Janice Hocker, and Thomas S. Frentz. *Projecting the Shadow: The Cyborg Hero in American Film.* Chicago: University of Chicago Press, 1995.

Russo, Anthony and Joe Russo, dirs. *Avengers: Infinity War.* Burbank, Calif.: Walt Disney Studios Motion Pictures, 2018.

———, dirs. *Captain America: Civil War.* Burbank, Calif.: Walt Disney Studios Motion Pictures, 2016.

Sagan, Carl. *Cosmos.* New York: Ballantine, 1980.

Sanders, Steven M. *The Philosophy of Science Fiction Film.* Lexington: University of Kentucky Press, 2008.

Schipper, Kristopher. *The Taoist Body.* Translated by Karen C. Duval. Berkeley: University of California Press, 1993.

Schlesinger, Arthur M., Jr. *Robert Kennedy and His Times.* Boston: Houghton Mifflin, 1978.

Scott, Brian. *The Reality Revolution.* YouTube Channel.

Scott, Ridely, dir. *Bladerunner.* Burbank, Calif.: Warner Bros., 1982.

Seifer, Mark J. *Wizard: The Life and Times of Nikola Tesla—Biography of a Genius.* Secaucus: Carol, 1996.

Selig, Paul. *The Book of Freedom.* New York: Penguin, 2018.

———. *The Book of Mastery.* New York: Penguin, 2016.

———. *The Book of Truth.* New York: Penguin, 2017.

Sheldrake, Rupert. *Morphic Resonance: The Nature of Formative Causation.* Toronto: Park Street Press, 2009.

———. *Science and Spiritual Practices. Transformative Experiences and Their Effects on Our Bodies, Brains, and Health.* Berkeley: Counterpoint, 2017.

Simons, Geoff. *Is Man a Robot?* Chichester: Wiley, 1986.

Smith, E. E. *The Skylark of Space: Commemorative Edition.* Lincoln: University of Nebraska Press, 2001.

St. Germain, Maureen J. *Waking Up in 5D: A Practical Guide to Multidimensional Transformations.* Rochester, Vt.: Bear & Company, 2017.

Standage, Tom. *The Turk: The Life and Times of the Famous Eighteenth-Century Chess-Playing Machine.* New York: Walker, 2002.

Stearns, R. P. *Science in the British Colonies of North America.* Urbana: University of Illinois Press, 1970.

Stent, Gunther. *The Double Helix.* Berkeley: University of California Press, 1980.

Stevenson, Jay. *Philosophy.* New York: Penguin, 2005.

Starr, Gabrielle G. *Feeling Beauty: The Neuroscience of Aesthetic Experience.* Cambridge: MIT Press, 2015.

Stork, David G. *HAL's Legacy: 2001's Computer Dream as Reality.* Cambridge: MIT Press, 1997.

Strieber, Whitley and Jeffrey J. Kripal. *The Super Natural: A New Vision of the Unexplained.* New York: Random House, 2016.

Struik, Dirk J. *Yankee Science in the Making.* New York: Dover, 1948.

Taibi, Matt. *The Divide: American Injustice in the Age of the Wealth Gap.* New York: Spiegel & Grau: New York, 2014.

Talbot, Michael. *The Holographic Universe: The Revolutionary Theory of Reality.* New York: HarperCollins, 1991.

Three Initiates. *The Kybalion: The Study of the Hermetic Philosophy of Ancient Egypt and Greece.* Mansfield: Martino Publishing, 2016,

Tegmark, Max. *Life 3.0: Being Human in the Age of Artificial Intelligence.* New York: Vintage, 2017.

Telotte, J. P. *Replications: A Robotic History of the Science Fiction Film.* Urbana: University of Illinois Press, 1995.

———. *Science Fiction Film.* Cambridge: Cambridge Press, 2001.

Tesla, Nikola. *My Inventions: The Autobiography of Nikola Tesla.* Momence: Baker & Taylor, 2006.

———. Method of and apparatus for controlling mechanisms of moving vessels or vehicles. U.S. Patent 613,809, filed July 1, 1898, issued November 8, 1898.

Tresch, John. "'The Potent Magic of Verisimilitude': Edgar Allan Poe within the Mechanical Age." *British Journal for the History of Science* 30, no. 3 (1997): 275–90.

Tuerk, Richard. *Oz in Perspective: Magic and Myth in the L. Frank Baum Books*. London: MacFarland & Company, 2007.

Turner, Victor. *The Ritual Process*. Chicago: University of Chicago Press, 1969.

Todorov, Tzvetan. *The Fantastic: A Structural Approach to a Literary Genre*. Ithaca: Cornell University Press, 1975.

Torode, Sam. *Everyday Emerson: The Wisdom of Ralph Waldo Emerson*. Nashville: American Renaissance, 2017.

Tye, Larry. *Bobby Kennedy: The Making of a Liberal Icon*. New York: Random House, 2016.

Vance, Ashlee. *Elon Musk: Tesla, SpaceX, and the Quest for the Fantastic Future*. New York: NY: HarperCollins, 2015.

Veyne, Paul. *Did the Greeks Believe in Their Myths?: An Essay on the Constitutive Imagination*. Chicago: University of Chicago Press, 1988.

Villiers de l'Isle-Adam, Auguste. *Eve of the Future Eden*. Lawrence: Coronado, 1981. Originally published as *l'Eve Future* in 1886.

Virk, Rizwan. *The Simulation Hypothesis: An MIT Computer Scientist Shows Why AI, Quantum Physics and Eastern Mystics Agree We Are in a Video Game*. New York: Bayview Books, 2019.

Wachowski, Lana, and Lilly Wachowski, dirs. *The Matrix*. Burbank, Calif.: Warner Bros, 1999.

Walsh, Toby. *Machines That Think: The Future of Artificial Intelligence*. Amherst: Prometheus, 2018.

Watts, Alan. *Psychotherapy: East and West*. Novato: New World, 1989.

Weiskel, Thomas. *The Romantic Sublime: Studies in the Structure and Psychology of Transcendence*. Baltimore: Johns Hopkins University Press, 1976.

Wegner, Daniel. *The Illusion of Conscious Will*. Cambridge: MIT Press, 2002.

Weightman, Gavin. *Signor Marconi's Magic Box*. New York: HarperCollins, 2003.

White, Mark D. *Iron Man and Philosophy: Facing the Stark Reality*. Sussex: Wiley, 2010.

White, T. G. "Mr. Tesla and the Universe: Human Energy and How to Increase It—His Philosophizing Questioned." *Science* 12, no. 299 (September 12, 1900): 447–51.

Wiebe, Robert H. *The Search for Order 1877-1920.* New York: Hill & Wang, 1967.

Wiener, Norbert. *Cybernetics: Or Control and Communication in the Animal and the Machine.* New York: Wiley, 1948.

———. *God and Golem, Inc. A Comment on Certain Points Where Cybernetics Impinges on Religion.* Cambridge: MIT Press, 1964.

———. *The Human Use of Human Beings: Cybernetics and Society.* New York: Double Day Anchor Books, 1950.

Wilcox, Fred M., dir. *Forbidden Planet.* Beverly Hills, Calif.: Metro-Goldwyn-Mayer, 1956.

Wilson, Colin. *Super Consciousness: The Quest for Peak Experience.* London: Watkins, 2009.

Wilson, E. O. *The Diversity of Life.* New York: Norton, 2002.

Wimsatt, W. K., Jr. "Poe and the Chess Automaton." *American Literature* 11, no. 2 (1939): 138–51.

Wise, Robert, dir. *The Day the Earth Stood Still.* Los Angeles, Calif.: 20th Century Fox, 1951.

Wolfe, Gary K. *Critical Terms of Science Fiction and Fantasy: A Glossary Guide to Scholarship.* New York: Greenwood Press, 1986.

Wood, Gaby. *Edison's Eve.* New York: Knopf, 2002.

Woolley, Benjamin. *Virtual Worlds.* New York: Penguin, 1993.

Yhnell, Diane J. "The Power of Myth in American Films." Ph.D. diss., University of California at Fresno Graduate School, 2002.

Index